AF885490

Cover Page Designed by:

Ms. Gowri Vemuri

First Edition

May, 2019

Author:

MATH-KNOTS

Index

Preface

The word "Science"brings to mind many different pictures: a fat textbook, bunch of equations, space shuttle, medicines and microscopes etc., simply discovering and learning about the nature around us. There are many branches of science and physics the most basic area of sciences, which deals with energy, matter, space, time and their interactions. The word physics is derived from the Latin word *physica*, which means "natural thing."

Physics as a fundamental science helps to uncover the laws of nature. The language of expression is mathematics. During ancient times, humans studied Stars and celestial bodies that lead to the development of disciplines like astronomy and mathematics. Based on size and speed, Physics as a subject is categorized several categories, which you can find from the following picture.

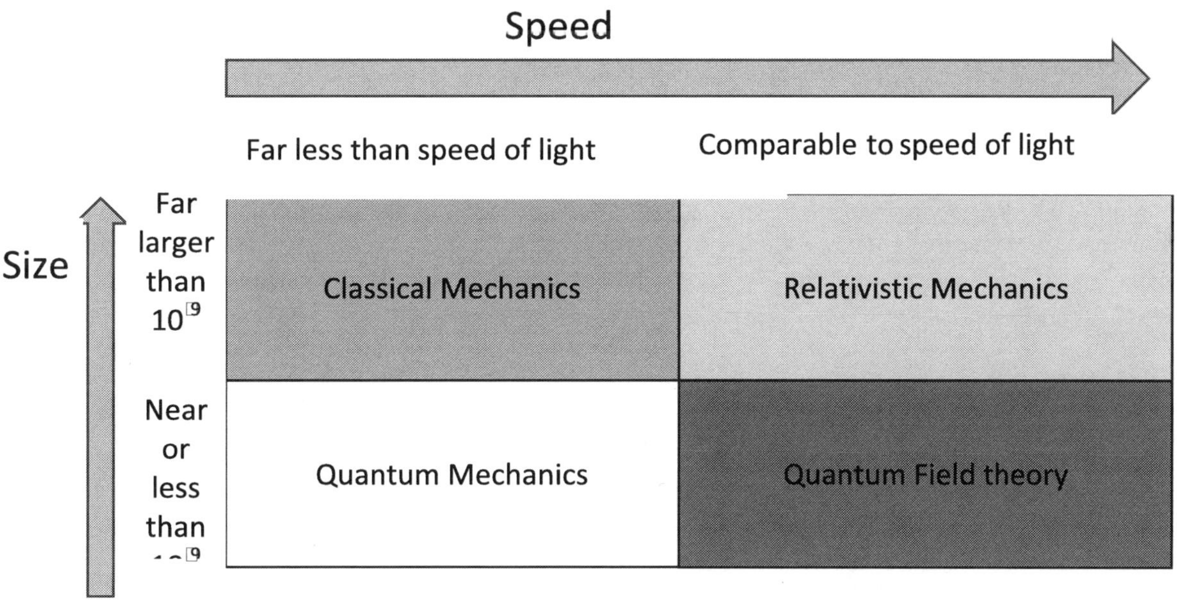

There are many ways to learn physics, but often students face difficulty in solving a problem by the concept they have learnt in the class. Our book is designed to help students in this area. Each chapter in the book, provides a summary of a concept, along with many practice problems.

Chapter - 1

ATOMS

ATOMS

SYNOPSIS

- At ordinary pressures and ordinary voltages gases are bad conductors of electricity.
- The p.d. required to start a spark through a gas is known as sparking potential. It is much greater than the p.d. required to maintain the spark once started.
- At ordinary pressures for moderately great spark lengths, the sparking potential is nearly proportional to the spark length, being about 30,000 V per cm in air between spherical electrodes of 1 cm diameter and less in case of pointed electrodes.
- According to Paschen's law, the sparking potential is directly proportional to the pressure (p) of the gas between the electrodes and the distance (d) between them.
- At ordinary pressures, the current is due to the presence of positive and negative ions as in the case of electrolytic conductivity of a solution.
- At low pressures, discharge (the passage of current through gases) occurs at much lower voltages than at high pressures.
- At lower pressures negative ion throws off its attendant atom or molecule and the resultant negatively charged particle, the electron travels free and faster than the positive ions.
- At low pressures in the discharge tube, as the pressure is reduced different phenomena such as sparking, positive column, glow of cathode, Faraday's dark space, Crooke's dark space and the striations of positive column are gradually observed.
- Finally at a pressure of about 0.01mm of Hg pressure, Crooke's dark space completely fills the discharge tube with the walls of the tube glowing with light i.e., producing fluorescence on the walls of the tube. This is due to some radiation emitted from the cathode surface to which the name cathode rays has been given.
- Cathode rays were first observed by Plucker. J J Thomson after studying their properties called them "Streams of negative corpuscles", while Johnson stoney who having found from the electrolysis that electricity was atomic in nature, suggested the name "electron".

J.J.Thomson's experimental determination of specific charge of an electron:

- Cathode rays can move undeflected through mutually perpendicular electric and magnetic fields acting in same region.

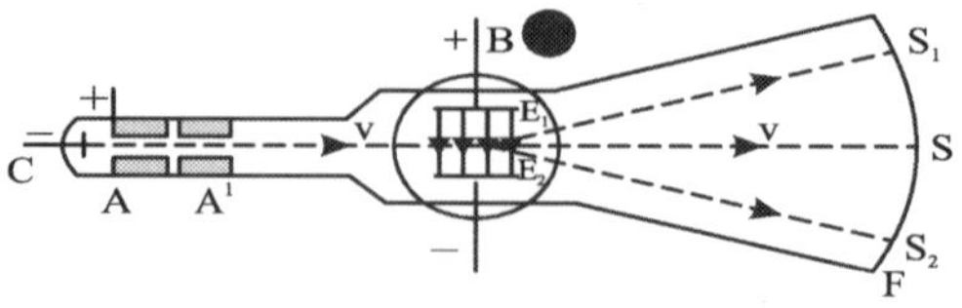

- If plate E_1 is connected to the positive terminal and E_2 to the negative terminal of a battery, vertically downward electric field (E) is produced perpendicular to the direction of motion of electrons. Then electrons will be deflected upwards striking the screen at S_1.
- If the magnetic field (B) alone is applied by using clockwise current carrying circular coil the direction of magnetic field is directed into the plane of the paper. Then by Fleming's left hand rule electrons will be deflected downwards striking the screen at S_2.
- Then both electric and magnetic fields are applied simultaneously and their values are adjusted so that the electron beam can move undeflected and strikes the screen at S.
- When the electron beam enters the region of crossed electric and magnetic fields with velocity v horizontally,

- If only electric field is switched on,
  Electric force on the electron which is vertically upwards is, $F_e = eE \rightarrow (1)$

  Where Electric field $E = \frac{V^1}{d}$

  V^1 = potential difference between the plates E_1 and E_2
  d = distance between the plates E_1 and E_2.
- If only magnetic field is switched on,

  Magnetic force acting on the electron, which is vertically downwards is, $F_m = Bev \rightarrow (2)$
  Where B = magnetic induction field
  e = charge of electron, v = velocity of electron
- If both electric and magnetic fields are applied simultaneously and the electrons moves undeflected, then magnetic and electric forces acting on the electrons are equal and opposite.
  $\therefore F_m = F_e$ $\qquad$ $Bev = eE$
  $\therefore$ Velocity of electron $v = \frac{E}{B} \rightarrow (3)$
  As the electron is accelerated from cathode to anode, the workdone on the electron by the accelerating voltage is equal to the increase in kinetic energy of the electron, K.E of electron at anode = work done on the electron $\frac{1}{2}mv^2 = eV_0 \rightarrow (4)$; $\therefore \frac{e}{m} = \frac{v^2}{2V_0} \rightarrow (5)$
  Where V_0 = potential difference between cathode and Anode.
  m = mass of electron, e = charge of electron
  v = velocity of electron.
  substituting equation (3) in (5) we get

  $$\frac{e}{m} = \frac{E^2}{2V_0B^2} \rightarrow (6)$$

  Since we know E, B and V_0, specific charge of electron can be calculated.
- The value of specific charge of electron as measured by J.J.Thomson is
  1.7588×10^{11} C kg^{-1}

  Note :

(a) Specific charge of electron is found to be independent of nature of the gas.
(b) Specific charge of electron is found to be independent of nature of the electrodes.
(c) Specific charge of electron decreases with increase in velocity of electron, because charge is independent of velocity and mass increases with increase in velocity, since $m = \frac{m_0}{\sqrt{1-\frac{v^2}{c^2}}}$

Where m_0 = rest mass of electron,
m = Relativistic mass of electron and
v = Velocity of electron; c = Velocity of light

Note : If cathode rays are moving along +ve x-axis and electric field is applied along +ve y-axis then magnetic field should be along +ve z-axis so that cathode rays moves undeflected. This is because net force on the electron

$$\vec{F} = e(\vec{E} + \vec{v} \times \vec{B}) = \vec{0} \Rightarrow \vec{E} = -(\vec{v} \times \vec{B})$$

W.E-1: ***An electron beam moving with a speed of 2.5×10^{7} ms^{-1} enters into the magnetic field 4×10^{-3} Wb/m^2 directed perpendicular to its direction of motion. Find the intensity of the electric field required so that the electron moves undeflected.***

Sol. When electron goes undeflected, force due to electric field is equal and opposite to force due to magnetic field i.e., $Ee = Bev$

$B = 4 \times 10^{-3}\ Wb/m^2$ and $v = 2.5 \times 10^{7}\ ms^{-1}$

$E = Bv = 4 \times 10^{-3} \times 2.5 \times 10^{7} = 10^{5}\ V/m$

W.E-2: ***A mono energetic electron beam with a speed of 5.2×10^{6} ms^{-1} enters into a magnetic field of induction 3×10^{-4} T, directed normal to the beam. Find the radius of the circle traced by the beam (Take $e/m = 1.76 \times 10^{11} Ckg^{-1}$)***

Sol. From the equation, $Bev = \frac{mv^2}{r}$

Radius of circle $r = \frac{mv}{Bq} = \frac{v}{B(e/m)}$

$$\therefore r = \frac{5.2 \times 10^{6}}{3 \times 10^{-4}\left(1.76 \times 10^{11}\right)} = 0.1m$$

W.E-3 : ***The deflecting plates in a Thomson's setup are x meters long. Intensity of electric field applied between the plates is E. The plates are maintained at a p.d of V volts. Electrons accelerated through a p.d of V volts enter from one edge of the plates midway in a direction parallel to the plates. Find the deflection at the other edge of the plates.***

Sol. Motion of a charged particle projected perpendicular to a uniform electric field is a parabola. Horizontal distance travelled by the electron in time t is $x = vt \Rightarrow t = \frac{x}{v}$

Deflection of the electron in y - direction is $y = \frac{1}{2}at^2$, $F = ma = Ee$; $a = \frac{Ee}{m}$; $y = \frac{1}{2}\left(\frac{Ee}{m}\right)\frac{x^2}{v^2}$

(since $u_y = 0$) $\Rightarrow y = \frac{Ee}{4 \times \left(\frac{1}{2}mv^2\right)}x^2$ ...... (1)

We know that, $\frac{1}{2}mv^2 = eV$ ...... (2)

From (1) and (2) $\Rightarrow y = \frac{E}{4V}x^2$

W.E-4: ***The specific charge of a proton is 9.6×10^{7} C/ Kg. The specific charge of an alpha particle is***

Sol: Specific change $= \frac{q}{m}$; $\frac{S_\alpha}{S_p} = \frac{q_\alpha}{q_p}\frac{m_p}{m_\alpha}$

$$\frac{S_\alpha}{9.6 \times 10^{7}} = \frac{2}{1} \cdot \frac{1}{4} = \frac{1}{2};\ S_\alpha = 4.8 \times 10^{7}\ C/kg$$

W.E-5: ***A beam of electrons enter at right angles to an uniform electric field with a velocity 3×10^{7} m/s, E= 1800 v/m while travelling through a distance of 10cm, the beam deflected by 2mm then specific charge of electron is***

Sol: $t = \frac{x}{u} = \frac{0.1}{3 \times 10^{7}} = \frac{1}{3} \times 10^{-8} s$; $y = \frac{1}{2}at^2$ $a = E\frac{q}{m}$

$$2\times10^{-3}=\frac{1}{2}\left(1800\,\frac{q}{m}\right).\left(\frac{1}{3}\times10^{-8}\right)^2;\frac{q}{m}=2\times10^{11}\,C/kg$$

W.E-6: ***In Thomson's experiment for determining e/m the potential difference between the cathode and the anode (in accelerating column) is same as that between the dflecting plates (in the region of crossed fields). If the p.d is doubled, by what factor should the magnetic field be increased to ensure that the electron beam remains undeflected***

Sol: $\frac{1}{2}mv^2=V.q$; $\frac{1}{2}m\left(\frac{E}{B}\right)^2=V.q$; $\frac{1}{2}m\frac{V^2}{d^2.B^2}=V.q$ $\quad \frac{q}{m}=\frac{V}{2d^2B^2}$; $B^2\alpha\ V\rightarrow B\alpha\ \sqrt{V}$;

$B^1\alpha\ \sqrt{2V}$

$B^1=\sqrt{2}B$

W.E-7: ***In a cathode ray tube, a p.d of 3000V is maintained between the deflector plates whose seperation is 2cm. A magnetic field of 2.5 x 10⁻³ T at right angles to the electric field gives no deflection of electron beam, which recieved an initial acceleration by a p.d of 10,000V. The e/m of electron is***

Sol: V_1 is p.d between the deflector plates and V_2 be the potential diference through which electrons are accelerated

$$\frac{1}{2}mv^2=V_2q \qquad \frac{1}{2}m.\frac{V_1^2}{d^2B^2}=V_2q;\ \frac{q}{m}=\frac{V_1^2}{2d^2\,B^2V_2}$$

$$\frac{q}{m}=\frac{9\times10^6}{2\times4\times10^{-4}\times6.25\times10^{-6}\times10^4}=1.8\times10^{11}c/kg$$

Determination of charge of an electron : Millikan's oil drop method.[Additional]

- Millikan's method to find charge of an oil drop is based on the measurement of terminal velocity of the charged oil drop under the action of gravity and under the combined action of gravity and electric field.

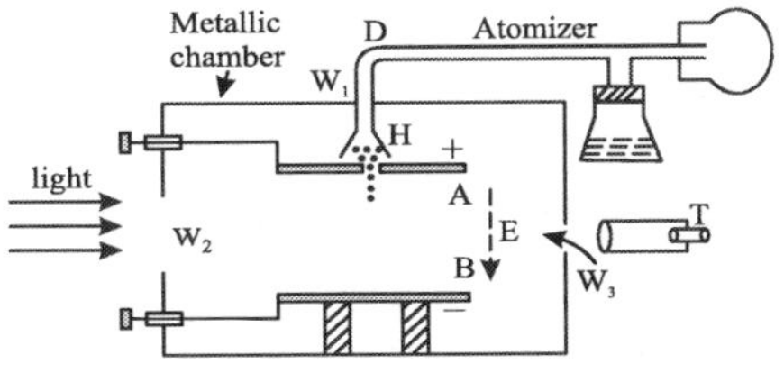

- Electric field can be produced between the plates A and B. Small droplets of non volatile oil (clock oil or apiezon oil) are sprayed using an atomizier above the hole in plate A. Some of the droplets acquire charge due to friction. The chamber is illuminated by sending light horizontally through it. The drops can be seen by using a telescope placed perpendicular to the light beam. A drop looks like a bright star moving upwards or downwards.
- An electric field is produced between the plates A and B by connecting upper plate to the positive terminal and lower plate to the negative terminal of a battery. The field is adjusted so that an oil drop slowly moves upwards. The drop experiences electric force and buoyancy in the upward direction where as weight and viscous force in the downward direction.
- The uniform velocity attained by the drop V_e is measured by measuring the distance travelled and time taken for it. Now electric field is switched off and the drop is allowed to fall freely. The oil drop now experiences viscous force and buoyancy directed upwards and weight downwards. The drop soon moves with terminal velocity V_g downwards. V_g can be measured by measuring distance travelled by the drop and time taken for it.

- Charge on the drop can be calculated using
  the formula $q = \frac{6\pi\eta(V_g + V_e)}{E}\left(\frac{9\eta V_g}{2(d-d^1)g}\right)^{1/2}$
  where η = Coefficient of viscosity of air
  d = density of oil; d^1 = density of air
  E = Intensity of electric field between the plates
  g = acceleration due to gravity
  V_e = terminal velocity of the oil drop in electric field
  V_g = terminal velocity of the oil drop in gravitational field
- After large number of observations Millikan found that the value of 'q' is an integral multiple of a common value 'e' such that q = ne. Where n is an integer and e is the charge of an electron. It is found that charge of electron, $e = 1.6\times10^{-19}$ C.
- When the electric field is applied between the plates A and B, suppose a negatively charged drop is moving upwards with constant velocity V_e.
  upward forces are electric force (F_e) and Buoyancy (F_b) while downward forces are viscous force (F_v) and weight (F_g).
  Net force on the drop is zero.

$$\therefore F_e + F_b = F_g + F_v$$

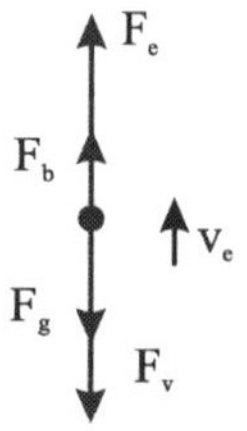

$$qE + \frac{4\pi}{3}R^3 d^1 g = \frac{4\pi}{3}R^3 dg + 6\pi\eta R V_e \rightarrow (1)$$

R - Radius of the drop

If electric field is switched off, suppose the drop falls downwards with terminal velocity V_g.

upward forces are Buoyancy and viscous force while downward force is weight.

Net force on the drop is zero.

$$\therefore F_b + F_v = F_g$$

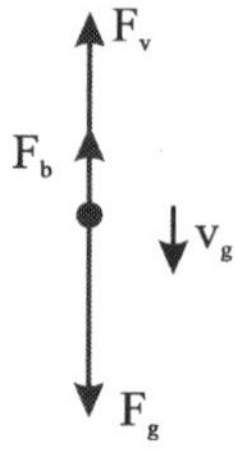

$$\frac{4\pi}{3}R^3 d^1 g + 6\pi\eta R V_g = \frac{4\pi}{3}R^3 dg \rightarrow (2)$$

subtracting (2) from (1)

$$qE - 6\pi\eta R V_g = 6\pi\eta R V_e$$

$$q = \frac{6\pi\eta R(V_e + V_g)}{E} \rightarrow (3)$$

from equation (2) $\frac{4\pi}{3}R^3(d - d^1)g = 6\pi\eta R V_g$

$$\therefore R^2 = \frac{9\eta V_g}{2(d-d^1)g}\ ;\ R = \sqrt{\frac{9\eta V_g}{2(d-d^1)g}} \rightarrow (4)$$

$$\therefore q = \frac{6\pi\eta(V_g + V_e)}{E}\left(\frac{9\eta V_g}{2(d-d^1)g}\right)^{1/2} \rightarrow (5)$$

Case-1 : In Millikan's oil drop method, if the drop is falling under gravity in the absence of electric field then, upward forces are Buoyancy (F_b) and viscous force (F_v) while downward force is weight (F_g). Net force on the drop is zero.

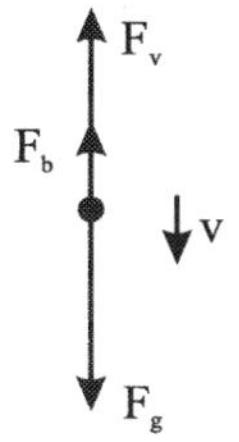

$$\therefore F_b + F_v = F_g \qquad \frac{4\pi}{3}R^3 d^1 g + 6\pi\eta RV = \frac{4\pi}{3}R^3 dg$$

Case-2: In Millikan's oil drop method if a positively charged drop is at rest in the presence of electric field directed upwards then, upward forces are Buoyancy (F_b) and electric force (F_e) while downward force is weight (F_g).
Net force on the drop is zero.

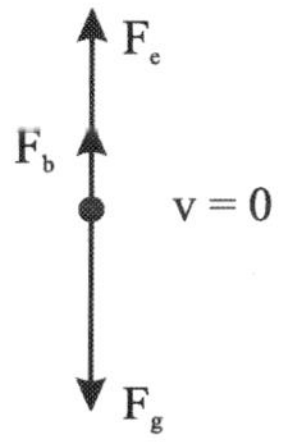

$$\therefore F_e + F_b = F_g$$

$$qE + \frac{4\pi}{3}R^3 d^1 g = \frac{4\pi}{3}R^3 dg$$

Case-3: In Millikan's oil drop method, if a positively charged drop is moving upwards with constant velocity V in the presence of electric field directed upwards then, upward forces are buoyancy (F_b) and electric force (F_e) while downward forces are weight (F_g) and viscous force (F_v). Net force on the drop is zero.

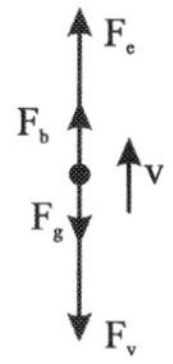

$$\therefore F_e + F_b = F_g + F_v$$

$$qE + \frac{4\pi}{3}R^3 d^1 g = \frac{4\pi}{3}R^3 dg + 6\pi\eta RV$$

Case-4: In Millikan's oil drop method if a positively charged drop is moving downwards with constant velocity V in the presence of electric field directed upwards then, upward forces are Buoyancy (F_b), electric force (F_e) and viscous force (F_v) while downward force is weight (F_g) .
Net force on the drop is zero.

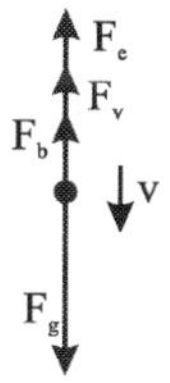

$$\therefore F_e + F_b + F_v = F_g$$

$$qE + \frac{4\pi}{3}R^3d^1g + 6\pi\eta RV = \frac{4\pi}{3}R^3dg$$

Case-5 : **A charged oil drop having charge q_1 is falling freely under gravity in the absence of electric field with a velocity 'V_1'. It is held stationary in an electric field. As it acquires a charge it moves up with a velocity V_2. Then**

i) The new charge on the drop is $q_2 = (k+1)q_1$

ii) Additional charge aquired is $\Delta q = kq_1$

Where $k = \frac{V_2}{V_1}$

In the absence of electric field, drop is falling downwords with velocity V_1.

$$F_g = F_b + F_v$$
$$F_g - F_b = F_v$$
$$F_g - F_b = 6\pi\eta RV_1 \quad -(1)$$

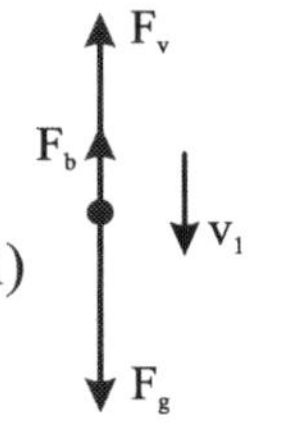

In the presence of electric field when the drop is at rest.

$$F_g = F_b + F_e$$
$$F_g - F_b = q_1E \quad -(2)$$

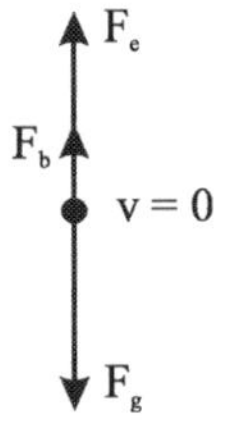

In the electric field, after acquiring final charge q_2 drop is moving upwards with velocity V_2. Then

$$F_g + F_v = F_b + F_e$$
$$F_g - F_b = q_2E - 6\pi\eta RV_2 \quad but, V_2 = kV_1$$
$$F_g - F_b = q_2E - k6\pi\eta RV_1 \quad -(3)$$

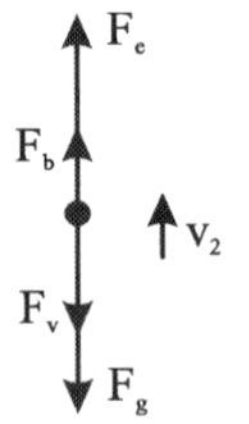

$$q_1E = q_2E - kq_1E$$

So, the final charge on the drop is $q_2 = (k+1)q_1$

Additional charge aquired is $\Delta q = kq_1$

Case-6 : **In the Millikan's oil drop experiment the oil drop is subjected to a horizontal electric field E and the drop moves with a constant velocity making an angle θ to the horizontal. If the weight of the drop is W, neglecting buoyancy, the charge on the drop is $\frac{W\cot\theta}{E}$**

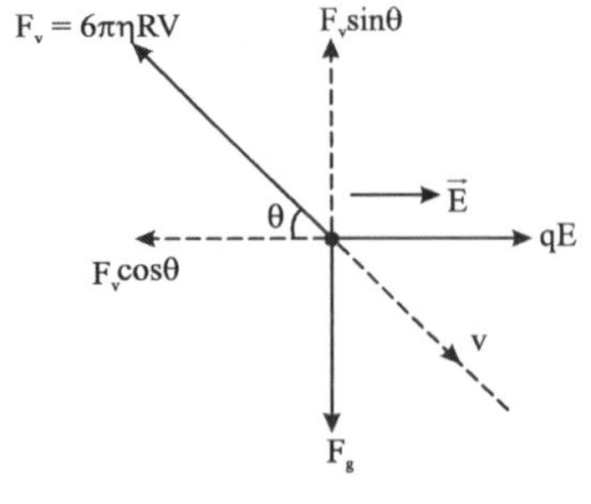

$$F_V\cos\theta = qE$$
$$F_V\sin\theta = F_g = W$$
$$\therefore q = \frac{W\cot\theta}{E}$$

W.E-8: ***In Millikan's oil drop experiment an oil drop of radius r and charge Q is held in equilibrium between the plates of a charged parallel plate capacitor when the potential difference is V. To keep another drop of same oil whose radius is 2r and carrying charge 2Q in equilibrium between the plates, find the new potential difference required.***

Sol. Since drop is at rest QE = mg

$$Q\frac{V}{d}=\frac{4\pi}{3}r^3\rho g \; ; \; V\propto\frac{r^3}{Q} \; ; \; \frac{V_2}{V_1}=\left(\frac{r_2}{r_1}\right)^3\frac{Q_1}{Q_2}$$

$$\frac{V_2}{V}=8\times\frac{1}{2} \quad ; \quad \therefore V_2=4V$$

W.E-9: ***A charged oil drop of charge q is falling under gravity with terminal velocity v in the absence of electric field. A electric field can keep the oil drop stationary. If the drop acquires an additional charge, it moves up with velocity 3v in that field. Find the new charge on the drop.***

Sol. In the absence of electric field

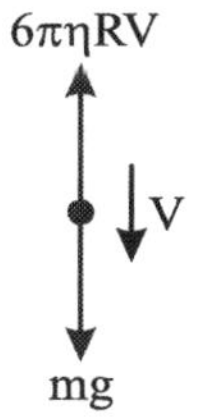

$$mg=6\pi\eta RV\rightarrow \quad (1)$$

when electric field is applied

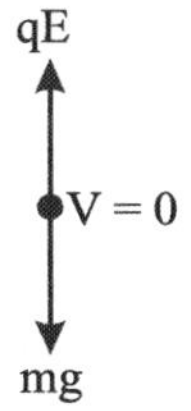

$$mg=qE\rightarrow \quad (2)$$

If q^1 = new charge and drop is moving up then

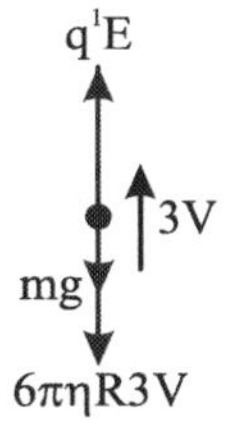

$$q^1E=mg+3\times6\pi\eta RV\rightarrow \quad (3)$$

$$q^1E=4mg=4qE \;\; \therefore q^1=4q$$

W.E-10: ***In Millikan's method of determining the charge of an electron, the terminal velocities of oil drop in the presence and in the absence of an electric field are xcm/s upwards and y cm/s downwards respectively. Find the ratio of electric force to gravitational force on the oil drop. (Neglect Buoyancy)***

Sol. In Gravitational field, weight = viscous force

$W = 6\pi\eta r y$........(1)

In electric field,

Electric force = weight + viscous force

$Eq = W + 6\pi\eta r x$ ....(2)

Substitute (1) in (2) then $Eq = 6\pi\eta r(y+x)$ ....(3)

$$\frac{Electric\ force}{Gravitational\ force} = \frac{x+y}{y}$$

W.E-11: ***In a Millikan's experiment an oil drop of radius 1.5 x 10^{-6} m and density 890 kg/m^3 is held stationary between two condenser plates 1.2 cm apart and kept at a p.d of 2.3 kV. If upthrust due to air is ignored, then the number of excess electrons carries by the drop will be***

Sol: Eq = mg; $\frac{V}{d}.ne = \frac{4}{3}\pi r^3.\rho g$; $n = \frac{4}{3}\pi \times r^3 \rho g \frac{d}{Ve}$

$$n = \frac{4}{3} \times \frac{\pi \times 3.375 \times 10^{-18} \times 890 \times 9.8 \times 1.2 \times 10^{-2}}{2.3 \times 10^3 \times 1.6 \times 10^{-19}}$$

$n \simeq 4$

W.E-12: ***A charged oil drop is of charge q is falling freely under gravity in the absence of electric field with a velocity 'v'. It is held stationary in an electric field, as it acquires a charge it moves up with a velocity '3v. Now the charge on the drop is***

Sol: $mg = 6\pi\eta r\upsilon$, Eq = mg

$Eq^1 = mg + 6\pi\eta r(3\upsilon)$, $Eq^1 = Eq + 3Eq$

$Eq^1 = 4Eq \Rightarrow q^1 = 4q$

W.E-13: ***A charged oil drop falls with a terminal velocity V in the absence of electric field. An electric field E keeps the oil drop stationary in it. When the drop acquires a charge 'q' it moves up with same velocity. Find the initial charge on the drop.***

Sol: $mg = 6\pi\eta r\upsilon$,Eq = mg

$E(Q+q) = mg + 6\pi\eta r\upsilon$,$E(Q+q) = 2EQ$

$Q + q = 2Q \Rightarrow q = Q$

W.E-14: ***Two oil drops in Millikan's experiment are falling with terminal velocities in the ratio 1:4. The ratio of their de-Broglie wave length is***

Sol: $\upsilon_T \propto r^2$ $\qquad r \propto \sqrt{V_T}$

$$\frac{r_1}{r_2} = \sqrt{\frac{V_1}{V_2}} = \frac{1}{2} \qquad \frac{m_1}{m_2} = \frac{1}{8}$$

$$\therefore \frac{\lambda_1}{\lambda_2} = \frac{m_2}{m_1}.\frac{\upsilon_2}{\upsilon_1} = \frac{8}{1}.\frac{4}{1} = 32:1$$

Rutherford's α -particle Scattering Experiment:

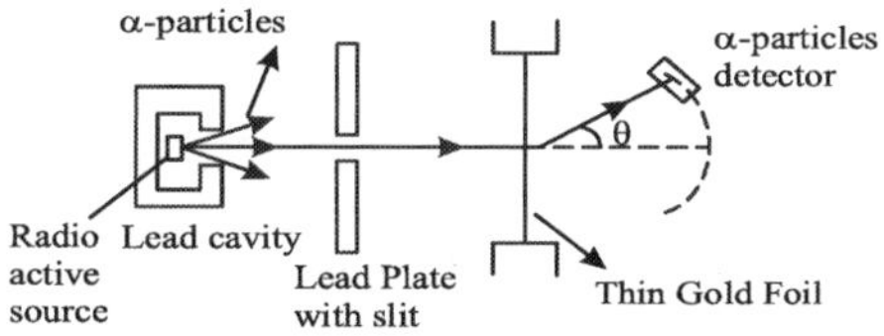

Experimental Observations:

a) Most of the α-particles were found to pass through the gold- foil without being deviated from their paths.

b) Some α-particles were found to be deflected through small angles $\theta < 90°$.

c) Few α-particles were found to be scattered at fairly large angles from their initial path $\theta > 90°$

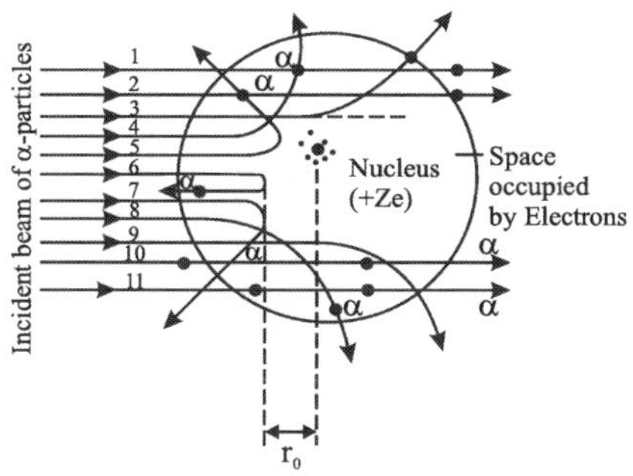

d) A very small number of α-particles about 1 in 8000 practically retracted their paths or suffered deflections of nearly 180°.

➢ The observation (a) indicates that most of the portion of the atom is hollow inside.

➢ Because α-particle is positively charged, from the observations (b), (c) and (d) atom also have positive charge and the whole positive charge of the atom must be concentrated in small space which is at the centre of the atom is called nucleus. The remaining part of the atom and electrons are revolving around the nucleus in circular objects of all possible radii. The positive charge present in the nuclei of different metals is different. Higher the positive charge in the nucleus, larger will be the angle of scattering of α-particle.

Distance of Closest Approach :

➢ An α-particle which moves straight towards the nucleus in head on direction reaches the nucleus i.e, it moves close to a distance r_0 as shown the figure.

➢ As the α-particle approaches the nucleus, the electrostatic repulsive force due to the nucleus increases and kinetic energy of the alpha particle goes on converting into the electrostatic potential energy. When whole of the kinetic energy is converted into electrostatic potential energy, the α-particle cannot further move towards the nucleus but returns back on its initial path i.e α-particle is scattered through an angle of 180°. The distance of α-particle from the nucleus in this stage is called as the distance of closest approach and is represented by $r_{0.}$

➢ Let m_α and v_α be the mass and velocity of the α-particle directed towards the centre of the nucleus. Then kinetic energy of the α-particle $K = \left(\frac{1}{2}\right) m_\alpha v^2_\alpha$

Because the positive charge on the nucleus is Ze and that on the α -particle 2e, hence the electrostatic potential energy of the α-particle, when at a distance r_0 from the centre of the nucleus, is given by

$$U = \frac{1}{4\pi \in_0} \cdot \frac{(2e)(Ze)}{r_0}$$

Because at $r = r_0$ kinetic energy of the α-particle appears as its potential energy, hence, K=U

$$\frac{1}{4\pi \in_0} \cdot \frac{(2e)(Ze)}{r_0} = \frac{1}{2} m_\alpha v_\alpha^2$$

$$r_0 = \frac{1}{4\pi \in_0} \frac{4Ze^2}{m_\alpha {v_\alpha}^2}$$

Note:

- When a mono energitic beam of α particles is projected towards a thin metal foil, some of the particles are found to deviate from their original path. This phenomenon is called α ray scattering
- It is caused by coulomb repulsive force between α particles and positive charges in atom.
- The number of α -particles scattered at an angle θ is given by $N = \dfrac{Qntz^2e^4}{(8\pi\varepsilon_0)^2 r^2 E^2 \sin^4\left(\frac{\theta}{2}\right)}$

where

$Q \rightarrow$ Total number of α particles striking the foil

$n \rightarrow$ number of atoms per unit volume of the foil

$r \rightarrow$ distance of screen from the foil

$t \rightarrow$ thickness of the foil

$z \rightarrow$ Atomic number of the foil atoms

$\theta \rightarrow$ angle of scettering

$E \rightarrow$ kinetic energy of α particles

$N \alpha\, t$; $N \alpha\, z^2$; $N\alpha \dfrac{1}{\sin^4\frac{\theta}{2}}$

$N \alpha \dfrac{1}{E^2}$ or $N \alpha \dfrac{1}{\upsilon^4}$

where ϑ is the velocity of α particles falling on the foil.

- **Impact Parameter(b):**The perpendicular distance of the initial velocity vector of the α-particle from centre of the nucleus is called "impact parameter".

$$b = \frac{Ze^2 \cot\left(\frac{\theta}{2}\right)}{4\pi \in_0 \times \frac{1}{2} mv^2}$$

Bohr's model of Hydrogen like atoms :

- Electron can revolve round the nucleus only in certain allowed orbits called stationary orbits and the Coulomb's force of attraction between electron and the positively charged nucleus provides necessary centripetal force.

$$\frac{k(Ze)e}{r^2} = \frac{mV^2}{r}$$

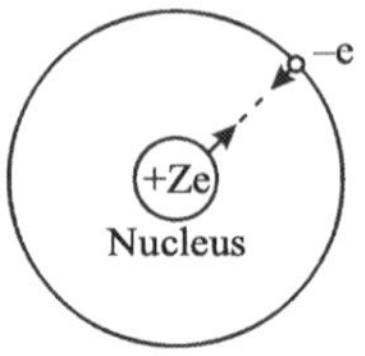

- Suppose m is the mass of electron, V is the velocity and 'r' is the radius of the orbit, then in stationary orbits the angular momentum of the electron is an integral multiple of $\frac{h}{2\pi}$, where h is the Planck's constant. The angular momentum $L = I\omega = mVr = n\frac{h}{2\pi}$ where n is called principal quantum number.
- An electron in a stationary orbit has a definite amount of energy. It posses kinetic energy because of its motion and potential energy on account of the attraction of the nucleus. Each allowed orbit is therefore associated with a certain quantity of energy called the energy of the orbit, which equals the total energy of the electron in it. In these allowed orbits electrons revolve without radiating energy.
- Energy is radiated or absorbed when an electron jumps from one stationary orbit to another stationary orbit. This energy is equal to the energy difference between these two orbits and emitted or absorbed as one quantum of radiation of frequency ν given by Planck's equation $E_2 - E_1 = h\nu = \frac{hc}{\lambda}$. This is called Bohr's frequency condition.
- **Radius of Bohr's orbit :** When mass of the nucleus is large compared to revolving electron, then electron revolves around the nucleus in circular orbit. According to first postulate

$$\frac{k(Ze)e}{r^2} = \frac{mV^2}{r} \left(\text{where } k = \frac{1}{4\pi \epsilon_0}\right) \text{......(1)}$$

According to second postulate

$$mVr = n\frac{h}{2\pi} \text{ where } n = 1,2,3,4\text{............}$$

$$\text{(or) } V = \frac{nh}{2\pi mr} \text{ .... (2)}$$

After solving the equations, radius of the orbit $r = \frac{n^2h^2}{4\pi^2 kZme^2}$

$$\text{For } n^{th} \text{ orbit } r_n = \frac{h^2}{4\pi^2 ke^2}\cdot\left(\frac{n^2}{mZ}\right) \text{ ....(3)}$$

For hydrogen atom Z = 1, radius of the first orbit (n = 1) is given by $r_1 = 0.529 \times 10^{-10} m \simeq 0.53Å$

This value is called as Bohr's radius and the orbit is called Bohr's orbit. In general, the radius of the nth orbit of a hydrogen like atom is given by $r_n = 0.53\left(\frac{n^2}{Z}\right)Å$ where n = 1,2,3,....... (4)

Velocity of the Electron in the orbit :

The velocity of an electron in nth orbit $V_n = \frac{nh}{2\pi mr_n}$ hence

$$V_n = \frac{2\pi ke^2}{h}\cdot\left(\frac{Z}{n}\right)\left(\because r_n = \frac{n^2h^2}{4\pi^2 kZme^2}\right) \text{..(5)}$$

i.e the velocity of electron in any orbit is independent of the mass of electron. The above equation can also be written as

$$\therefore V_n = \left(\frac{c}{137}\right)\cdot\frac{Z}{n} \text{ m/s .... (6)}$$

Where 'c' is the speed of light in vacuum.

Time period of electron in the orbit :

Angular velocity of electron in nth orbit

$\omega_n = \frac{V_n}{r_n} = \frac{\omega_0 Z^2}{n^3}$ where $\omega_0 = \frac{8\pi^3 k^2 e^4 m}{h^3}$ ......(7) is the angular velocity of electron in first Bohr's orbit. The time period of rotation of electron in n^{th} orbit $T=\frac{2\pi}{\omega_n}=\frac{n^3}{2\pi\omega_0 Z^2}$ .... (8) i.e $T\propto\frac{n^3}{Z^2}$.

The time period of rotation increases as n increases and is independent on the mass of the electron.

Energy of the electron in the orbit :

➢ The kinetic energy of the electron revolving round the nucleus in n^{th} orbit is given by

$$K_n=\frac{1}{2}mV^2=\frac{1}{2}m\left[\frac{2\pi ke^2}{h}.\frac{Z}{n}\right]^2$$

$$K_n=\frac{2\pi^2k^2e^4}{h^2}.\left(\frac{mZ^2}{n^2}\right) ....(9);\ K_n\,\alpha\,\frac{mZ^2}{n^2}$$

If the reference level (zero potential energy level) is at infinity then the electrostatic potential energy is given by

$$U_n=-\frac{k(Ze)e}{r_n}=-kZe^2\left[\frac{4\pi^2kmZe^2}{n^2h^2}\right]$$

$$U_n=-\frac{4\pi^2k^2e^4}{h^2}\left(\frac{mZ^2}{n^2}\right) .... (10)$$

Total energy of the electron in n^{th} orbit

$$E_n=K_n+U_n=-\frac{2\pi^2k^2mZ^2e^4}{n^2h^2}$$

$$E_n=-\frac{2\pi^2k^2e^4}{h^2}\left(\frac{mZ^2}{n^2}\right) .... (11)$$

The expression of total energy for hydrogen like atom may be simplified as

$$E_n=-13.6\frac{Z^2}{n^2}\quad eV,\ n=1,2,3.....\ (12)$$

Where -13.6 eV is the total energy of the electron in the ground state of an hydrogen atom.

From the equations (9),(10)&(11) it is clear that

PE : K.E : T.E = -2 : 1 : -1

i.e $U=-2K$ and $E=-K$ (or) $E=-K=U/2$

➢ The state $n=1$ is called ground state and $n>1$ states are called excited states. When electron go from lower orbit to higher orbit speed and hence kinetic energy decrease, but both potential energy and total energy increases. $E\alpha\frac{1}{n^2}$ tells us that the energy gap between the two successive levels decreases as the value of n increases. At infinity level the total energy of the atom becomes zero.

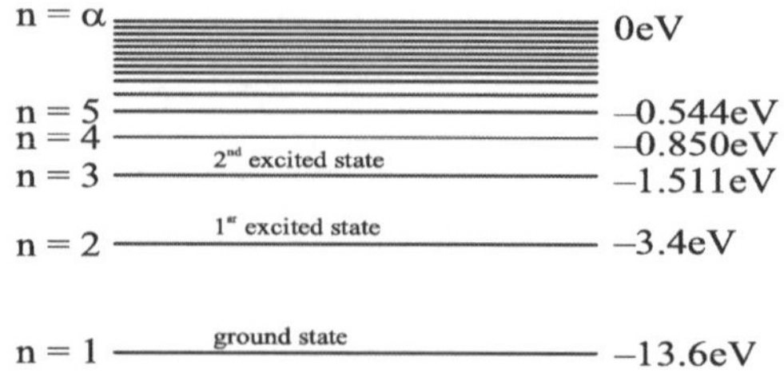

Energy level diagram of hydrogen atom (Z = 1) for normal and excited states as shown the figure.

The energy level diagram of hydrogen like atom with atomic number Z for normal and excited states as shown in Figure.

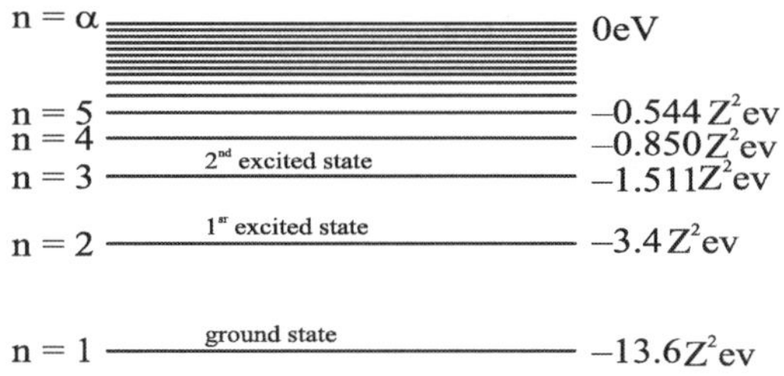

The total energy of the electron is negative implies the atomic electron is bound to the nucleus. To remove the electron from its orbit beyond the attraction of the nucleus , energy must be required.

The minimum energy required to remove an electron from the ground state of an atom is called its ionization energy and it is $13.6\ Z^2$ eV.

- In hydrogen atom the ground state energy of electron is –13.6 eV, so 13.6 eV is the ionization energy of the Hydrogen atom.

Emission of radiation:

- When an electron jumps from higher energy level n_2 to a lower energy level n_1 in stationary atom, the difference in energy is radiated as a photon whose frequency ν is given by Planck's formula.

$$E_{n_2} - E_{n_1} = h\nu$$

(or) $$h\nu = E_2 - E_1 = 13.6Z^2\left[\frac{1}{n_1^2} - \frac{1}{n_2^2}\right] \text{ e.V}$$

$$\left(\therefore E_n = -\frac{13.6Z^2}{n^2}\text{ e.V}\right) \text{ since } 1\text{eV} = 1.6\times10^{-19}\text{J} \quad \text{hence } h\frac{c}{\lambda} = (12.8\times10^{-18})Z^2\left[\frac{1}{n_1^2} - \frac{1}{n_2^2}\right] \text{ J}$$

(or) wave number $$\bar{\nu} = \frac{1}{\lambda} = RZ^2.\left[\frac{1}{n_1^2} - \frac{1}{n_2^2}\right] \text{ m}^{-1}$$

where R is called for "Rydberg constant", when the nucleus is infinitely massive as compared to the revolving electron. In other words the nucleus is considered to be stationary. The numerical value of R is 1.097×10^7 m^{-1}.

Emission Spectrum of Hydrogen atom :

Electron in hydrogen atom, can be in excited state for very small time of the order of 10^{-8} second. This is because in the presence of conservative force system particles always try to occupy stable equilibrium position and hence minimum potential energy, which is least in ground state. Because of instability, when an electron in excited state makes a transition to lower energy state, a photon is emitted. Collection of such emitted photon frequencies is called an emission spectrum. This is as showing in figure.

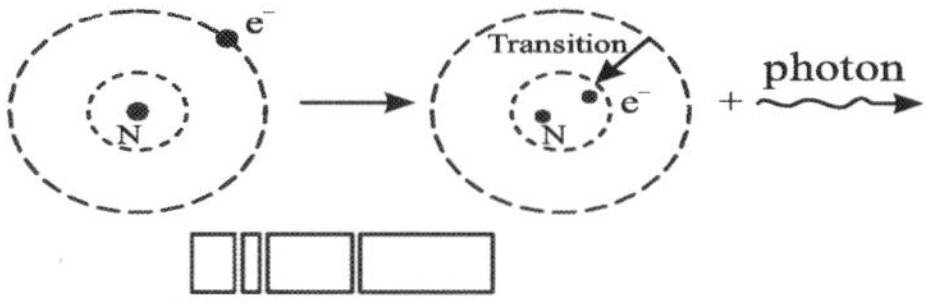

Emission spectrum

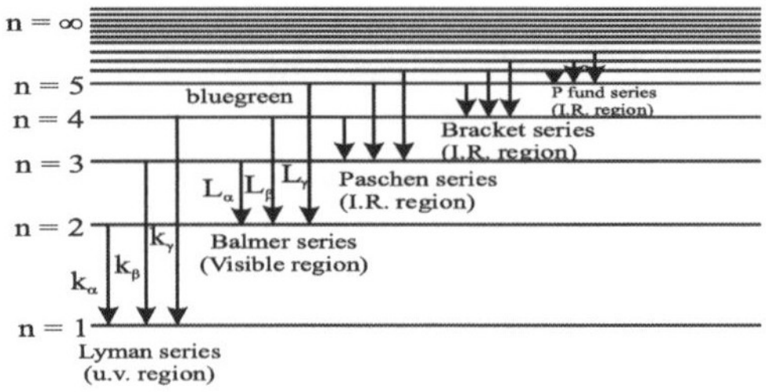

The Spectral Series of Hydrogen Atom as shown in figure, are explained below.

(a) **Lyman Series :** Lines corresponding to transition from outer energy levels $n_2 = 2,3,4,..........\infty$ to first orbit (n_1= 1) constitute Lyman series. The wave numbers of different lines are given by, $\bar{v} = \frac{1}{\lambda} = R\left[\frac{1}{1^2} - \frac{1}{n_2^{\,2}}\right]$

➢ Line corresponding to transition from $n_2 = 2$ to $n_1 = 1$ is first line; its wavelength is maximum.

$\frac{1}{\lambda_{max}} = R\left[\frac{1}{1^2} - \frac{1}{2^2}\right] = 1.1\times10^7\left[\frac{1}{1} - \frac{1}{4}\right] \therefore \lambda_{max} = 1212Å$ Similarly transition from $n_2 = \infty$ to $n_1 = 1$ gives line of minimum wavelength. $\frac{1}{\lambda_{min}} = R\left[\frac{1}{1^2} - \frac{1}{\infty^2}\right] = 1.1\times10^7 \quad \therefore \lambda_{min} = 912Å$

➢ Lyman series lies in ultraviolet region of electro magnetic spectrum.

➢ Lyman series is obtained in emission as well as in absorption spectrum.

(b) **Balmer Series:** Lines corresponding to $n_2 = 3,4,5,........\infty$ to $n_1 = 2$ constitute Balmer series. The wave numbers of different lines are given by, $\bar{v} = \frac{1}{\lambda} = R\left[\frac{1}{2^2} - \frac{1}{n_2^2}\right]$

➢ Line corresponding to transition n_2=3 to n_1=2 is first line, wavelength corresponding to this transition is maximum. Line corresponding to transition $n_2 = \infty$ to $n_1 = 2$ is last line; wavelength of last line is minimum.

$\frac{1}{\lambda_{max}} = R\left[\frac{1}{2^2} - \frac{1}{3^2}\right] \quad \therefore \lambda_{max} = 6568Å \qquad \frac{1}{\lambda_{min}} = R\left[\frac{1}{2^2} - \frac{1}{\infty^2}\right] \therefore \lambda_{min} = 3636Å$

➢ Balmer series lies in the visible region of electromagnetic spectrum. The wavelength of L_α line is 656.8 nm (red). The wavelength of L_β line is 486 nm (blue green). The wavelength of L_γ line is 434 nm (violet). The remaining lines of Balmer series closest to violet light wavelength. The speciality of these lines is that in going from one end to other, the brightness and the separation between them decreases regularly.

➢ This series is obtained only in emission spectrum. Absorption lines corresponding to Balmer series do not exist, except extremely weakly, because very few electrons are normally in the state n = 2 and only a very few atoms are capable of having an electron knocked from the state n = 2 to higher states. Hence photons that correspond to these energies will not be strongly absorbed. In highly excited hydrogen gas there is possibility for detecting absorption at Balmer-line wavelengths.

(c) **Paschen Series:** Lines corresponding to $n_2 = 4,5,6,.......\infty$ to $n_1 = 3$ constitute Paschen series. The wave number of different lines are given by $\bar{v} = R\left[\frac{1}{3^2} - \frac{1}{n_2^2}\right]$

➢ Line corresponding to transition $n_2 = 4$ to $n_1 = 3$ is first line, having maximum wavelength. Line corresponding to transition $n_2 = \infty$ to $n_1 = 3$ is last line, having minimum wavelength

$\frac{1}{\lambda_{max}} = R\left[\frac{1}{3^2} - \frac{1}{4^2}\right] \therefore \lambda_{max} = 18747Å$

$$\frac{1}{\lambda_{min}} = R\left[\frac{1}{3^2} - \frac{1}{\infty}\right] = 1.1 \times 10^7 \times \left[\frac{1}{9} - 0\right]$$

$\therefore \lambda_{min} = 8202Å$

- Paschen series lies in the infrared region of electromagnetic spectrum.
- This series is obtained only in the emission spectrum.

(d) **Bracket Series:** The series corresponds to transitions from $n_2 = 5,6,7......., \infty$ to $n_1 = 4$. The wave number are given by, $\bar{v} = \frac{1}{\lambda} = R\left[\frac{1}{4^2} - \frac{1}{n_2^2}\right]$

- Line corresponding to transition from n_2=5 to n_1=4 has maximum wavelength and $n_2 = \infty$ to $n_1 = 4$ has minimum wavelength.

$$\frac{1}{\lambda_{max}} = R\left[\frac{1}{4^2} - \frac{1}{5^2}\right] \therefore \lambda_{max} = 40477Å$$

$$\frac{1}{\lambda_{min}} = R\left[\frac{1}{4^2} - \frac{1}{\infty^2}\right] \therefore \lambda_{min} = 14572Å$$

- This series lies in the infrared region of electromagnetic spectrum.

(e) **Pfund Series:** This series corresponds to transitions from $n_2 = 6,7,8,....,\infty$ to $n_1 = 5$. The wave numbers are given by $\bar{v} = \frac{1}{\lambda} = R\left[\frac{1}{5^2} - \frac{1}{n_2^2}\right]$

- Line corresponding to transition from $n_2 = 6$ to $n_1 = 5$ has maximum wavelength and $n_2 = \infty$ to $n_1 = 4$ has minimum wavelength. $\frac{1}{\lambda_{max}} = R\left[\frac{1}{5^2} - \frac{1}{6^2}\right]$

$\therefore \lambda_{max} = 74563Å$ $\frac{1}{\lambda_{min}} = R\left[\frac{1}{5^2} - \frac{1}{\infty^2}\right]$

$\therefore \lambda_{min} = 22768Å$

- This series lies in infrared region of electromagnetic spectrum.

Note **:** In an atom emission transition may start from any higher energy level and end at any energy level below of it. Hence in emission spectrum the total possible number of emission lines from some excited state n_2 to another energy state $n_1 (< n_2)$ is $\frac{(n_2 - n_1)(n_2 - n_1 + 1)}{2}$

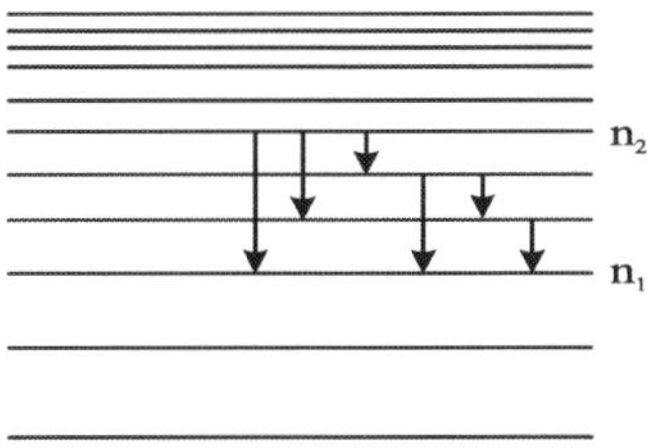

Note 1: for $n_2 = 4$, and $n_1 = 1$, the number of possible lines are 6.

Note 2 : If ΔE is the energy difference between two given energy states, then due to transition between these two states wavelength of emitted photon is $\lambda(Å) = \frac{12400}{\Delta E(eV)}$

Limitation of Bohr's model :

Despite its considerable achievements, the Bohr's model has certain short coming.

- It could not interpret the details of optical spectra of atoms containing more than one electron.
- It involves the concept of orbit which could not be checked experimentally
- It could be successfully applied only to single-electron atoms (e.g., H, He^{+} , Li^{2+}, etc.)
- Bohr's model could not explain the binding of atoms into molecules.
- No justification was given for the "principle of quantization of angular momentum".
- Bohr's model could not explain the reason why atoms should combine to form chemical bonds and why do the molecules become more stable on such combinations.
- Bohr had assumed that an electron in the atom is located at definite distance from the nucleus and is revolving with a definite velocity around it. This is against the Heisenberg uncertainty principle. With the advancements in quantum mechanics, it be came clear that there are no well defined orbits; rather there are clouds of negative charge.

S.No	Name of the series	Final State (n_1)	Initial State (n_2)	Formula	Series limit	Maximum wavelength	Region
1.	Lyman	$n_1 = 1$	$2, 3, 4, ...\infty$	$\frac{1}{\lambda} = R\left(\frac{1}{1^2} - \frac{1}{n_2^2}\right)$	$\lambda = \frac{1}{R} = 911 A^\circ$	$\lambda = \frac{4}{3R}$	UV
2.	Balmer	$n_1 = 2$	$3, 4, 5....\infty$	$\frac{1}{\lambda} = R\left(\frac{1}{2^2} - \frac{1}{n_2^2}\right)$	$\lambda = \frac{4}{R}$	$\lambda = \frac{36}{5R}$	Visible
3.	Paschen	$n_1 = 3$	$4, 5, 6....\infty$	$\frac{1}{\lambda} = R\left(\frac{1}{3^2} - \frac{1}{n_2^2}\right)$	$\lambda = \frac{9}{R}$	$\lambda = \frac{144}{7R}$	Near IR
4.	Brackett	$n_1 = 4$	$5, 6, 7....\infty$	$\frac{1}{\lambda} = R\left(\frac{1}{4^2} - \frac{1}{n_2^2}\right)$	$\lambda = \frac{16}{R}$	$\lambda = \frac{400}{9R}$	Middle IR
5.	Pfund	$n_1 = 5$	$6, 7, 8....\infty$	$\frac{1}{\lambda} = R\left(\frac{1}{5^2} - \frac{1}{n_2^2}\right)$	$\lambda = \frac{25}{R}$	$\lambda = \frac{9000}{11R}$	Far IR

W.E-15: ***The electron in a hydrogen atom makes a transition $n_1 \to n_2$ where n_1 and n_2 are the principal quantum numbers of the two states. Assume the Bohr model to be valid. The time period of the electron in the initial state is eight times that in the final state. What are the possible values of n_1 and n_2?***

Sol. Since, $T \propto n^3$; $\frac{T_1}{T_2} = \frac{n_1^3}{n_2^3}$, As $T_1 = 8T_2$, $\left(\frac{n_1}{n_2}\right)^3 = 8$ (or) $n_1 = 2n_2$. Thus the possible values of n_1 and n_2 are $n_1 = 2, n_2 = 1, n_1 = 4, n_2 = 2, n_1 = 6, n_2 = 3$; and so on.

W.E-16: ***Find the kinetic energy, potential energy and total energy in first and second orbit of hydrogen atom if potential energy in first orbit is taken to be zero.***

Sol. $E_1 = -13.60eV$; $K_1 = -E_1 = 13.60eV$

$U_1 = 2E_1 = -27.20eV$ $E_2 = -3.40eV$ $K_2 = 3.40eV$ and $U_2 = -6.80eV$

Now, $U_1 = 0$, i.e., potential energy has been increased by 27.20eV. So, we will increase U and E in all energy states by 27.20eV while kinetic energy will remain unchanged.

Hence	K(eV),	U(eV),	E(eV)
First orbit are	13.6,	0,	13.6
in Second orbit	3.40,	20.40,	23.80

W.E-17: ***A small particle of mass m moves in such a way that the potential energy $U = ar^2$ where a is a constant and r is the distance of the particle from the origin. Assuming Bohr's model of quantization of angular momentum and circular orbits, find the radius of nth allowed orbit.***

Sol. The force at a distance r is, $F = -\frac{dU}{dr} = -2ar$ Suppose r be the radius of nth orbit. Then the necessary centripetal force is provided by the above force. Thus, $\frac{mv^2}{r} = 2ar$ -------- (i)

Further, the quantization of angular momentum gives, $mvr = \frac{nh}{2\pi}$ -------- (ii)

Solving Eqs. (i) and (ii) for r, we get $r = \left(\frac{n^2h^2}{8am\pi^2}\right)^{1/4}$

W.E-18: ***Consider a hydrogen-like atom whose energy in n^{th} excited state is given by $E_n = -\frac{13.6Z^2}{n^2}$ when this excited atom makes transition from excited state to ground state most energetic photons have energy $E_{max} = 52.224eV$ and least energetic photons have energy $E_{min} = 1.224eV$. Find the atomic number of atom and the state of excitation.***

Sol. Maximum energy is liberated for transition $E_n \to 1$ and minimum energy for $E_n \to E_{n-1}$

Hence, $\frac{E_1}{n^2} - E_1 = 52.224eV$ ... (1)

and $\frac{E_1}{n^2} - \frac{E_1}{(n-1)^2} = 1.224\,eV$ ...... (2)

Solving above equations simultaneously, we get

$E_1 = -54.4eV$ and n = 5 Now $E_1 = -\frac{13.6Z^2}{1^2} = -54.4eV$. Hence, Z = 2 i.e, gas is helium originally excited to n = 5 energy state.

W.E-19: ***A hydrogen-like atom (atomic number Z) is in a higher excited state of quantum number n. This excited atom can make a transition to the first excited state by successively emitting two photons of energies 10.20 eV and 17.00 eV respectively. Alternatively the atom from the same excited state can make a transition to the second excited state by successively emitting two photons of energies 4.25 eV and 5.95 eV respectively. Determine the values of n and Z (ionization energy of hydrogen atom = 13.6 eV)***

Sol. The electronic transitions in a hydrogen-like atom from a state n_2 to a lower state n_1 are given by

$\Delta E = 13.6Z^2\left(\frac{1}{n_1^2} - \frac{1}{n_2^2}\right)$. For the transition from a higher state n to the first excited state n_1 = 2, the total energy released is (10.2 + 17.0) eV or 27.2eV. Thus $\Delta E = 27.2$ eV, $n_1 = 2$ and $n_2 = n$.

We have $27.2 = 13.6Z^2\left[\frac{1}{4} - \frac{1}{n^2}\right]$ ...... (1)

For the eventual transition to the second excited state $n_1 = 3$, the total energy released is (4.25 + 5.95) eV or 10.2eV.

Thus $10.2 = 13.6Z^2\left[\frac{1}{9} - \frac{1}{n^2}\right]$ .....(2)

Dividing the Eq. (1) by Eq. (2) we get $\frac{27.2}{10.2} = \frac{9n^2 - 36}{4n^2 - 36}$.

Solving we get $n^2 = 36$ or n = 6

Substituting n = 6 in any one of the above equations, we obtain $Z^2 = 9$ (or) Z = 3, Thus n=6 and Z=3.

W.E-20: ***A doubly ionized lithium atom is hydrogen like with atomic number Z = 3. Find the wavelength of the radiation required to excite the electron in Li^{2+} from the first to the third Bohr orbit. Given the ionization energy of hydrogen atom as 13.6 eV.***

Sol. The energy of nth orbit of a hydrogen-like atom is given as $E_n = -\frac{13.6Z^2}{n^2}$ Thus for Li^{2+} atom, as Z = 3, the electron energies of the first and third Bohr orbits are For n = 1, $E_1 = -122.4eV$, for n = 3, $E_3 = -13.6eV$. Thus the energy required to transfer an electron from E_1 level to E_3 level is, $E = E_3 - E_1 = -13.6 - (-122.4) = 108.8eV$. Therefore, the radiation needed to cause this transition should have photons of this energy. hv = 108.8 eV. The wavelength of this radiation is or $\lambda = \frac{hc}{108.8eV} =$ 114.25 Å

W.E-21: ***A hydrogen atom in a state of binding energy 0.85 eV makes a transition to a state of excitation energy of 10.2 eV.***
(i) What is the initial state of hydrogen atom?
(ii) What is the final state of hydrogen atom ?
(iii) What is the wavelength of the photon emitted ?

Sol. (i) Let n_1 be initial state of electron. Then $E_1 = -\frac{13.6}{n_1^2}eV$ Here E_1 = - 0.85 eV, therefore $-0.85 = -\frac{13.6}{n_1^2}$ or $n_1 = 4$

(ii) Let n_2 be the final excitation state of the electron. Since excitation energy is always measured with respect to the ground state, therefore

$\Delta E = 13.6\left[1 - \frac{1}{n_2^2}\right]$ here $\Delta E = 10.2eV$, therefore,

$10.2 = 13.6\left[1 - \frac{1}{n_2^2}\right]$ or $n_2 = 2$ Thus, the electron jumps from $n_1 = 4$ to $n_2 = 2$.

(iii) The wavelength of the photon emitted for a transition between $n_1 = 4$ to $n_2 = 2$, is given by

$$\frac{1}{\lambda} = R_\infty\left[\frac{1}{n_2^2} - \frac{1}{n_1^2}\right] (\text{or}) \frac{1}{\lambda} = 1.09 \times 10^7\left[\frac{1}{2^2} - \frac{1}{4^2}\right] = 4860\text{Å}.$$

W.E-22: ***A hydrogen atom initially in the ground level absorbs a photon, which excites it to the n = 4 level. Determine the wavelength and frequency of photon. To find the wavelength and frequency of photon use the relation of energy of electron in hydrogen atom is*** $E_n = -\frac{13.6}{n^2}eV$.

Sol. For ground state $n_1 = 1$ to $n_2 = 4$.
Energy absorbed by photon, $E = E_2 - E_1$

$$= +13.6\left(\frac{1}{n_1^2} - \frac{1}{n_2^2}\right) \times 1.6 \times 10^{-19} \text{ J}$$

$$= 13.6\left(\frac{1}{1} - \frac{1}{4^2}\right) \times 1.6 \times 10^{-19}$$

$= 13.6 \times 1.6 \times 10^{-19}\left(\frac{15}{16}\right) = 20.4 \times 10^{-19}$.

or $E = h\nu = 20.4 \times 10^{-19}$

Frequency $\nu = \frac{20.4 \times 10^{-19}}{h} = \frac{20.4 \times 10^{-19}}{6.63 \times 10^{-34}}$

$= 3.076 \times 10^{15} = 3.1 \times 10^{15}$ Hz.

Wavelength of photon $\lambda = \frac{c}{\nu} = \frac{3 \times 10^8}{3.076 \times 10^{15}} = 9.74 \times 10^{-8}$ m. Thus, the wavelength is 9.7×10^{-8} m and frequency is 3.1×10^{15} Hz.

W.E-23: ***(a) Using the Bohr's model calculate the speed of the electron in a hydrogen atom in the n = 1, 2 and 3 levels.***

(b) Calculate the orbital period in each of these levels.

Sol. **(a)** Speed of the electron in Bohr's nth orbit

$$\nu = \frac{c}{v}\alpha \text{ where, } \alpha = \frac{2\pi K e^2}{ch}$$

$$\alpha = 0.0073 \qquad \therefore \quad v = \frac{c}{n} \times 0.0073$$

For $n = 1$, $\quad v_1 = \frac{c}{1} \times 0.0073$

$= 3 \times 10^8 \times 0.0073 = 2.19 \times 10^6$ m/s

For $n = 2$ $\quad v_2 = \frac{c}{2} \times 0.0073$

$= \frac{3 \times 10^8 \times 0.0073}{2} = 1.095 \times 10^6$ m/s

For $n = 3$ $\quad v_3 = \frac{c}{3} \times 0.0073$

$= \frac{3 \times 10^8 \times 0.0073}{3} = 7.3 \times 10^5$ m/s.

(b) Orbital period of electron is given by $T = \frac{2\pi r}{v}$

Radius of nth orbit $r_n = \left(\frac{n^2}{m}\right)\left(\frac{h}{2\pi}\right)^2 \frac{4\pi\varepsilon_0}{e^2}$

$$\therefore r_1 = \frac{(1)^2 \times (6.63 \times 10^{-34})^2}{4 \times 9.87 \times (9 \times 10)^9 \times 9 \times 10^{-31} \times 6.6 \times 10^{-19})} = 0.53 \times 10^{-10} \text{ m.}$$

For n = 1 $T_1 = \frac{2\pi r_1}{v_1}$

$= \frac{2\times3.14\times0.53\times10^{-10}}{2.19\times10^{6}} = 1.52 \times 10^{-16}$ s

For n = 2, radius $r_n = n^2 r_1$
$r_2 = 2^2 .r_1 = 4\times0.53\times10^{-10}$ m and

velocity $v_n = \frac{v_1}{n}$

$\therefore\ v_2 = \frac{v_1}{2} = \frac{2.19\times10^{6}}{2}$

Time period $T_2 = \frac{2\times3.14\times4\times0.53\times10^{-10}\times2}{2.19\times10^{6}}$

$= 1.216 \times 10^{-15}$ s.
For n = 3, radius $r_3 = 3^2$
$r_1 = 9\ r_1 = 9 \times 0.53 \times 10^{-10}$ m

and velocity $v_3 = \frac{v_1}{3} = \frac{2.19\times10^{6}}{3}$ m/s

Time period $T_3 = \frac{2\pi r_3}{v_3}$

$= \frac{2\times3.14\times9\times0.53\times10^{-10}\times3}{2.19\times10^{6}} = 4.1 \times 10^{-15}$ s.

W.E-24: ***The radius of the innermost electron orbit of a hydrogen atom is 5.3 × 10^{-11} m. What are the radii of the n = 2 and n = 3 orbits?***

Sol. Given, the radius of the innermost electron orbit of a hydrogen $r_1 = 5.3 \times 10^{-11}$ m.
As we know that $r_n = n^2 r_1$
For n = 2, radius $r_2 = 2^2$
$r_1 = 4 \times 5.3 \times 10^{-11} = 2.12 \times 10^{-10}$ m.
For n = 3, radius $r_3 = 3^2$
$r_1 = 9 \times 5.3 \times 10^{-11} = 4.77 \times 10^{-10}$ m.

W.E-25: ***A 12.5 eV electron beam is used to bombard gaseous hydrogen at room temperature. What series of wavelength will be emitted?***

Sol. Energy of electron beam E=12.5 eV $= 12.5 \times 1.6 \times 10^{-10}$ J
Planck's constant h=6.63 × 10^{-34} J-s
Velocity of light c=3 × 10^{8} m/s

Using the relation $E = \frac{hc}{\lambda} = \frac{6.62\times10^{-34}\times3\times10^{8}}{12.5\times1.6\times10^{-19}}$

$= 0.993 \times 10^{-7}\,m = 993 \times 10^{-10}\,m = 993\,A^0$

This wavelength falls in the range of Lyman series ($912\,A^0$ to $1216A^0$)

thus, we conclude that Lyman series of wavelength $993\,A^0$ is emitted.

W.E-26:*In accordance with the Bohr's model, find the quantum number that characterises the earth's revolution around the sun in an orbit of radius 1.5×10^{11} m with orbital speed 3×10^4 m/s. (Mass of earth = 6.0×10^{24} kg).*

A. Given, radius of orbit $r = 1.5 \times 10^{11}\,m$

Orbital speed $v = 3\times 10^4 m/s$;

Mass of earth $M = 6\times 10^{24} kg$

Angulalr momentum, $mvr = \frac{nh}{2\pi}$ or $n = \frac{2\pi vrm}{h}$

[where, n is the quantum number of the orbit]

$$= \frac{2\times 3.14\times 3\times 10^4\times 1.5\times 10^{11}\times 6\times 10^{24}}{6.63\times 10^{-34}}$$

$= 2.57 \times 10^{74}$ or $n = 2.6 \times 10^{74}$. Thus, the quantum number is 2.6×10^{74} which is too large. The electron would jump from n=1 to n=3

$$E_3 = \frac{-13.6}{3^2} = -1.5\,eV.$$

So, they belong to Lyman series.

CONCEPTUAL

THOMSON'S METHOD

1. **If 'C' denotes the velocity of light, velocity of cathode rays is**
   1) equal to C 2) greater than C 3) less than C
   4) either greater or less than C
2. **An electron has no initial velocity in a direction different from that of an electric field. The path of the electron is**
   1) a straight line 2) a circle 3) a parabola 4) an ellipse
3. **Cathode rays enter an electric field normal to the field. Then their path in the electric field is**
   1) a parabola 2) a circle 3) a straight line 4) an ellipse
4. **When the electron in the discharge tube is accelerated to high speeds**
   1) The charge on the electron will decrease 2) The specific charge will decrease
   3) The charge of the electron will increase 4) The value of e/m will increase
5. **The discovery of electron was a consequence of study of the**
   1) discharge of electricity through atmosphere. 2) discharge of electricity through rarefied gases.
   3) photoelectric effect. 4) nuclear fission.
6. **If a proton and an electron are accelerated through the same potential difference**
   1) both the proton and electron have same K.E
   2) both the proton and electron have same momentum
   3) both the proton and electron have same velocity
   4) both the proton and electron have same temperature

7. **A cathode ray tube has a potential difference of V between the cathode and anode. The speed of the cathode rays is given by**

1) $v \propto V$ 2) $v \propto V^{-1}$ 3) $v \propto \sqrt{V}$ 4) $v \propto V^2$

8. **In a region of space, cathode rays move along positive z-axis and a uniform magnetic field is applied along x-axis. If cathode rays pass undeviated, the direction of electric field will be along**

1) Negative x-axis 2) Positive y-axis 3) Negative y-axis 4) Positive z-axis

9. **A cathode ray particle is accelerated from rest through a potential difference of V volt. The speed of the particle is**

1) $\sqrt{\frac{2eV}{m}}$ 2) $\frac{m}{eV}$ 3) $\sqrt{meV}$ 4) $\sqrt{\frac{4eV}{m}}$

10. **In J.J.Thomson's method, electric field of intensity E, magnetic field of induction B and velocity V of the electrons were in mutually perpendicular directions. The condition for velocity is**

1) $V = E / B$ 2) $V = B / E$ 3) $V = B E$ 4) $V = \sqrt{B / E}$

11. **A positive ion of mass M kg and charge e coulomb travels from rest through a p.d. of V volt. The final K.E. is**

1) e V joule 2) MeV joule 3) Me/V joule 4) eV/M joule

12. **The increasing order of specific charge for electron(e), proton(p), neutron(n) and alpha particle(α) is**

1) e, p, n, α 2) n, α, p, e 3) n, p, e, α 4) n, p, α, e

13. **A uniform electric field and a uniform magnetic field are produced pointing in the same direction. An electron is projected with its velocity pointed in the same direction.**

1) the electron will turn to its right
2) the electron will turn to its left
3) the electron velocity will increase in magnitude
4) the electron velocity will decrease in magnitude

14. **When a positively charged particle enters a uniform magnetic field with uniform velocity its trajectory can be**

a) straight line b) a circle c) a helix

1) a only 2) a or b 3) a or c 4) any one of a,b and c

15. **The ratio e/m of electron is independent of**

a) Nature of cathode, anode
b) Nature of gas in discharge tube
c) Applied voltage
d) Size of discharge tube

1) a ,b,c are only correct
2) a,b,d are only correct
3) b,c,d are only correct
4) a,b,c,d are correct

16. **An electron is not deflected on passing through a certain region, because**

1) there is magnetic field in that region and the electron enters into it in any direction.
2) there may be magnetic field but the velocity of electron may be parallel to the direction of magnetic field
3) electron is a chargeless particle
4) there is electric field and the electron enters into it in any direction.

17. **In Thomson's experiment, when the electron strikes the undeflected spot, then it moves with**

1) constant acceleration
2)non uniform velocity
3) constant velocity
4)constant retardation

18. **An electron enters a magnetic field with a speed of 10^8 cm/s. The particle experiences a force due to the magnetic field and the speed of the electron**

1) will decrease 2) will increase
3) will remain constant 4) may increase or decrease

19. **A neutron, a proton, an electron and an alpha particle enter a region of constant magnetic field with equal velocities. The magnetic field is along the inward normal to the plane of the paper. The tracks of the particles are labled in the figure. The electron follows ____ track and alpha particle follows track ______**

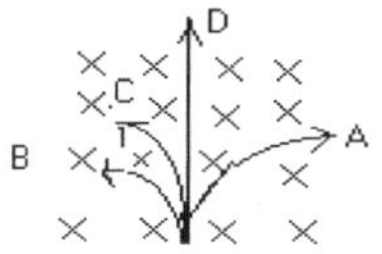

1) C,D 2) B, A 3) A,C 4) A,D

20. **When a charged particle moves through a magnetic field, the quantity which is not affected in the magnetic field is**

1) particle velocity 2)particle acceleration
3) linear momentum of the particle 4) Kinetic energy of the particle

21. **A negatively charged electroscope with zinc disc discharges when irradiated by an ultraviolet lamp. What caused this?**

1) α - particles from the source combine with electrons of the disc
2) electrons escape from the disc when ultraviolet radiation falls on it
3) ultraviolet rays ionize the air surrounding the electroscope
4) the disc becomes hot and thermionic emission takes place

22. **The force felt by an electron on entering into a magnetic field is independent of its**

1) Charge 2) Strength of the field
3) Mass 4) Direction of its velocity

23. **When an electron moves through a magnetic field, its speed will**

1) decrease 2) increase
3) remain the same 4) increase first and then decrease

24. **The direction of deflection of a cathode ray particle passing through a magnetic field can be found by**

1) Fleming's left hand rule 2) Laplace's law
3) Maxwell's cork screw rule 4) Ampere's rule

25. **Which of the following particles cannot be deflected by magnetic field**

1) electrons 2) neutrons 3) α - particles 4) protons

26. **A β - particle enters a magnetic field making an angle of 45^0 with the field lines. The path of the particle is**

1) circular 2) elliptical 3) spiral 4) a straight line

27. **An electron and a proton are injected into a uniform magnetic field at right angle to its direction with the same momentum. Then**

1) electron's path is less curved than proton's path
2) proton's path will be less curved than electron's path
3) the paths of both will be equally curved 4) both the trajectories will be straight

28. A proton and an electron simultaneously enter into a region in which a uniform magnetic field acts normal to the motion of both the particles. The frequency of revolution of

1) the proton is greater than that of the electron
2) the electron is greater than that of the proton
3) the proton is equal to that of the electron
4) both are having same frequency, but revolve in opposite direction

29. You are sitting in a room in which uniform magnetic field is present in vertically downward direction. When an electron is projected in horizontal direction, it will be moving in circular path with constant speed

1) clockwise in vertical plane
2) clockwise in horizontal plane
3) anticlockwise in vertical plane
4) anticlockwise in horizontal plane

30. A charged particle of charge Q and mass m moves with velocity v in a circular path due to transverse magnetic field, B, then its frequency is

1) $\frac{QB}{2\pi m}$
2) $\frac{QvB}{2\pi m}$
3) $\frac{QmvB}{2\pi}$
4) $\frac{vB}{2\pi Qm}$

31. A proton and an electron simultaneously enter a region in which a uniform magnetic field acts normal to the motion of both the particles. The frequency of revolution of

1) proton is greater than that of electron
2) electron is greater than that of proton
3) proton is equal to that of electron
4) proton depends on its velocity

32. Imagine that you are sitting in a room with your back to one wall and that an electron beam traveling horizontally from the back wall towards the front one is deflected towards the right. What is the direction of the magnetic induction field that exists in the room?

1) vertically upwards
2) vertically downwards
3) horizontal and perpendicular to the direction of motion of the electron beam
4) horizontal and parallel to the direction of motion of the electron beam

33. Cathode rays are made to pass between the poles of a magnet as shown in figure. The effect of magnetic field is

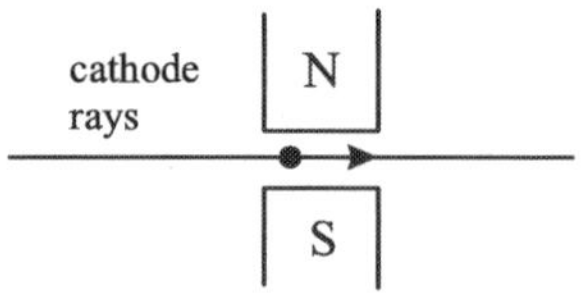

1) to deflect them towards the south pole
2) to deflect them perpendicular to the plane of the paper and towards the observer
3) to deflect them towards the north pole
4) to increase the velocity of the rays

34. In which of the following fields cathode rays show minimum deflection

1) Electric field
2) Magnetic field
3) Plasma field
4) Gravitational field

MILLIKAN'S OIL DROP EXPERIMENT

35. An oil drop of mass m and charge +q is balanced in vacuum by a uniform electric field of intensity E. The direction of this field should be

1) vertically up
2) vertically down
3) horizontal
4) inclined at 45^0 to the horizontal

36. An oil drop of mass m falls through a medium that offers a viscous drag force F. If the velocity of the drop is constant it means that

1) F > mg
2) F < mg
3) F > mg
4) F = mg

37. An oil drop of mass m falls through a viscous medium. The viscous drag force (F) is proportional to the velocity of the drop. At the instant it begins to fall the net force that acts on the oil drop is (neglect buoyancy)

1) mg 2) mg-F 3) F-mg 4) F

38. An oil drop of mass m falls through air with a terminal velocity in the presence of upward electric field of intensity E. The drop carries a charge +q. R is the viscous drag force and F is the buoyancy force. Then for the motion of the drop.

1) there is a net force directed upwards
2) there is a net force directed downwards.
3) mg = Eq + F + R
4) mg + Eq = F + R

39. If g_E and g_M are the accelerations due to gravity on the surfaces of the earth and the moon respectively and if Millikan's oil drop experiment could be performed on the two surfaces, one will find the ratio $\frac{\text{electronic charge on the moon}}{\text{electronic charge on the earth}}$ to be

1) g_E / g_M 2) 1 3) 0 4) g_M / g_E

40. In Millikan's oil drop method, electron charge is accurately measured by conducting an experiment in which

1) Apparent weight of drop is balanced by electric force on it
2) Apparent weight of drop is balanced by magnetic force
3) Electric force of drop is equal to magnetic force
4) Apparent weight of drop is balanced by force due to surface tension

41. In Millikan's experiment, the oil drops acquire charge by

1) induction 2) Friction 3) electric field 4) magnetic field

42. According to Stoke's law, the viscous drag force on an oil drop is proportional to

1) $\sqrt{V}$ 2) V^2 3) V^{-1} 4) V

43. An oil drop of radius r carrying a charge q remains stationary in the presence of electric field of intensity E. If the density of oil is ρ, then

1) $E = \frac{4}{3}\pi r^3 \rho g q$ 2) $E = \frac{4}{3}\pi r^3 \rho g$ 3) $E = \frac{4}{3}\pi r^3 \rho g / q$ 4) $E = \frac{4}{3}\pi r^3 \rho / g^3$

44. The important conclusion given by Millikan's experiment about the charge is

1) Charge is never quantised
2) Charge has no definite value
3) Charge is quantised
4) Charge on an oil drop always increases

ALPHA RAY SCATTERING

45. 'Coulomb's law correctly describes the electric force is that (pick the wrong statement)

1) binds the electrons and neutrons in the nucleus of an atom.
2) binds electrons to nucleus
3) binds atoms together to form molecules
4) binds atoms and molecules to from solids

46. Is the probability of backward scattering (i.e.scattering of α - particals at angles greater than 90^0) predicted by Thomson's model much less, about the same, or much greater than that predicted by Ruthorford's model?

1) much greater 2) much less 3) same 4) slightly greater

47. An electron with KE 6 eV is incident on a hydrogen atom in its ground state.The collision.

1) must be elastic
2) may be partially elastic
3) must be completly inelastic
4) may be partially inelastic.

48. The angular momentum of the α -particles which are scattered through large angles by the heavier nuclei,is conserved because of the

1) nature of repulsive forces
2) conservation of kinetic energy
3) conservation of potential energy
4) there is no external torque

49. The Incorrect statement regarding Rutherford's atomic model is

1) Atom contains nucleus
2) Size of nucleus is very small in comparison to that of atom
3) Nucleus contains about 90 % mass of the atom
4) Electrons revolve round the nucleus with uniform speed

50. α -particles are

1) helium nucleii
2) sodium nucleii
3) ionised nucleii
4) hydrogen nucleii

51. In scattering experiment,the force that scatters particles is

1) nuclear force
2) coulomb force
3) Both(1)and(2)
4) gravitational force

BOHR'S THEORY

52. Bohr's atom model assumes

1) the nucleus is of infinite mass and is at rest
2)electron in a quantized orbit will not radiate energy.
3) mass of the electron remains constant.
4) 1,2,3 are correct

53. The ratio of magnetic dipole moment to angular momentum in a hydrogen like atom is

1) $\frac{e}{m}$
2) $\frac{e}{2m}$
3) $\frac{e}{3m}$
4) $\frac{2e}{m}$

54. The radius of hydrogen atom,when it is in its second excited state, becomes:

1) half
2) double
3) four times
4) nine times

55. For electron moving in n^{th} orbit of the atom, the angular velocity is proportional to:

1) n
2) $\frac{1}{n}$
3) n^3
4) $\frac{1}{n^3}$

56. The order of size of nucleus and Bohr's radius of an atom respectively are:

1) $10^{-14}m, 10^{-10}m$
2) $10^{-10}m, 10^{-8}m$
3) $10^{-20}m, 10^{-16}m$
4) $10^{-8}m, 10^{-6}m$

57. In the lowest energy level of hydrogen atom,electron has the angular momentum:

1) $\frac{\pi}{h}$
2) $\frac{h}{\pi}$
3) $\frac{h}{2\pi}$
4) $\frac{2\pi}{h}$

58. Atomic hydrogen is excited to the nth energy level.The maximum number of spectral lines which it can emit while returning to the ground state ,is:

1) $\frac{1}{2}n(n-1)$
2) $\frac{1}{2}n(n+1)$
3) $n(n+1)$
4) $n(n+1)$

59. As the orbit number increases ,the distance between two consecutive orbits in an atom or ion having single electron:

1) increases
2)decreases
3) remains the same
4) first increases and then becomes constant

60. Ionization energy of a hydrogen -like ion A is greater than that of another hydrogen like ion B. Let $r, u,$ E and L represents the radius of the orbit, speed of the electron, energy of the atom and orbital angular momentum of the electron respectively in ground state, then

1) $r_A > r_B$ 2) $u_A > u_B$ 3) $E_A > E_B$ 4) $L_A > L_B$

61. The classification of discrete energy levels in atoms was first given experimentally by

1) Thomson's experiment 2) Millikan's oil drop experiment.
3) Frank - Hertz experiment 4) Leonard experiment

62. An atomic nucleus contains

1) only electrons 2) only protons 3) only neutrons
4) both protons and neutrons

63. On decreasing principal quantum number n, the values of r and v will

1) decrease 2) increase
3) r will increase but v will decrease. 4) r will decrease but v will increase

64. The possible values of principal quantum number can be

1) 1, 2, 3...8. 2) 0, 1, 2...8. 3) Only zero 4) only odd numbers

65. According to Sommerfeld, an electron revolves round a nucleus in

1) Circular orbits. 2) Elliptical orbits. 3) Hyperbolic orbits 4) parabolic

66. The main defect of Bohr atom model is

1) mixing of classical and quantum theories 2) exclusion of nuclear motion
3) failed to explain the fine structure of spectral lines 4) failed to explain other atoms

67. According to Bohr hypothesis, discrete quantity is

1) Momentum 2) Angular velocity 3) Potential energy 4) Angular momentum

68. If E_n and are the magnitude of total energy and angular momentum of electron in the nth Bohr orbit respectively, then

1) $E_n \propto J_n^2$ 2) $E_n \propto \frac{1}{J_n^2}$ 3) $E_n \propto J_n$ 4) $E_n \propto \frac{1}{J_n}$

69. If the radius of first Bohr orbit is r, then radius of second orbit will be

1) 2r 2) $\frac{r}{2}$ 3) 4r 4) $\sqrt{2r}$

ATOMIC SPECTRA

70. According to quantum mechanics, one of the following is wrong about spin of electron

1) it is related to intrinsic angular momentum 2) spin is rotation of electron about its own axis
3) value of spin quantum number must not be 1
4) +1/2 value of spin quantum number represents up spin

71. According to classical theory, the path of electron in Rutherford atom nuclei is

1) straight line 2) spiral 3) circular 4) parabolic

72. One of the following radiations are not emitted by electron transition in atoms, choose the option

1) ultra violet 2) infrared radiations
3) visible rays 4) α -rays

73. **The energy emitted by a source is in the form of**

1)Photons 2)electrons 3)protons 4)neutrons

74. **The unit of Planck's constant is equivalent to that of**

1) energy 2) angular momentum 3) velocity 4) force

75. **Hydrogen atom will be in its ground state,if its electron is in**

1) any energy level
2) the lowest energy state
3) the highest energy state
4) the intermediate state

76. **The wavelengths involved in the spectrum of deuterium ($_1D^2$) are slightly different from that of hydrogen spectrum, because**

1) the size of the nuclei are different
2) the nucelar forces are different in two cases
3) the masses of the two nuclei are different
4) the attraction between electron and the nucelus is different in the two cases

77. **When an electron jumps from n_1 th orbit to n_2 th orbit then the formula for energy radiated out is**

1) $E_1 - E_2 = h\nu$ 2) $E_2 - E_1 = h\nu$ 3) $E_2 + E_1 = h\nu$ 4) $E_2 - 2E_1 = h\nu$

78. **Which of the following parameters are the same for all hydrogen -like atoms and ions in their ground states?**

1) radius of the orbit
2) speed of the electron
3) energy of the atom
4) orbital angular momentum of the electron

79. **The fine structure of hydrogen spectrum can be explained by**

1) the presence of neutrons in the nucleus
2) the finite size of nucleus.
3) the orbital angular momentum of electrons
4) the spin angular momentum of electrons

80. **The visible region of hydrogen spectrum was first studied by**

1) Lyman 2) Balmer 3) Pfund 4) Brackett

81. **With the increase in quantum number the energy difference between consecutive energy levels**

1) remains constant 2) decreases. 3) increases
4)sometimes increases sometimes decreases

82. **Hydrogen atom does not emit X-rays because:**

1) its energy levels are too close to each other
2) its energy levels are too far apart
3) it has a very small mass
4) it has a single electron

83. **An electron makes transition from n= 3 , n=1 state in a hydrogen atom. The maximum possible number of photons emitted will be**

1) 1 2) 2 3) 3 4) 6

ASSERTION & REASONING

Choose the correct answer :

1) A and R are true and R is the correct explanation of A
2) A and R are true and R is not the correct explanation of A
3) A is true and R is false. **4) A is false but R is true**

84. **Assertion (A) :** Rydberg's constant varies with mass number of a given element.

Reason (R): The reduced mass of the electron is dependent on the mass of the nucleus only

85. **Assertion (A) :** Study of discharge of electricity through gases at low pressures resulted in the discovery of cathode rays.

Reason (R): Cathode rays are deflected by both magnetic and electric fields and the direction of deflection shows that they are negatively charged.

86. **Assertion (A) :** If electrons travel undeflected by the electric field E and magnetic field B then the velocity of electrons is given by $V = \frac{E}{B}$.

Reason (R) : When both electric and magnetic fields are applied simultaneously on electron beam. If force due to electric field is equal and opposite to force due to magnetic field then they travel undeflected.

87. **Assertion (A) :** In the Thomsons e/m experiment of electrons, the specific charge of electrons is independent of nature of the gas in the discharge tube.

Reason (R): Charge of a body is quantized.

88. **Assertion (A):** Between any two given energy levels, the number of absorption transition is always less than number of emission transition.

Reason (R): Absorption transitions start from the lowest energy level only and may end at any higher energy level. But emission transitions may start from any higher energy level and end at any energy level below it.

MATCH THE FOLLOWING

89. **In each situation of Column I, a physical quantity related to orbiting electron in hydrogen-like atom is given.The terms 'Z'and 'n'given in Column II have usual meaning in Bohr's theory. Match the quantities in Column I with the terms they depend on it Column II**

Column-I	**Column-II**
a. Frequency of orbiting electron	p. is directly proportional to n
b. Angular momentum of electron	q. is inversely proportional to n
c. Magnetic moment of electron	r. is inversely proportional to n^3
d.The average current due to orbiting of electron	s.is independent of Z

1)a -p,r ; b -q,s; c -q,s; d -q,r
2) a -q,s ; b -p,r; c-q,s; d -q,r
3) a-q,r ; b-q,s; c-q,s; a-p,r
4) a-p,r; b-q,r; c-q,s; d-q,s

90. **Column-I**

a. Radius of orbit is related with atomic number(Z)

b. Current associated due to orbital motion of electron with atomic number(Z)

c. Magnetic field at the center due to orbital motion of electron related with Z

d. Velocity of an electron related with atomic number(Z)

Column-II

p. is proportional to Z
q. is inversely proportion to Z
r. is proportional to Z^2
s. is proportional to Z^3

1) a-q b-r c-r d-p 2) a-q b-s c-r d-p 3)a-p b-r c-s d-q 4) a-p b-s c-s d-q

91. **The spectral lines of hydrogen-like atom fall within the wavelength range from 950A^0 to 1350A^0 Then, match the following.**

Column-I

a. If it is atomic hydrogen atom and energy $E-=0.85\ eV$

b. If it is atomic hydrogen atom and energy atom and energy $E-=3.4\ eV$

c. If it is double ionized lithium atom, then

d. If it is singly ionized helium atom, then

Column-II

p. λ =1212A^0 and it corresponds to transition from 2 to 1

q. λ =134A^0 and it corresponds to transition from 2 to 1

r. λ =303A^0 and it corresponds to transition from 2 to 1

s. λ =970A^0 and it corresponds to transition from 4 to 1

1) a-s b-p c-q d-r 2) a-s b-q c-p d-r 3)a-p b-q c-r d-s 4) a-q b-r c-s d-p

92. Excitation energy of hydrogen atom is 13.6 eV. Match the following.

Column-I	Column-II
a. Energy of second excited state of hydrogen	p. – 3.4 eV
b. Energy of fourth state	q – 13.6eV of He^+
c. Energy of first excited	r. – 1.5 eV state of Li^{+2}
d. Energy of third excited state of Be^{+3}	s.None

1) a-r b-p c-s d-q 2)a-r b-s c-p d-q 3)a-s b-p c-q d-r 4)a-q b-p c-s d-r

93. Match the column–I with column–II

Column – I	Column – II
(a) Radius of orbit depends on principal quantum number as	(p) Increase
(b) Due to orbital motion of electrons, magnetic field arises at the centre of nucleus is proportional to principal quantum number as	(q) Decrease
(c) If electron is going from lower energy level to higher energy level than velocity of electron will	(r) Is proportional to $\frac{1}{n^2}$
(d) If electron is going from lower energy level to higher energy level than total energy of electron will	(s) Is proportional to n^2
	(t) Is proportional to $\frac{1}{n^5}$

1) a-s b-t c-q d-p 2) a-t b-s c-p d-q 3)a-p b-q c-t d-s 4) a-p b-q c-s d-t

94. Match the column–I with column–II

A certain amount of He+ ions are subjected to excitation and subsequently the emission spectrum for the same is observed. Column – I shows four different series to which the spectral lines belong to and column – II shows five different energies of the photons emitted during the process

Column – I	Column – II
(a) Lyman series	(p) 1.33 eV
(b) Balmer series	(q) 12.09 eV
(c) Paschen series	(r) 51 eV
(d) pfund sereis	(s) 5.5 eV
	(t) 3.87 eV

1) a-r b-q c-s,t d-p 2) a-q b-r c-p,t d-s 3)a-r b-p c-q d-t,s 4) a-s b-q c-p d-t

CONCEPTUAL - KEY

1) 3	2) 1	3) 1	4) 2	5) 2	6) 1	7) 3	8) 3	9) 1	10) 1
11) 1	12) 2	13) 4	14) 4	15) 4	16) 2	17) 3	18) 3	19) 3	20) 4
21) 2	22) 3	23) 3	24) 1	25) 2	26) 3	27) 3	28) 2	29) 2	30) 1
31) 2	32) 2	33) 2	34) 4	35) 1	36) 2	37) 1	38) 3	39) 2	40) 1
41) 2	42) 4	43) 3	44) 3	45) 1	46) 2	47) 1	48) 4	49) 3	50) 1
51) 2	52) 4	53) 2	54) 4	55) 4	56) 1	57) 3	58) 1	59) 1	60) 2
61) 3	62) 4	63) 4	64) 1	65) 2	66) 1	67) 4	68) 2	69) 3	70) 2
71) 2	72) 4	73) 1	74) 2	75) 2	76) 3	77) 1	78) 4	79) 4	80) 2
81) 2	82) 1	83) 3	84) 4	85) 2	86) 1	87) 2	88) 1	89) 1	90) 1
91) 1	92) 1	93) 1	94) 1						

LEVEL - I -1

J.J.THOMSONS METHOD

1. **An electron passes undeflected through perpendicular electric and magnetic fields of intensity 3.4×10^3 V/m and 2×10^{-3} Wb/m^2 respectively. Then its velocity in m/s is**
   1) 1.7×10^6 2) 6.8×10^6 3) 6.8 4) 1.7×10^8 m/s

2. **The ratio of specific charges of an electron to that of a hydrogen ion is**
   1) 2:1 2) 1:1 3) 1: 1840 4) 1840:1

3. **An α - particle and a proton are subjected to the same electric field, then the ratio of the forces acting on them is**
   1) 2: 3 2) 1: 2 3) 3 :2 4) 2 :1

4. **An electron is accelerated in an electric field of 40 V cm^{-1}. If e/m of electron is 1.76×10^{11} Ckg^{-1}, the acceleration is**
   1) 14.0×10^{14}ms^{-2} 2) 14.0×10^{10}ms^{-2} 3) 7.0×10^{10}ms^{-2} 4) 7.04×10^{14}ms^{-2}

5. **An electron beam moving with a speed of 2.5×10^7 ms^{-1} enters into the magnetic field directed perpendicular to its direction of motion. The magnetic induction of the field is 4×10^{-3} wb/m^2. The intensity of the electric field applied so that the electron is undeflected due to the magnetic field is.**
   1) 10^4N/C 2) 10^5N/C 3) 10^7N/C 4) 10^3N/C

6. **A particle carrying a charge moves perpendicular to a uniform magnetic field of induction B with a momentum p, then the radius of the circular path is**
   1) Be/p 2) pe/B 3) p/Be 4) Bep

MILLIKAN OIL DROP EXPERIMENT

7. **A water drop of mass 3.2×10^{-18} kg and carrying a charge of $1.6\times10^{-19}C$ is suspended stationary between two plates of an electric field. Given g=10 m/s^2, the intensity of the electric field required is**
   1) 2 V/m 2) 200 V/m 3) 20 V/m 4) 2000 V/m

8. **In a Millikan's oil drop experiment, an oil drop of mass 0.64×10^{-14} kg, carrying a charge 1.6×10^{-19} C remains stationary between two plates seperated by a distance of 5 mm. Given g=9.8 m/s^2; the voltage that must be applied between the plates being**
   1) 980 V 2) 1960 V 3) 3920V 4) 2880V

ALPHA RAY SCATTERING

9. α – particles are projected towards the nuclei of the different metals, with the same kinetic energy. The distance of closest approach is minimum for

1) Cu(Z=29) 2) Ag(Z=47) 3) Au(Z=79) 4) Pd(Z=46)

10. In Rutherford experments on α – ray scattering the number of particles scattered at 90^0 be 28 per minute. Then the number of particles scattered per minute by the same foil but at 60^0 are

1) 56 2) 112 3) 60 4) 120

11. For a given impact parameter (b), if the energy increase then the scattering angle (θ) will

1) Decrease 2) increase 3) become zero 4) become

BOHR'S MODEL OF ATOM

12. Find the frequency of revolution of the electron in the first stationary orbit of H-atom

1) $6\times10^{14}\,Hz$ 2) $6.6\times10^{10}\,Hz$ 3) $6.6\times10^{-10}\,Hz$ 4) $6.6\times10^{15}\,Hz$

13. Let the potential energy of a hydrogen atom in the ground state be zero. Then its energy in the first excited state will be

1) 10.2eV 2) 13.6eV 3) 23.8eV 4) 27.2eV

14. According to bohr model, the diameter of first orbit of hydrogen atom will be

1) $1.A^0$ 2) $0.529A^0$ 3) $2.25A^0$ 4) $0.725A^0$

15. The angular momentum of electron is J.its magnetic moment will be

1) $\frac{mJ}{2e}$ 2) $\frac{eJ}{2m}$ 3) $\frac{2m}{eJ}$ 4) $\frac{emJ}{2}$

16. The radius of shortest orbit in one electron system is 18 pm.It may be.

1) ${}_1^1H$ 2) ${}_1^2H$ 3) He^+ 4) Li^+

17. In the Bohr model of a hydrogen atom, the centripetal force is furnished by the coulomb attraction between the proton and the electron. If a_0 is the radius of the ground state orbit, m is the mass, e is the charge on the elctron and ε_0 is the vacuum permittivity, the speed of the electron is: [CBSE 1998]

1) Zero 2) $\frac{e}{\sqrt{\varepsilon_0 a_0 m}}$ 3) $\frac{e}{\sqrt{4\pi\varepsilon_0 a_0 m}}$ 4) $\frac{\sqrt{4\pi\varepsilon_0 a_0 m}}{e}$

18. The energy necessary to remove the electron from n=10 state in hydrogen atom will be

1)1.36 eV 2)0.0136 eV 3)13.6 eV 4) 0.136 eV

19. The ratio of energies of first two excited states hydrogen atom is

1) 3/1 2) 1/4 3) 4/9 4) 9/4

ATOMIC SPECTRA

20. The number of different wavelengths may be observed in the spectrum from a hydrogen sample if the atoms are excited to third excited state is

1) 3 2)4 3)5 4)6

21. The ratio of the frequencies of the long wavelength limits of the Balmer and Lyman series of hydrogen is

1)27:5 2)5:27 3)4:1 4)1:4

22. **When an electron jumps from higher orbit to the second orbit in hydrogen ,the radiation emitted out will be in** $(R = 1.09\times10^{7} m^{-1})$:

1)ultraviolet 2) visible region 3)infrared region 4)X-ray region

23. **The energy required to seperate a hydrogen atom into a proton and an electron is 13.6eV. Then the velocity of electron in a hydrogen atom is**

1) $2.2\times10^{4}\,\mathrm{m/s}$ 2) $2.2\times10^{2}\,\mathrm{m/s}$ 3) $2.2\times10^{6}\,\mathrm{m/s}$ 4) $2.2\times10^{10}\,\mathrm{m/s}$

LEVEL-I -1 - KEY

1) 1	2) 4	3) 4	4) 4	5) 2	6) 3	7) 2	8) 2	9) 1	10) 2
11) 1	12) 4	13) 3	14) 1	15) 2	16) 4	17) 3	18) 4	19) 4	20) 4
21) 1	22) 2	23) 3							

LEVEL-I -1 - HINTS

1. $V = \frac{E}{B}$

2. $\frac{specific\ charge\ of\ electron}{specific\ charge\ of\ hydrogen\ ion} = \frac{q_1}{q_2}\times\frac{m_2}{m_1}$

3. F = Ee

4. $a = \frac{Ee}{m}$

5. $V = \frac{E}{B}$

6. $P = mv, r = \frac{mv}{Be}$

7. mg = qE

8. $Eq = mg \Rightarrow \left(\frac{V}{d}\right)q = mg$

9. $r = \frac{1}{4\pi\varepsilon_0}\frac{q_1q_2}{(KE)} \Rightarrow r\alpha q_2$

10. No. of particles scattered at angle θ is $N\alpha\frac{1}{\sin^4\left(\frac{\theta}{2}\right)}$

12. $T = \frac{2\pi r}{V}$ where $V = \frac{3\times10^{8}}{137}$; $r = 0.53\times10^{-10} m$

13. $P.E_2 = -\frac{27.2}{4} + 27.2 = 20.4$ eV

$K.E_1 = 13.6 eV, \quad K.E_2 = 3.4 eV$

$\therefore T.e_2 = 20.4 + 3.4 \quad = 23.8 eV$

14. Diameter $= 2r_0 = 2 \times 0.529 = 1.058 \overset{0}{A}$

15. $\frac{\text{magnetic moment}}{\text{angular moment}} = \frac{e}{2m}$

16. $r_n = \frac{n^2 r_B}{Z} \Rightarrow \frac{0.53A^\circ}{Z} = 0.18A^\circ \Rightarrow Z = 3$

17. $\frac{1}{4\pi\varepsilon_0}\frac{e^2}{a_0^2} = \frac{mv^2}{a_0}$

18. Energy requried $= E_\alpha - E_{10} = \frac{13.6}{10^2} = 0.136eV$

19. $E_n = -13.6Z^2 / n^2$

20. N$=\frac{n(n-1)}{2}$: where n=4

21. $\lambda_B = \frac{36}{5R} : \lambda_2 = \frac{4}{3R}$, $\frac{\lambda_B}{\lambda_c} = \frac{27}{5}$:

22. *For* $\lambda_{\max}$:
$n_1 = 2 \ \& n_2 = 3;$

23. Velocity of revolving electron is given as

$$\frac{mv^2}{r} = \frac{1}{4\pi\varepsilon_0}\frac{e^2}{r^2}, \qquad \therefore v = \frac{e}{\sqrt{4\pi\varepsilon_0 mr}}$$

LEVEL - I -2

J.J.THOMSONS METHOD

1. **A cathode emits 1.8 x 10^{17} electrons/s and all the electrons reach the anode when it is given a positive potential of 400 V. Given e=1.6 x 10^{-19}C, the maximum anode current is**
   1) 2.88 mA 2) 28.8 mA 3) 7.2 mA 4) 6.4 mA
2. **An electron of mass 9 x 10^{-31} kg moves with a speed of 10^7 m/s. It acquires a K.E. of**
   1) 562.50 eV 2) 1125 eV 3) 1250 eV 4) 281.25 eV
3. **Two electron beams having velocities in the ratio 1:2 are subjected to the same transverse magnetic field. The ratio of the deflections is**
   1) 1:2 2) 2:1 3) 4:1 4) 1:4
4. **The velocity of electrons accelerated by potential difference of 1×10^4 V (The charge of the elctron is 1.6×10^{-19} C and mass is 9.11×10^{-31}kg) is**
   1) 5.93×10^7 ms^{-1} 2) 2.94×10^7 ms^{-1} 3) 6.87×10^7 ms^{-1} 4) 3.98×10^7 ms^{-1}
5. **Cathode ray tube is operating at 5 KV. Then the K.E. acquired by the electrons is**
   1) 5 eV 2) 5 MeV 3) 5 KeV 4) 5 V

6. A steam of similar negatively charged particals enters an electrical field normal to the electric lines of force with a velocity of $3\times10^{7}\,m/s$. The electric intensity is 1800 V/m. While through a distance of 100 m, the electron beam is deflected by 2mm. Then the specific charge value of in c/kg is

1) 2×10^{10} 2) 2×10^{7} 3) 2×10^{11} 4) 2×10^{4}

MILLIKAN OIL DROP EXPERIMENT

7. An oil drop having a mass 4.8×10^{-10}g. and charge 2.4×10^{-18}C stands still between two charged horizantal plates seperated by a distance of 1cm. If now the polarity of the plates is changed the instantaneous acceleration of the drop is (in ms^{-2}) ($g=10ms^{-2}$)

1) 5 2)10 3) 20 4) 40

8. In millkan's oil drop experiment a charged drop of mass 1.8 X 10^{-14}kg is stationary between the plates. The distance between the plates is 0.9cm and potential difference is 2000 V. The number of electrons in the drop is ($g = 10ms^{-2}$)

1) 2 2) 4 3) 5 4) 1

ALPHA RAY SCATTERING

9. An $\alpha-$particle accelerated through V volt is fired towards a nucleus.Its distance of closest approach is r.If a proton is accelerated through the same potential and fired towards the same nucleus,the distance of closest approach of proton will be:

1) r 2) 2r 3) r/2 4) r/4

10. In $\alpha-$ray scattering, the scattering angle (θ) for impact parameter (b) to become zero is

1) 0° 2) 90° 3) 180° 4) 45°

11. The impact parameter at which the scattering angle is 90°, $z=79$ and initial energy 10 MeV is

1) $1.137\times10^{-14}m$ 2) $1.137\times10^{-16}m$ 3) $2.24\times10^{-17}m$ 4) $2.24\times10^{-18}m$

BOHR'S MODEL OF ATOM

12. When a hydrogen atom emits a photon of energy 12.1 eV,its orbital angular momentum changes by

1) $1.05\times10^{-34}J-s$ 2) $2.11\times10^{-34}J-s$ 3) $3.16\times10^{-34}J-s$ 4) $4.22\times10^{-34}J-s$

13. An electron is in an excited state in a hydrogen like atom. It has a total energy -3.4 eV. The kinetic energy of the electron is E and its de broglie wavelength is λ.

1) $E=6.8eV.\lambda=6.6\times10^{-10}m$ 2) $E=3.4eV,\lambda=6.6\times10^{-10}m$

3) $E=3.4eV,\lambda=6.6\times10^{-11}m$ 4) $E=6.8eV,\lambda=6.6\times10^{-11}m$

14. The ionisation potential of hydrogen atom is

1) $12.97V$ 2) $10.2V$ 3) $13.6V$ 4) $27.2V$

15. The energy of electron in an excited hydrogen atom is -3.4 eV. Its angular momentum according to bohr's theory will be

1) $\frac{h}{\pi}$ 2) $\frac{h}{2\pi}$ 3) $\frac{3h}{2\pi}$ 4) $\frac{3}{2\pi h}$

16. **The velocity of an electron in its fifth orbit,if the velocity of an electron in the second orbit of sodium atom (atomicnumber=11)is v,will be:**

1) v 2) $\frac{22v}{5}$ 3) $\frac{5}{2}v$ 4) $\frac{2}{5}v$

17. **The ratio of the kinetic energy and the potential energy of electron in the hydrogen atom will be**

1)1:2 2) -1:2 3) 2:1 4) -2:1

18. **If the potential energy of a H-atom in the ground state be zero then its potential energy in the first excited state will be**

1) 10.2 ev 2)20.4 eV 3) 23.8 eV 4)27.2 eV

ATOMIC SPECTRA

19. **The value of wavelength radiation emitted due to transition of electrons from $n = 4$ to $n = 2$ state in hydrogen atom will be**

1) $\frac{5R}{36}$ 2) $\frac{16}{3R}$ 3) $\frac{36}{5R}$ 4) $\frac{3R}{16}$

20. **The maximum number of photons emitted by an H-atom,if atom is excited to states with principal quantum number four is:**

1)4 2) 6 3) 2 4) 1

21. **For certain atom,there are energy levels A,B,C corresponding to energy values $E_A < E_B < E_c$.**

Choose the correct option if $\lambda_1, \lambda_2, \lambda_3$ are the wavelength of radiations corresponding to the transition from C to B,B to A and C to A respectively:

1) $\lambda_3 = \lambda_1 + \lambda_2$ 2) $\lambda_3 = \frac{\lambda_1\lambda_2}{\lambda_1 + \lambda_2}$ 3) $\lambda_1 + \lambda_2 + \lambda_3 = 0$ 4) $3\lambda_2 = \lambda_3 + 2\lambda_2$

22. **If 13.6 eV is the energy required to saperate a hydrogen atom into a proton and an electron then its orbital radius is**

1) 5.3×10^{-11}m 2) 5.3×10^{-12}m 3) 7.6×10^{-13}m 4) 7.6×10^{-14}m

LEVEL - I -2- KEY

1) 2	2) 4	3) 2	4) 1	5) 3	6) 1	7) 3	8) 3	9) 1	10) 3
11) 1	12) 2	13) 2	14) 3	15) 1	16) 4	17) 2	18) 2	19) 2	20) 2
21) 2	22) 1								

LEVEL - I -2- HINTS

1. $i = \frac{ne}{t}$

2. K.E. $= \left(\frac{\frac{1}{2}mv^2}{e}\right)$ in ev where e is charge of electron

3. $\frac{y_1}{y_2} = \frac{v_2}{v_1}$

4. $v=\sqrt{\frac{2V_0e}{m}}$ Where V_0 is the P.D

5. K.E. = Ve Where V is P.D

6. $y=\frac{1}{2}\frac{F}{m}t^2=\frac{1}{2}\left(\frac{Eq}{m}\right)\left(\frac{x}{v}\right)^2$

7. mg = Eq and F = (mg+Eq) $\Rightarrow$ ma = 2mg $\Rightarrow$ a=2g

8. $Eq=mg\Rightarrow\left(\frac{V}{d}\right)ne=mg$

9. KE =PE $\Rightarrow q_1V=\frac{1}{4\pi\varepsilon_0}\frac{q_1q_2}{r}$

Independent of charge of the particle projected

11. $b=\frac{ze^2\cot(\theta/2)}{4\pi\varepsilon_0\times E}$

12. $\Delta E=-13.6eV\left(\frac{1}{n_1^2}-\frac{1}{n_2^2}\right)$; $\Delta L=\frac{h}{2\pi}(n_1-n_2)$

13. $|T.E|=K.E;\lambda=\frac{h}{\sqrt{2mK.E}}$

14. $E_{ionisation}=E_\alpha-E_1$

15. $.E_n=-\frac{13.6}{n^2}=-3.4\Rightarrow n=2;L=\frac{nh}{2\pi}=h/\pi$

16. $\upsilon_n\alpha\frac{1}{n}$:

17. P.e =-2 K.E :

18. $P.E.=0=-27.2+27.2=0$

$P.E.=-\frac{27.2}{4}+27.2=20.4$

19. $\frac{1}{\lambda}=RZ^2\left(\frac{1}{n^2{}_1}-\frac{1}{n_2{}^2}\right)$:

20. For maximum number. $N=\frac{(n-1)}{2}n$

21. $\frac{1}{\lambda_3}=\frac{1}{\lambda_1}+\frac{1}{\lambda_2}$ $\Rightarrow\lambda_3=?$

22. Total energy of electron in hydrogan atom,

$E=-13.6eV=-13.6\times1.6\times10^{-19}J$

$=\frac{-e^2}{8\pi\varepsilon_0 r}$ $\therefore r=\frac{e^2}{8\pi\varepsilon_0 E}$

LEVEL - II -1

J.J.THOMSONS METHOD

1. **Two ions having masses in the ratio 1:1 and charges 1:2 are projected into uniform magnetic field perpendicular to the field with speeds in the ratio 2:3 .The ratio of the radii of circular paths along which the two particles move is**

1) 4:3 2) 2:3 3) 3:1 4) 1:4

2. **In Thomson's experiment, a magnetic field of induction 10^{-2} wb/m^2 is used. For an undeflected beam of cathode rays, a p.d. of 500 V is applied between the plates which are 0.5 cm apart. Then the velocity of the cathode ray beam is ...... m/s.**

1) 4×10^7 2) 2×10^7 3) 2×10^8 4) 10^7

3. **A cathode ray beam is bent into an arc of a circle of radius 0.02 m by a field of magnetic induction 4.55 milli tesla. The velocity of electrons is (given e = 1.6×10^{-19} c and m = 9.1×10^{-31} kg)**

1) 2×10^7m/s 2) 3×10^7m/s 3) 1.6×10^7 m/s 4) 3.2×10^7m/s

4. **When two electrons enter into a magnetic field with different velocities, they deflect in different circular parts, in such a way that the radius of one path is double that of the other. 1×10^7 ms^{-1} is the velocity of electron in smaller circle of radius 2×10^{-3} m. The velocity of electron in the other circular path is :**

1) 4×10^7 ms^{-1} 2) 4×10^6 ms^{-1} 3) 2×10^7 ms^{-1} 4) 2×10^6 ms^{-1}

MILLIKAN OIL DROP EXPERIMENT

5. **A charged oil drop falls with terminal velocity V_0 in the absence of electric field. An electric field E keeps it stationary. The drop acquires additional charge q and starts moving upwards with velocity V_0. The initial charge on the drop was**

1) 4q 2) 2q 3) q 4) q/2

6. **A charged oil drop is moving with a velocity V_1. As it acquires an additional charge it moves up with the velocity V_2 in the same electric filed. It fall freely with a Velocity 'V' in the absence of electric field . The ratio of the charges before and after acquiring additional charge is**

1) $\frac{V_1 + V}{V_2 - v_1}$ 2) $\frac{V_1 + V_2}{2V}$ 3) $\frac{V + V_1}{V + V_2}$ 4) $\frac{V_2 - 2V_1}{2V_1 + V}$

ALPHA SCATTERING

7. **A proton strikes another proton at rest with speed V_o . Assume impact parameter to be zero. Their closest distance of approach is (mass of proton is m)**

1) $\frac{e^2}{4\pi\varepsilon m\upsilon_o^2}$ 2) $\frac{e^2}{\pi\varepsilon_0 m\upsilon_o^2}$ 3) $\frac{e^2}{m\upsilon_0^2}$ 4) zero

8. **A closest distance of approach of an α particle travelling with a velocity v towards Al_{13} nucleus is d. The closest distance of approach of an alpha particle travelling with velocity 4V towards Fe_{26} nucleus is**

1) d/2 2) d/4 3) d/8 4) d/16

BOHR'S THEORY

9. **The energy required to excite an electron from n=2 to n=3 energy state is 47.2 e V. The charge number of the nucleus, around which the electron revolving will be**

1) 5 2) 10 3)15 4) 20

10. The debroglie wave length of an electron in the first Bohr orbit is

1) Equal to the circumference of the first orbit

2) 1/2 th circumference of the first orbit

3) 1/4 th circumference of the first orbit

4) 3/4th circumference of the first orbit

11. If the radius of first Bohr's orbit is x, then de-Broglie wavelength of electron in 3rd orbit is nearly:

1) $2\pi x$ 2) $6\pi x$ 3) $9x$ 4) $x/3$

12. The angular momentum of the electron in third orbit of hydrogen atom, if the angular momentum in the second orbit of hydrogen atom is L is

1) L 2) 3L 3) (3/2) L 4) 2/3 L

13. The de-Broglie wave length of the electron in the second Bohr orbit is (given r_0=0.53A^0)

1) 3.33A^o 2) 6.66A^o 3) 9.99A^o 4) 1.06A^o

ATOMIC SPECTRA

14. The maximum wavelength of Brackett series of hydrogen atom will be ______ A^0

1) 35,890 2) 14,440 3) 62,160 4) 40,477

15. When a hydrogen atom emits a photon in going from n = 5 to n = 1, its recoil speed is almost

1) $10^{-4} m/s$ 2) $2\times10^{-2} m/s$ 3) $4 m/s$ 4) $8\times10^{-2} m/s$

16. An orbital electron in the ground state of hydrogen has the magnetic moment μ_1. This orbital electron is excited to 3rd excited state by some energy transfer to the hydrogen atom. The new magnetic moment of the electron is μ_2, then

1) $\mu_1 = 2\mu_2$ 2) $2\mu_1 = \mu_2$ 3) $16\mu_1 = \mu_2$ 4) $4\mu_1 = \mu_2$

LEVEL-II-1-KEY

1) 1 2) 4 3) 3 4) 3 5) 3 6) 3 7) 2 8) 3 9) 1 10) 1
11) 2 12) 3 13) 2 14) 4 15) 3 16) 4

LEVEL-II-1-HINTS

1. $\frac{r_1}{r_2} = \frac{m_1}{m_2} \times \frac{v_1}{v_2} \times \frac{q_2}{q_1}$

2. $V = \frac{E}{B}$

3. $v = \frac{rBe}{m}$

4. $r = \frac{mv}{Bq}$

5. $6\pi\eta r V_0 = mg; Eq = mg$

$E\left(q^1 + q\right) = 6\pi\ \eta r V_0 + mg = 2mg = 2Eq$

6. $Eq_1 - mg = 6\pi\eta aV_1 \rightarrow 1$ $\qquad mg = 6\pi\eta aV$

$Eq_2 - mg = 6\pi\eta aV_2 \rightarrow 2$

7. Use:LCLM and conservation of energy.

ie: $m\upsilon_0 = 2m\upsilon - (1)$

$$\frac{m\upsilon_o^{\,2}}{2} = 2\times\frac{1}{2}m\left(\frac{\upsilon_o}{2}\right)^2 = \frac{e^2}{4\pi\varepsilon r}$$

8. $\frac{1}{2}mv^2 = \frac{1}{4\pi\varepsilon_0}\times\frac{q_1\,q_2}{r}$; $r\alpha\frac{z^2}{v^2}$

9. $E = E_3 - E_2 = \frac{13.6Z^2}{4} - \frac{13.6Z^2}{9} = 47.2$

$\Rightarrow Z = ?$

10. $mvr = \frac{nh}{2\pi}$; $2\pi = \frac{h}{mv}$; $2\pi r = \lambda$

11. $mvr = \frac{nh}{2\pi}\left(r = \frac{r_0 n^2}{Z}\right) r_0 = x$

$2\pi r = n\lambda \Rightarrow \lambda = \frac{2\pi r}{n} = 6\pi\, x$

12. $L = \frac{nh}{2\pi}$

13. $2\pi r = n\lambda$;

$\therefore \lambda = \frac{2\pi r}{n} = \frac{2\pi r_0 n^2}{n} = 2\pi r_0 n$

14. $RZ^2\left(\frac{1}{n_1^{\,2}} - \frac{1}{n_2^{\,2}}\right) = \frac{1}{\lambda}$ *Where* $n_1 = 4; n_2 = 5$

15. From law of conservation of momentum

$\therefore m\upsilon = p = \frac{h}{\lambda} \Rightarrow \upsilon = \frac{h}{m\lambda}$

m $\rightarrow$ mass of proton :

$\lambda \rightarrow$ wave length of emitted photon

16. $\mu = niA = iA$, $\quad = \frac{qv}{2\pi r}\times\pi r^2$

$M\alpha vr$ where $v\alpha\frac{z}{n}$ & $r\alpha\frac{n^2}{z}$

$\therefore \mu\alpha n$

LEVEL - II -2

J.J.THOMSONS METHOD

1. **A proton and an α-particle enter a magnetic field in a direction perpendicular to it. If the force acting on the proton is twice that acting on the α-particle, the ratio of their velocities is**

   1) 4 : 1 2) 1. : 4 3) 1 : 2 4) 2 : 1

2. **A proton, a deuteron and an α-particle are accelerated through the same p.d. of V volt. The velocities acquired by them are in the ratio**

   1) $1 : 1 : \sqrt{2}$ 2) $1 : \sqrt{2} : 1$ 3) $1 : 1 : 1$ 4) $\sqrt{2} : 1 : 1$

3. **An electron starts from rest and travels**

   0.9 m in an electric field of 200 V/m. After this, it enters a magnetic field at right angles to its direction of motion. If the radius of circular path of the electron is 9 cm, the magnetic field induction is (Given e=1.6 x 10^{-19}C, m=9x10^{-31}kg)

   1) 5 x 10^{-4} wb/m^2 2) 5 x 10^{-5} wb/m^2 3) 5 x 10^{-3} wb/m^2 4) 5 x 10^{-2} wb/m^2

MILLIKAN OIL DROP EXPERIMENT

4. **In the Millikan's experiment, the oil drop is subjected to a horizontal electric field of 2 N/C and the drop moves with a constant velocity making an angle of 45^0 with the horizontal. If the weight of the drop is W, then the electric charge, in coulomb, on the drop is**

   1) W 2) W/2 3) W/4 4) W/8

ALPHA SCATTERING

5. **An α -Particle of energy 5 Mev is scattered through 180^0 by a fixed uranium nuclens.The closest distance is in the order of:**

   1) $1A^o$ 2) $10^{-10} cm$ 3) $10^{-12} cm$ 4) $10^{-16} cm$

6. **An alpha nucleus of energy $\frac{1}{2}mv^2$ bombards a heavy nuclear target of charge Ze. Then the distance of closest approach for the alpha nucleus will be proportional to (AIEEE-2006)**

   1) $\frac{1}{v}$ 2) $\frac{1}{ze}$ 3) v^2 4) $\frac{1}{m}$

BOHR'S THEORY

7. **In the Bohr model of hydrogen atom, the ratio of the kinetic energy and total energy of electron in the nth quantum state will be**

   1) 1 2) -1 3) 2 4) -12

8. **The number of revolutions done by an electron 'e' in one second in the first orbit of hydrogen atom is**

   1) 6.57x10^{15} 2) 6.57 x 10^{13} 3) 1000 4) 6.57 x 10^{14}

9. **If $\left(\frac{0.51\times10^{-10}}{4}\right)$ m is the radius of smallest electron orbit in hydrogen like atom ,then this atom is :**

   1) hydrogen atom 2) He^+ 3) Li^{2+} 4) Be^{3+}

10. **In Bohr's orbit of hydrogen atom m kg is mass of an electron and e coulomb is the charge on it .The ratio (in SI units) of magnetic dipole moment to that of the angular momentum of electron is:**

    1) e/2m 2) e/m 3) 2e/m 4)2e/3m

11. **In a sample of hydrogen like atoms all of which are in ground state, a photon beam contain photons of various energies is passed. In absorption spectrum, five dark line are observed. The number of bright lines in the emission spectrum will be(assume that all transitions take place)**

1) 21 2) 10 3) 15 4) 12

ATOMIC SPECTRA

12. **Let f_1 be the frequency of the series limit of the lyman series, f_2 be the frequency of the first line of the lyman series,and f_3 be the frequency of the series limit of the balmer series .**

1) $f_1 - f_2 = f_3$ 2) $f_2 - f_1 = f_3$ 3) $f_3 = \frac{1}{2}(f_1 + f_2)$ 4) $f_1 + f_2 = f_3$

13. **Ratio of difference of spacing between the energy levels with n=3 and n=4 and the spacing between the energy levels with n=8 and n=9 for a hydrogen like atom or ion is**

1) 0.71 2) 0.41 3) 2.43 4) 14.82

14. **A stationary hydrogen atom emits photon corresponding to the first line of Lyman series. If R is the Rydberg's constant and M is the mass of the atom, then the velocity acquired by the atom is (neglect energy absorbed by the photon)**

1) $\frac{3Rh}{4M}$ 2) $\frac{4M}{3Rh}$ 3) $\frac{Rh}{4M}$ 4) $\frac{4M}{Rh}$

LEVEL-II -2- KEY

1) 1 2) 4 3) 1 4) 2 5) 3 6) 4 7) 2 8) 1 9) 4 10) 1
11) 3 12) 1 13) 2 14) 1

LEVEL-II -2- HINTS

1. $F = Bqv$

2. $v = \sqrt{\frac{2eV}{m}} \Rightarrow v \propto \sqrt{\frac{e}{m}}$

3. $V^2 - U^2 = 2as$; $V^2 = 2\frac{Eq}{m}S \Rightarrow V = \sqrt{2\frac{Eq}{m}S}$

$r = \frac{mV}{Bq} \Rightarrow B = \frac{mV}{rq}$

4. $\text{Tan } 45^0 = \frac{Eq}{w}$ Eq = w

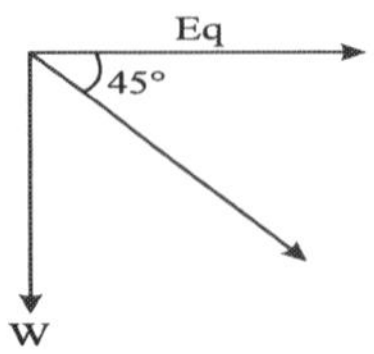

5. $\frac{1}{2}mv_o^2 = \frac{q_1 q_2}{4\pi\varepsilon_o r}$

6. $\frac{1}{2}mv^2 = \frac{1}{4\pi\varepsilon_0}\frac{(2e)(ze)}{r}$

7. $K.e = +\frac{13.6Z^2}{n^2}$ $\therefore K.e = |T.e|$ &

$T.e = -13.6z^2/n^2$ $T.e = -K.e$

8. $\upsilon_1(t) = n.2\pi r_1 \Rightarrow n = ?$

9. $r_n = \frac{n^2}{Z} r_o$:

10. Magnetic dipole moment $= \frac{er^2 w}{2}$

Angular momentum $= mr^2 w$

11. $\frac{n(n-1)}{2}$ Here $n = 6$

12.
f_1 relates $\Rightarrow n_1 = 1; n_2 = \alpha$
f_2 relates $\Rightarrow n_1 = 1; n_2 = 2$
f_3 relates $\Rightarrow n_1 = 2 : n_2 = \alpha$

13. $\frac{r_4 - r_3}{r_9 - r_8} = \frac{4^2 - 3^2}{9^2 - 8^2} = 0.41$

14. $p = -mv$; $\frac{h}{\lambda} = mv \Rightarrow v = \frac{h}{m}\left(\frac{1}{\lambda}\right)$ & $\frac{1}{\lambda} = \frac{3R}{4}$

LEVEL - III

1. **Magnetic moment due to the motion of the electron in nth energy state of hydrogen atom is proportional to**

1) n 2) n^0 3) n^5 4) n^3

2. **The ratio between total accelertion of the electron in singly ionized helium atom and hydrogen atom (both in ground state) is**

1) 1 2) 8 3) 4 4) 16

3. **The shortest wavelength of the Brackett series of a hydrogen like atom (atomic number =Z) is the same as the shortest wavelength of the Balmar series of hydrogen atom. The value of Z is**

1) 2 2) 3 3) 4 4) 6

4. **According to Bohr's theory of hydrogen atoms, the product of the binding energy of the electron in the nth orbit and its radius in the n the orbit**

1) is proportional to n^2
2) is inversely proportional to n^3
3) has a constant value of 10.2eV-A^0
4) has constant value 7.2 eV-A^0

5. **If an electron drops from 4 th orbit to 2nd orbit in an H-atom,then**

1) it gains 2.55 eV of potential energy
2) it gains 2.55 eV of total energy
3) it emits a 2.55 eV electron
4) it emits a 2.55 eV photon

6. **The energy of an atom or ion in the ground state is -54.4 eV .it may be:**

1) He^+ 2) Li^{2+} 3) hydrogen 4)deuterium

7. **An atom absorbs 2 eV energy and is excited to next energy state. The wavelength of light absorbed will be**

1) 2000 A° 2) 4000 A°. 3) 8000 A° 4) 6206 A°

8. **When an electron in the hydrogen atom in ground state absorbs a photon of energy 12.1 eV, its angular momenturm**

1) decreases by $2.11\times10^{-34} J-s$
2) decreases by $1.055\times10^{-34} J-s$
3) increases by $2.11\times10^{-34} J-s$
4) increases by $1.055\times10^{-34} J-s$

9. **Magnetic field at the centre (at nucleus) of the hydrogen like atoms (atomic number =Z) due to the motion of electron in n^{th} orbit is proportional to**

1) $\frac{n^3}{Z^5}$ 2) $\frac{n^4}{Z}$ 3) $\frac{Z^2}{n^3}$ 4) $\frac{Z^3}{n^5}$

10. **A neutron moving with a speed v makes a head on collision with a hydrogen atom in ground state kept at rest. The minimum kinetic energy of neutron for which inelastic collision will take place is (assume that mass of proton is nearly equal to the mass of neutron)**

1) 10.2 eV 2) 20.4eV 3) 12.1eV 4) 16.8ev

11. **A charged particle is moving in a uniform magnetic field in a circular path. The energy of the particle is doubled. If the initial radius of the circular path was R, the radius of the new circular path after the energy is doubled will be**

1) R/2 2) $\sqrt{2}R$ 3) 2R 4) R/ $\sqrt{2}$

12. **An electron in hydrogen atom after absorbing an energy photon jumps from energy state n_1 to n_2. Then it returns to ground state after emitting six different wavelengths in emission spectrum. The energy of emitted photons is either equal to, less than or greater than the absorbed photons. Then n_1 and n_2 are**

1) $n_2=4, n_1=3$ 2) $n_2=5, n_2=3$ 3) $n_2=4, n_1=2$ 4) $n_2=4, n_1=1$

13. **The photon radiated from hydrogen corresponding to 2nd line of Lyman series is absorbed by a hydrogen like atom 'X' in 2nd excited stae. As a result then , the hydrogen-like atom 'X' makes a transition of nth orbit.**

1) X=He^+, n=4 2) X=Li^{++},n=6 3) X=He^+, n=6 4) X=Li^{++},n=9

14. **In a hypothetical system, a particle of mass m and charge -3q is moving around a very heavy particle charge q. Assume that Bohr's model is applicable to this system.then velocity of mass m in first orbit is**

1) $\frac{3q^2}{2\in_0 h}$ 2) $\frac{3q^2}{4\in_0 h}$ 3) $\frac{3q}{2\pi\in_0 h}$ 4) $\frac{3q}{4\pi\varepsilon_0 h}$

15. **Consider a hydrogen-like atom whose energy in n^{th} excited state is given by** $E_n=\frac{13.6Z^2}{n^2}$

When this excited atom makes a transition from excited state to ground state, most energetic photons have energy E_{max}= 52.224 eV and least energetic photons have energy E_{min}= 1.224eV. Find the atomic number of atom and the state of excitation.

1) Z=2; n=5 2) Z=2, n=4 3) Z=3, n=6 4) Z=4, n =6

16. **29 electrons are removed from Zn-atom (Z=30) by certain means .The minimum energy needed to remove the 30th electron ,will be:**

1)12.24 keV 2)408 keV 3)0.45 keV 4)765keV

17. **The ionisation energy of Li^{2+} atom in ground state is :**

1) 13.6 x 9 eV 2) 13.6J 3) 13.6 erg 4) 13.6×10^{-19} J

18. **A photon of energy 15 eV collides with H-atom. Due to this collision, H-atom gets ionized .The maximum kinetic energy of emitted electron is :**

1) 1.4 eV 2) 5 eV 3) 15eV 4) 13.6eV

19. **Monochromatic radiation of wavelength λ is incident on a hydrogen sample in ground state.Hydrogen atoms absorb a fraction of light and subsequently emit radiation of six different wavelengths. Find the value of λ.**

1) 80 nm 2) 97.5 nm 3) 105 nm 4) 60 nm

20. **The figure shows energy levels of a certain atom, when the system moves from level 2E to E, a photon of wavelength λ is emitted. The wavelength of photon produced during its transition from level 4/3 E to E level is:**

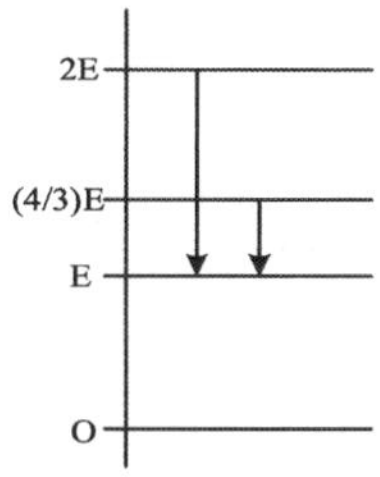

1) 3λ 2) $3/4\lambda$ 3) $\lambda/4$ 4) 2λ

21. **When the electron in hydrogen atom jumps from the second orbit to the first orbit, the wavelength of the radiation emitted is λ. When the electron jumps from the third to the first orbit, the wavelength of the radiation emitted is:**

1) $\frac{9}{4}\lambda$ 2) $\frac{4}{9}\lambda$ 3) $\frac{27}{32}\lambda$ 4) $\frac{32}{27}\lambda$

22. **The ratio of the largest to shortest wavelengths in Balmer series of hydrogen spectra is:**

1) $\frac{25}{9}$ 2) $\frac{17}{6}$ 3) $\frac{9}{5}$ 4) $\frac{5}{4}$

23. **An electron in a hydrogen atom makes a transition $n_1 \rightarrow n_2$, where n_1 and n_2 are principal quantum numbers of the states. Assume the Bohr's model to be valid. The time period of the elctron in the initial states is eight times to that of final state. What is ratio of $\frac{n_1}{n_2}$**

1) 8:1 2) 4:1 3) 2:1 4) 1:2

24. **Any radiation in the ultra violet region of Hydrogen spectrum is able to eject photoelectrons from a metal. Then the maximum value of threshold frequency of the metal is, nearly**

1) 3.3×10^{15} HZ 2) 2.5×10^{15} HZ 3) 4.6×10^{14} HZ 4) 8.2×10^{14} HZ

25. **A hydrogen atom emits a photon corresponding to an electron transition from n=5 to n=1. The recoil speed of hydrogen atom in nearly equal to**

1) 10^{-4} m/s 2) 2×10^{-2} m/s 3) 4 m/ 4) 8×10^{-2} m/s

26. **The wave number of energy emitted when electron jumps from fourth orbit to second orbit in hydrogen in 20,497 cm^{-1}. The wave number of energy for the same transition in He^{+} is**

1) 5,099 cm^{-1} 2) 20,497 cm^{-1} 3) 40,994 cm^{-1} 4) 81,988 cm^{-1}

27. **In a Bohr atom the electron is replaced by a particle of mass 150 times the mass of the electron and the same charge. If a_0 is the radius of the first Bohr orbit of the orbital atom, then that of the new atom will be**

1) 150 a_0 2) $\sqrt{150}\,a_0$ 3) $\frac{a_0}{\sqrt{150}}$ 4) $\frac{a_0}{150}$

28. **If the wavelength of first member of Balmer series of hydrogen spectrum is 6564 A^0, the wavelength of second member of Balmer series will be:**

1) 1215 A^0 2) 4848 A^0 3) 6050 A^0
4) data given is insufficient to calculate the value

29. **A hydrogen like atom (atomic number Z) is in a higher excited state of quantam number 'n'. This excited atom can make a transition to the first excited state by emitting a photon of energy 27.2eV. Alternatively the atom from the same excited state can make a transition to second excited state by emitting a photon of energy 10.20 eV. The value of n and z are given (Ionization energy of hydrogen atom is 13.6eV)**

1) n = 6 and z = 3 2) n = 3 and z = 6 3) n = 8 and z = 4 4) n = 4 and z = 8

30. **Photons from n=2 to n=1 in Hydrogen atom is made to fall on a metal surface with work function 1.2ev. The maximum velocity of photo electrons emitted is nearly equal to**

1) 6×10^5 m/s 2) 3×10^5 m/s 3) 2×10^5 m/s 4) 18×10^5 m/s

31. **Let ϑ_1 be the frequency of the series limit of the Lyman series and ϑ_2 be the frequency of the first line of the Lyman series and ϑ_3 be the frequency of the series limit of Balmar series, then**

1) $\vartheta_1 - \vartheta_2 = \vartheta_3$ 2) $\vartheta_2 - \vartheta_1 = \vartheta_3$ 3) $2\vartheta_3 = \vartheta_1 + \vartheta_2$ 4) $\vartheta_1 + \vartheta_2 = \vartheta_3$

32. **a)Find the wavelength of the radiation required to excite the electron in Li from the first to the third Bohr orbit,**

b) How many spectral lines are observed in the emission spectrum of the above excited system?

1) 108.8 eV, 3 2) 13.6 eV, 4 3) 54.4 eV, 2 4) 10.2 eV, 3

33. **Find the wavelengths in a hydrogen spectrum between the range 500nm to 700nm.**

1) 540 nm 2) 580 nm 3) 654 nm 4) 696 nm

34. **The largest wavelength in the ultraviolet region of the hydrogen spectrum is 122 nm. The smallest wavelength in the infrared region of the hydrogen spectrum (to the nearest integer) is**

1) 802 nm 2) 823 nm 3) 1882 nm 4)1648 nm

35. **The electric potential between a proton and an electron is given by $V = V_0 ln\frac{r}{r_0}$, where r_0 is a constant. Assuming Bohr's model to be applicable, write variation of r_n with *n*, *n* being the principal quantum number**

1) $r_n \propto n$ 2) $r_n \propto 1/n$ 3) $r_n \propto n^2$ 4) $r_n \propto 1/n^2$

36. **If elements of quantum number greater than *n* were not allowed, the number of possible elements in nature would be ?**

1) $\frac{1}{2}n(n+1)$ 2) $\left\{\frac{n(n+1)}{2}\right\}^2$ 3) $\frac{1}{2}n(n+1)(2n+1)$ 4) $\frac{1}{3}n(n+1)(2n+1)$

37. **Magnetic field at the centre (at nucleus) of the hydrogen like atoms (atomic number $= z$) due to the motion of electron in n^{th} orbit is praportional to**

1) $\frac{n^3}{z^5}$ 2) $\frac{n^4}{z}$ 3) $\frac{z^2}{n^3}$ 4) $\frac{z^3}{n^5}$

38. **The recoil speed of a hydrogen atom after it emits a photon in going from $n = 5$, state to $n = 1$ state is (in ms^{-1})**

1) 4.718 2) 7.418 3) 4.178 4) 7.148

39. **The binding energy of an electron in the ground state of He-atom is $E_0 = 24.7eV$. The energy required to remove both the electrons from the atom is**

1) 24.6eV 2) 79.0eV 3) 54.4eV 4) none of these

40. **In hydrogen atom, the radius of nth Bohr orbit is V_n. The graph between $\log\left(\frac{r_n}{r_1}\right)$ and $\log n$ will be**

1)
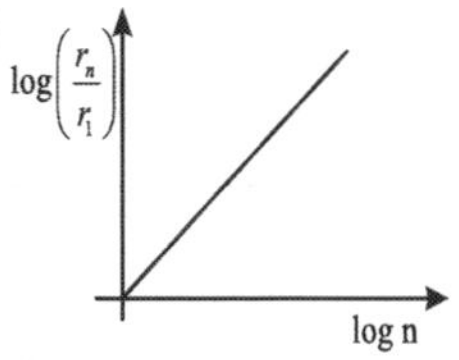

2)
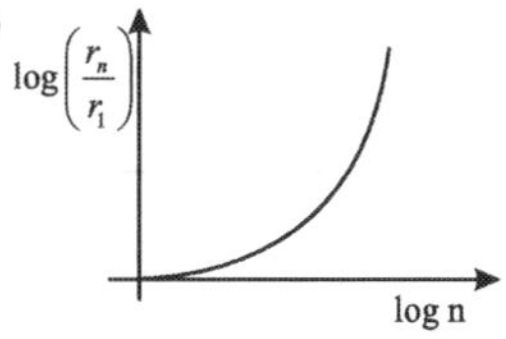

3)
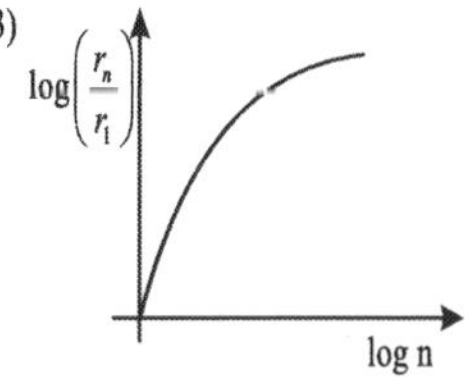

4)
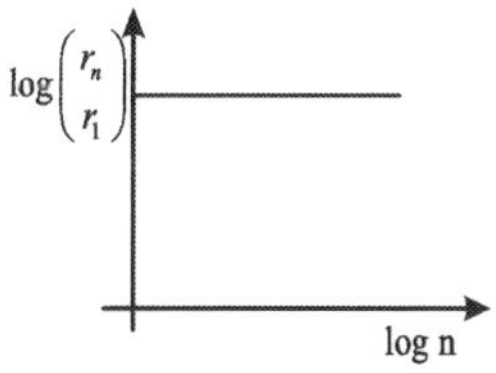

41. **In hydrogen atom, the area enclosed by n^{th} orbit is A_n. The graph between $\log\left(\frac{A_n}{A_1}\right)$ and $\log n$ will be**

1)
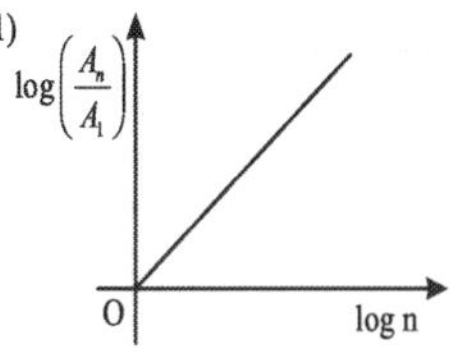

2)
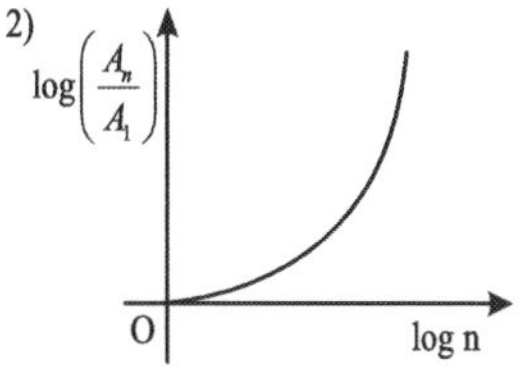

3)
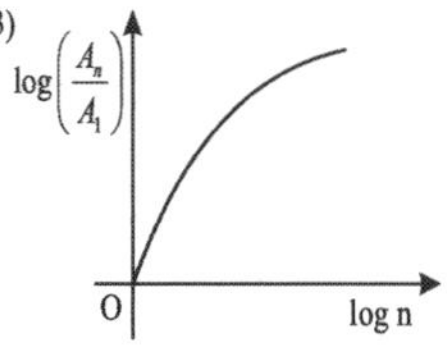

4)
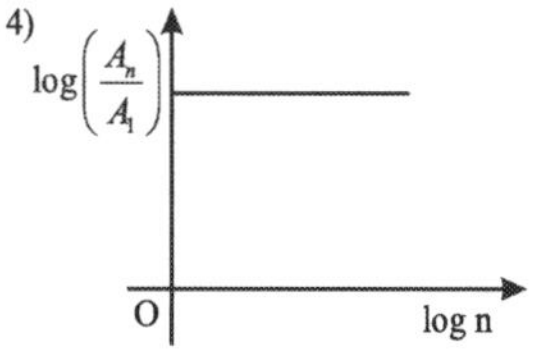

42. An electron in the ground state of hydrogen atom is revolving in anticlockwise direction in the cicular orbit of radius R. The atom is placed in a uniform magnetic induction B such that the plane normal of electron orbit makes an angle 30^0 with B, as shown in figure. The torque experienced by electron willbe

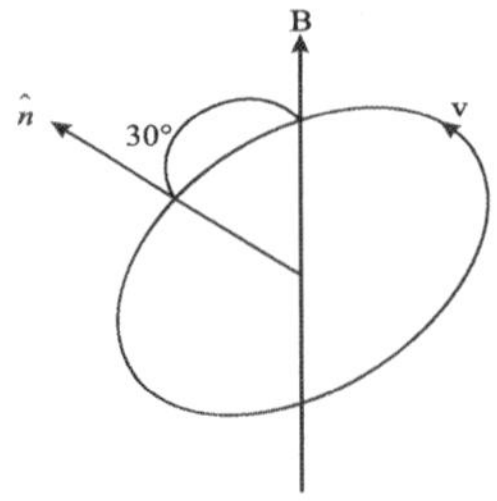

1) $\frac{ehB}{8\pi m}$ 2) $\frac{eh}{8\pi Bm}$ 3) $\frac{eB}{8\pi mh}$ 4) $\frac{hB}{8\pi em}$

43. If we assume only gravitational attraction between proton and electron in hydrogen atom and the Bohr's quantization rule to be followed, then the expression for the ground state energy of the atom will be (the mass of proton is M and that of electron is m)

1) $\frac{G^2M^2m^2}{h^2}$ 2) $\frac{2\pi^2G^2M^2m^3}{h^2}$ 3) $\frac{2\pi^2GM^2m^3}{h^2}$ 4) $\frac{h^2}{G^2M^2n^2}$

LEVEL - III - KEY

1) 1	2) 2	3) 1	4) 4	5) 4	6) 1	7) 4	8) 3	9) 4	10) 2
11) 2	12) 3	13) 4	14) 1	15) 1	16) 1	17) 1	18) 1	19) 2	20) 1
21) 3	22) 3	23) 3	24) 1	25) 3	26) 4	27) 4	28) 2	29) 1	30) 4
31) 1	32) 1	33) 3	34) 2	35) 1	36) 4	37) 4	38) 3	39) 2	40) 1
41) 1	42) 1	43) 2							

LEVEL - III - HINTS

1. $\frac{Magnetic\ Moment}{Angular\ Momentum} = \frac{e}{2m}$

$\therefore$ Magnetic Moment $\propto$ angular momentum

$M \propto n \left(L = \frac{nh}{2\pi} \right)$

2. $a = \frac{v^2}{r} \therefore a \propto \frac{(z)^2}{(1/z)}$ or $\therefore a \propto z^3 \therefore \frac{a_1}{a_2} = \left(\frac{2}{1}\right)^3 = 8$

3. $(z^2)\left(\frac{13.6}{16}\right) = \left(\frac{13.6}{4}\right)$; $Z^2 = 4$ or Z=2

4. $E_n \propto \frac{1}{n^2}$ and $r_n \propto n^2$; $\therefore E_n r_n$ is independent of n

Hence, $E_1 r_1 = (13.6eV)(0.53\overset{0}{A}) = 7.2eV - \overset{0}{A}$ =constant

5. $E_n = -\dfrac{13.6Z^2}{n^2} eV; \Delta E = E_4 - E_2$

6. $E = \dfrac{-13.6z^2}{n^2}$

7. $E = \dfrac{hc}{\lambda} \Rightarrow \lambda = ?$

8. After absorbing a photon of energy 12.1 eV electron jumps from ground state (n=1) to second excited state (n=3).Therefore change in angular momentum $\Delta L = L_3 - L_1$ $= 3\left(\dfrac{h}{2\pi}\right) - \dfrac{h}{2\pi} = \dfrac{h}{\pi}$

$= \dfrac{6.6 \times 10^{-34}}{3.14} J - s = 2.11 \times 10^{-34} J - s$

9. $B_n = \dfrac{\mu_0 I_n}{2r_n}$ (or) $B_n \propto \dfrac{I_n}{r_n} \propto \dfrac{(f_n)}{r_n}$

$B_n \propto \dfrac{(v_n / r_n)}{r_n} \propto \dfrac{v_n}{(r_n)^2}$; $\propto \dfrac{(Z/n)}{(n^2/Z)^2} \propto \dfrac{Z^3}{n^5}$

10. Let v=speed of neutron before collision
v_1= speed of neutron after collision
v_2= speed of proton or hydrogen atom after collision and ΔE=energy of exitation.
From conservation of linear momentum

$mv = mv_1 + mv_2$ ....(1)

From conservation of energy

$\dfrac{1}{2}mv^2 = \dfrac{1}{2}mv_1^2 + \dfrac{1}{2}mv_2^2 + \Delta E$ ....(2)

from 1 and 2 As $v_1 - v_2$ must be real

$v^2 - 4\dfrac{\Delta E}{m} \geq 0$

$\Rightarrow \dfrac{1}{2}mv^2 = 2\Delta E = 2 \times 10.2 = 20.4 eV$

11. $r \propto \sqrt{KE} \Rightarrow \dfrac{r_2}{r_1} = \sqrt{\dfrac{KE_2}{KE_1}}$

12. From $n_2 = 4$, six lines are obtained in emission spectrum. Now: $E_{4-2} = E_{absorbed}$

$E_{4 \to 3} < E_{absorbed}$ and $E_{4-1} E_{3-1}, E_{2-1} > E_{absorbed}$

Hence, $n_1 = 2$ and $n_2 = 4$

13. Energy of nth state in hydrogen is same as energy of 3n[th] state in Li^{++}.

$\therefore$ $3 \to 1$ transition in H would give same energy as the $9 \to 3$ transition in Li^{++}.

14. $\frac{mv^2}{r} = \frac{1}{4\pi \in_o} \frac{3q^2}{r^2}$; $mvr = \frac{nh}{2\pi}$

using above equations, putting n=1; $v = \frac{3q^2}{2\in_o h}$

15. Maximum energy is liberated for transition $E_n \rightarrow 1$ and minimum energy for $E_n \rightarrow E_{n-1}$

Hence $\Delta E = (Rhc)\, Z^2 \left[\frac{1}{n_1^2} - \frac{1}{n_2^2}\right]$

16. After removing 29 electrons,Zn atom will become Zn^{+29} $\therefore$ E= 13.6 Z^2 eV=12,240 eV =12.24KeV

17. (E)= Li^{+2} ; $E = Z^2(E)_H; E = 9\times 13.6\,eV$

18. Energy of photon ionization energy + $K.E_{max}$ =15ev - 13.6ev=1.4ev

19. The atom is excited to n=4 state

$$\frac{hc}{\lambda} = E_4 - E_1 = 12.75eV \Rightarrow \lambda = 97.5nm$$

20. $\frac{hc}{\lambda_1} = 2E - E = E \rightarrow 1$; $\frac{hc}{\lambda_2} = \frac{4}{3}E - E = \frac{E}{3} \rightarrow 2$

$$\frac{\lambda_2}{\lambda_1} = \frac{E}{\frac{E}{3}} = \frac{3}{1}$$

21. $\frac{1}{\lambda} = RZ^2\left(\frac{1}{n_1^2} - \frac{1}{n_2^2}\right)$

22. $\frac{1}{\lambda_{max}} = RZ^2\left(\frac{1}{n_1^2} - \frac{1}{n_2^2}\right)$ *Where* $n_1 = 2; n_2 = 3$

$\frac{1}{\lambda_{min}} = RZ^2\left(\frac{1}{n_1^2} - \frac{1}{n_2^2}\right)$ *Where* $n_1 = 2; n_2 = \alpha$

23. Time period of revolution of electron in the nth orbit is $T \alpha n^3$

24. $\therefore \upsilon = \frac{c}{\lambda}$

25. $v = \frac{h}{m\lambda}$; $\frac{1}{\lambda} = R\left[1 - \frac{1}{25}\right]$

26. $\bar{v} = RZ^2\left(\frac{1}{n_1^2} - \frac{1}{n_2^2}\right); \bar{v}\alpha Z^2$

27. $r_n \alpha \frac{1}{m}$

28. $\frac{1}{\lambda} = RZ^2\left(\frac{1}{n_1^2} - \frac{1}{n^2_2}\right)$ for balmer series

n_1=2 and n_2= 3,4,5....

29. $\Delta E_1 = 27.2 = (13.6)\left[\frac{1}{4} - \frac{1}{n^2}\right]z^2$

$\Delta E_2 = 10.2 = (13.6)\left[\frac{1}{9} - \frac{1}{n^2}\right]z^2$

30. E= E_2-E_1; $E_n = -\frac{13.6}{n^2}; \frac{1}{2}mv^2 = E - W$

$\frac{1}{\lambda} = R\left(\frac{1}{1^2} - \frac{1}{\alpha^2}\right)$ series limit for lyman series

$\frac{1}{\lambda} = R\left(\frac{1}{1^2} - \frac{1}{2^2}\right)$ series for lyman first line

31. $\frac{1}{\lambda} = R\left(\frac{1}{2^2} - \frac{1}{\alpha^2}\right)$ series limit for balmer series

32. (a) The energy in the first orbit=E_1=Z^2E_0 where E_0=-13.6eV is the energy of a hydrogen atom in ground state. Thus for Li^{++},

$E_1 = 9E_0 = 9\times(-13.6eV)$

The energy in the third orbit is

$$E_3 = \frac{E_1}{n^2} = \frac{E_1}{9} = -13.6eV$$

Thus, $E_3 - E_1 = 108.8\ eV$; $\frac{hc}{\lambda} = E_a - E_1$

33. The energy of a photon of wavelength 500 nm is

$$\frac{hc}{\lambda} = \frac{1242eV - nm}{500nm} = 2.48eV$$

The energy of a photon of wavelength 700nm is

$$\frac{hc}{\lambda} = \frac{1242eV - nm}{700nm} = 1.77eV$$

The energy difference between the states involved in the transition should, therefore, be between 1.77 eV and 2.44 eV. $n = 3 \rightarrow n = 2$ falls in this range $\lambda = \frac{hc}{\Delta E} = \frac{1242eV - nm}{1.9eV} = 654nm$

34. $\frac{1}{\lambda} = RZ^2\left(\frac{1}{n_1^2} - \frac{1}{n_2^2}\right)$

35. $U = eV = eV_0 \ln\left(\frac{r}{r_0}\right)$

$$F = \frac{-dU}{dr} \Rightarrow F = -\frac{eV_0 r_0}{r} \Rightarrow F = -\frac{K}{r}$$

$$\frac{mv^2}{r_n} = \frac{K}{r_n} \qquad \Rightarrow mv = \sqrt{mK}\ ;\ mvr_n = \frac{nh}{2\pi}$$

$$\sqrt{mK}\, r_n = \frac{nh}{2\pi} \quad \Rightarrow r_n \propto n$$

36. For each principal quantum number n, number of electrons permitted equals the number of elements corresponding to the quantum number.

$$\begin{pmatrix} Total\ number \\ of\ electrons \end{pmatrix} = \Sigma 2n^2 = \frac{n(n+1)(2n+1)}{3}$$

37. $B_n = \frac{\mu_0 I_n}{2r_n}$ or $B_n \propto \frac{I_n}{r_n}; \propto \frac{(f_n)}{r_n}$

$$\therefore\ B_n \propto \frac{(v_n / r_n)}{r_n} \propto \frac{v_n}{(r_n)^2} \propto \frac{(z/n)}{(n^2/z)^2} \propto \frac{z^3}{n^5}$$

38. $E = R_\infty hc\left(1 - \frac{1}{25}\right)$ Momentum of photon emitted is $p = \frac{E}{c} = R_\infty h\left(\frac{24}{25}\right)$ Recoil momentum of H atom will also be p. $mv = p$

$$v = \frac{p}{m} = \frac{(1.097\times10^7)(6.626\times10^{-34})24}{(25)(1.67\times10^{-27})}$$

$\therefore v = 4.178\,ms^{-1}$

39. The total energy required to make the electron free from nucleus is the sum of the energy required to separate the electrons from the influence of each
other and the energy required to separate the electrons from the influence of nucleus, i.e., Total required energy = BE of electron in He atom + ionization energy of He atom

$$= (24.6 + 13.6\times2^2)eV = (24.6 + 54.4)eV = 79eV$$

40. $r_n \propto n^2$

41. $A_n \propto n^4$

42. $\tau = M \times B$ and $M = iA$

Where $i = \frac{ev}{2\pi r}$ $\quad A = \pi r^2$

$$M = \frac{ev}{2\pi r}\times\pi r^2 = \frac{(mvr)e}{2m} = \frac{(nh)e}{2\pi(2m)}$$

$$\therefore M = \frac{eh}{4\pi m} \quad \tau = MB\sin 30^{\circ}$$

43. $mvr = \frac{nh}{2\pi}$; $\frac{mv^2}{r} = \frac{GMm}{r^2}$

Energy $E = \frac{1}{2}mv^2 - \frac{GMm}{r}$

LEVEL - IV

COMPREHENSION TYPE

Passage-I:

A particle of charge equal to that of an electron, -e, and mass 208 times the mass of electron (called μ -meson)moves in a circular orbit around a nucleus of charge+3e.(Take the mass of the nucleus to be infinite). Assuming that Bohr model of the atom is applicable to this system:

1. **Derive an expression for the radius of the nth Bohr orbit**

1) $\frac{\varepsilon_0 n^2 h^2}{208\pi m_e e^2}$ 2) $\frac{\varepsilon_0 n^2 h^2}{3\pi m_e e^2}$ 3) $\frac{\varepsilon_0 n^2 h^2}{624\pi m_e e^2}$ 4) $\frac{\varepsilon_0 n^2 h^2}{64\pi m_e e^2}$

2. **Find the value of n for which the radius of the orbit is approximately the same as that of the first Bohr orbit for the hydrogen atom.**

1) 10 2)15 3) 25 4) 30

3. **Find the wavelength of the radiation emitted when the μ – meson jumps from the third orbit to the first**

1) 0.4500 A^0 2) 0.5500 A^0 3) 0.6500 A^0 4) None of these

(Rydberg's constant=$1.097 \times 10^7 m^{-1}$)

Passage-II

Phtoelectrons are emitted when 4000 A^0 radiation is incident on a surface of work function 1.9eV. These photoelectrons pass through a region has a-particles to form He^+ ion, emitting a single photon in this process He^+ ions thus formed are in their fourth excited state.

4. **Energy of the fourth excited state is approx**

1) -4.2eV 2) -2.2eV 3) -3.2eV 4) -1.2eV

5. **Energy released during the combination of He^+ ions is.**

1) 5.38 eV 2) 3.38eV 3) 2.38eV 4) 1.38eV

6. **Energy of emitted photon in range of 3eV& 4eV after combination is**

1) 3.86eV 2) 3.24eV 3) 3.29eV 4) 5.24eV

Passage - III:

A sample of hydrogen gas in its ground state is irradiated with photons of 10.2eV energies. The radiation from the above sample is used to irradiate two other samples of excited ionized He^+ and excited ionized Li^{2+}, respectively. Both the ionized samples absorb the incident radiation.

7. **How many spectral lines are obtained in the spectra of Li^{2+}**

1) 10 2) 15 3) 20 4) 17

8. **Which is the smallest wavelength that will be observed in spectra of He^+ ion?**

1) 24.4nm 2) 28.8nm 3) 22.2nm 4) 30.6nm

9. **How many spectral lines are observed in spectra of He^+ ion?**

1) 2 2) 4 3) 6 4)8

10. **Which is the smallest wavelength observed in the spectra of Li^{2+}?**
1) 8.6nm 2) 10.4nm 3) 12.8nm 4) 14.6nm

11. **Consider the spectral line resulting from the transition n = 2 to n = 1 in the atoms and the ions given below. The shortest wavelength is produced by**
1) hydrogen atom 2) deuterium atom
3) singly ionized helium 4) doubly ionized lithium

12. **A gas of monoatomic hydrogen is bombared with a stream of electron that have been accelerated from rest through a potential difference of 12.75 volt. In the emission spectrum, one cannot observe any line of**
1) Lyman series 2) Balmer series 3) Paschen series 4) Pfund series

13. **An electron jumps from the 4th orbit to 2nd orbit of hydrogen atom. Given the Rydberg's constant $r = 10^5\ cm^{-1}$, the frequency in hertz of the emitted radiation will be**
1) $\frac{3}{16}\times 10^5$ 2) $\frac{3}{16}\times 10^{15}$ 3) $\frac{9}{16}\times 10^{15}$ 4) $\frac{3}{4}\times 10^{15}$

14. **With increasing quantum number, the energy difference between adjacent energy levels in atoms**
1) decreases 2) increases 3) remains constant
4) decreases for low z and increases for high z

15. **Which of the following statements are true regarding Bohr's model of hydrogen atom?**
(I) orbiting speed of electron decreases as it shifts to discrete orbits away from the nucleus
(II) radii of allowed orbits of electron are proportional to the principal quantum number
(III) frequency with which electron orbits around the nucleus in discrete orbits is inversely proportional to the principal quantum number.
(IV) binding force with which the electrons are bound to the nucleus increases as it shifts to outer orbits.
Select the correct answer
1) I and III 2) II and IV 3) I, II and III 4) II, III and IV

16. A Hydrogen atom and a Li^{++} **ion are both in the second excited state. If l_H and l_{Li} are their respective electronic angular momenta, and E_H and E_{Li} their respective energies, then**
1) $l_H > l_{Li}$ and $|E_H| > |E_{Li}|$ 2) $l_H = l_{Li}$ and $|E_H| < |E_{Li}|$
3) $l_H = l_{Li}$ and $|E_H| > |E_{Li}|$ 4) $l_H < l_{Li}$ and $|E_H| < |E_{Li}|$

17. **Consider the spectral line resulting from the transition n = 2 $\rightarrow$ n = 1 in the atoms and ions given below, the shortest wavelength is produced by**
1) Hydrogen atom 2) deuterium atom
3) Singly ionized helium 4) doubly ionized lithium

18. **For hydrogen like system,**
(I) Ratio of magnetic moment to angular momentum is e/2m.
(II) energy of the electron is directly proportional to $1/n^2$.
(III) angular momentum varies inversely with n.
(IV) radius of an orbit is inversely related to n.
1) I, II 2) II, IV 3) I, II, IV 4) III, IV

19. **As per Bohr model,**

(I) minimum energy required to remove an electron from ground state of doubly ionized Li atom(z = 3) is 122.4 eV.

(II) energy of transition n = 3 to m = 2 is less than that of m = 2 to n = 1.

(III) minimum energy required to remove an electron from ground state of singly-ionised He atom (z = 2) is 27.2 eV.

(IV) A transition from state n = 3 to n = 2 in a hydrogen atom results in U-V radiations

1) II, III, IV 2) I, II 3) II, III 4) I, III, IV

20. **Assertion: A discharge tube appears dark, when evacuated to very low pressures.**
Reasoning: No colour is left at such low pressures.

1) If both assertion and reason are true and reason is the correct explanation of assertion.
2) If both assertion and reason are true but reason is not the correct explanation of assertion.
3) If assertion is true and reason is false. 4) If assertion is false but reason is true

21. **Assertion: An electron in hydrogen atom passes from n = 4 to n = 1 level. The maximum number of photons that can be emitted is 6.**
Reasoning: maximum number of photons omitted can only be 4.

1) If both assertion and reason are true and reason is the correct explanation of assertion.
2)If both assertion and reason are true but reason is not the correct explanation of assertion.
3)If assertion is true and reason is false. 4)If assertion is false but reason is true

LEVEL - IV - KEY

1) 3	2) 3	3) 2	4) 3	5) 2	6) 2	7) 2	8) 1	9)3	10) 2
11) 4	12) 4	13) 3	14) 1	15) 1	16) 2	17) 4	18) 1	19) 2	20) 3
21) 3									

LEVEL - IV - HINTS

1. $r_n = \dfrac{n^2 h^2 \in_o}{\pi m e^2 Z}$; Put $Z = 3$; $m = 208\, m_e$

2. $r_n = a_o \Rightarrow \dfrac{n^2 h^2 \in_o}{624\pi\, m_e e^2} = \dfrac{\in_o h^2}{\pi\, m_e e^2} \Rightarrow$ n= 25

3. $\dfrac{hc}{\lambda} = \Delta E$

4. $K_{max} = \dfrac{hc}{\lambda} - \phi = 1.18eV$; $E = -(13.6eV)\dfrac{Z^2}{n^2}$

5. Difference of energy when photoelectron recombines with α particle =1.18-(-2.2)=3.38eV
6. For de-excitation from n = 5 to n = 3; E= 3.86 eV
7. Li^{+2} is excited from $n = 3$ *to* $n = 6$ So, $N = 15$
8. He^{+} is excited from n = 2 to n = 4

$$\lambda_{min} = \frac{hc}{\Delta E} = 24.4\, nm$$

9. $N = \dfrac{4x(4-1)}{2} = 6$

10. $\lambda_{min} = \dfrac{hc}{\Delta E} = 10.4\, nm$

Chapter - 2

NUCLEI

NUCLEI

SYNOPSIS

NUCLEUS:

- The nucleus of an atom is at the centre. Most of the mass of an atom is at the centre. The entire positive charge of an atom lies in the nucleus.
- All atomic nuclei are made up of elementary particles called protons and neutrons. Proton is the nucleus of the hydrogen atom. It has a positive charge of 1.6×10^{-19} C having a mass of 1.6726×10^{-27} kg. This is nearly equal to 1836 times the electron mass. Neutron is electrically neutral (i.e. neutron carries no charge). Mass of neutron is slightly greater than that of the proton (1.6750×10^{-27} kg). Both the proton and neutron together constitute the nucleus. They are called nucleons.
- Generally, atomic number is denoted by Z and mass number is denoted by A and (A-Z) gives number of neutrons (N) in the nucleus.

  $\therefore N = A - Z$; $A = Z + N$
- Nucleus is positively charged and its shape is considered as spherical.

TYPES OF NUCLEI:

- **ISOTOPES:** Atomic nuclei having same atomic number but different mass numbers are known as isotopes. They occupy same position in the periodic table and possess identical chemical properties. They have same proton number.

  Ex: 1) $_3Li^6, _3Li^7$ 2) $_1H^1, _1H^2, _1H^3$
- **ISOTONES :** Atomic nuclei having same number of neutrons are called isotones.

  Ex: 1) $_{17}Cl^{37}, _{19}K^{39}$, 2) $_7N^{17}, _8O^{18}, _9F^{19}$
- **ISOBARS:** Atomic nuclei having same mass number but different atomic numbers are called Isobars. They have same number of nucleons.

  Ex:–1) $_{18}Ar^{40}, _{20}Ca^{40}$, 2) $_{32}Ge^{76}, _{34}Se^{76}$
- **ISOMERS:** Atomic nuclei having same mass number and same atomic number but different nulear properties are called isomers.

  Ex:- $m\,_{35}Br^{80}$ metastable Bromine and $g_{35}Br^{80}$ ground state Bromine are two isomers with different half lives
- **ISODIAPHERS:** Nuclei having different Atomic number (Z) and mass number (A) but with same excess number of neutrons over protons (A-2Z) are called isodiaphers. Ex:- $_{11}Na^{23}$, $_{13}Al^{27}$

SIZE OF THE NUCLEUS:

- Nuclear sizes are very small and are measured in fermi (or) femtometer. 1 fermi=10^{-15}m
- Radius of the nucleus depends on number of nucleons. $R = R_0 A^{1/3}$

  above equation does not apply to heavy nucleides

  Value of $R_o = 1.4 \times 10^{-15}$m
- Radius of the nucleus is in the order of 10^{-15} m.
- Size of an atom is in the order of 10^{-10}m.
- If an α -particle with an initial kinetic energy E

approaches a target of atomic number Z, if the distance of closest approach is "d" then $\frac{1}{4\pi \in_\circ}\frac{2Ze^2}{d} = E$

(Where 'e' is charge of an electron) If "v" represents the initial velocity of α particle, (m is mass of "α" particle) then $\frac{1}{4\pi \in_\circ}\frac{2Ze^2}{d} = \frac{1}{2}mv^2$

Note : If a particle of charge q, mass m is projected towards a nucleus of charge Q with velocity v from infinity then the distance of closest approach d is give by $\frac{1}{4\pi \in_\circ}\frac{qQ}{d} = \frac{1}{2}mv^2$

Note : If R, S and V be the Radius, surface area and volume of a nucleus with mass number A then

$$R \propto A^{\frac{1}{3}} \Rightarrow \frac{R_1}{R_2} = \left(\frac{A_1}{A_2}\right)^{\frac{1}{3}}; S \propto R^2 \propto A^{\frac{2}{3}} \Rightarrow \frac{S_1}{S_2} = \left(\frac{A_1}{A_2}\right)^{\frac{2}{3}}$$

$$V \propto R^3 \propto A \Rightarrow \frac{V_1}{V_2} = \frac{A_1}{A_2}$$

Note : If a stationary nucleus splits in to two lighter nuclei with mass numbers A_1 and A_2 then according to law of conservation of linear momentum, the two lighter nuclei move in opposite directions with equal momenta hence $\mathbf{m_1v_1 = m_2v_2}$

Ratio of velocities of the two nuclei

$$\frac{v_1}{v_2} = \frac{m_2}{m_1} = \frac{A_2}{A_1} = \left(\frac{R_2}{R_1}\right)^3 \quad (\because m \propto A \propto R^3)$$

Ratio of kinetic energy of the two nuclei

$$\frac{KE_1}{KE_2} = \frac{m_2}{m_1} = \frac{A_2}{A_1} = \left(\frac{R_2}{R_1}\right)^3$$

$$\left(\because KE = \frac{p^2}{2m} \;\&\; KE \propto \frac{1}{m} \text{ when p is constant}\right)$$

DENSITY OF THE NUCLEUS:

- Density of nucleus is independent of mass number of the atom.
- Density of the nucleus is 1.45×10^{17} Kgm$^{-3.}$
- The density is maximum at the centre and gradually falls to zero as we move radially out wards.
- Radius of the nucleus is taken as the distance between the centre and the point where the density falls to half of its value at the centre.
- Density of nucleus is of the order of $10^{14}\, gm/cc = 10^{17}\, kg/m^3$

W.E-1: ***Compare the radii of the nuclei of mass numbers 27 and 64.***

A. The ratio of the radii of the nuclei is

$$\frac{R_1}{R_2} = \left(\frac{A_1}{A_2}\right)^{\frac{1}{3}} = \left(\frac{27}{64}\right)^{\frac{1}{3}} (\because R = R_0A^{1/3}) = \frac{3}{4}$$

W. E-2 **: *The radius of the oxygen nucleus $^{16}_{8}O$ is 2.8×10^{-15}m. Find the radius of Lead nucleus***

A. $R_0 = 2.8 \times 10^{-15}$ m, $A_0 = 16$, $A_{Pb} = 205$ $R \propto A^{1/3}$

$$\frac{R_0}{R_{Pb}} = \left(\frac{A_0}{A_{Pb}}\right)^{1/3} = \frac{2.8 \times 10^{-15}}{R_{Pb}} = \left(\frac{16}{205}\right)^{1/3}$$

$R_{Pb} = 6.55 \times 10^{-15}$m.

ATOMIC MASS UNIT (A.M.U):

i) The masses of atoms, nuclei, sub atomic particles are very small. Hence, a small unit is used to express these masses. This unit is called as atomic mass unit (amu). ***1 amu is equal to one twelth part of the mass of carbon ($_6C^{12}$) isotope.***

Mass of $_6C^{12}$ is exactly 12 amu

ii) Now, the mass of 1 gm -mole of carbon is 12 gm and according to Avogadro's Hypothesis it has N (Avogadro's Number) atoms. Thus, the mass of one atom of carbon is (12/N) gm. According to the definition.

$$1amu = 1u = \frac{1}{12} \times (mass\ of\ one\ carbon\ atom)$$

$$= \frac{1}{12} \times \frac{12}{N} = \frac{1}{N} gm = \frac{1}{6.023 \times 10^{23}} gm$$

$$= 1.660565 \times 10^{-24} gm = 1.660565 \times 10^{-27} Kg$$

MASS - ENERGY EQUIVALENCE : According to Einstein's mass-energy equivalence principle, mass is another form of energy. Mass can be converted into energy & energy can be converted into mass according to the equation $E = mC^2$

Here m is the mass that disappears and E is the energy liberated. C is the velocity of light in vacuum.

When 1 amu of mass is converted in to energy

Energy liberated is given by

$E = (1.660565 \times 10^{-27}) \times 9 \times 10^{16}$ J = 931.5 MeV

hence 1 amu of mass is equivalent to 931.5 MeV of energy $\therefore$ $\boxed{1\ amu = 931.5\ MeV/C^2}$

The masses of electron, proton and neutron in terms of various units are :

Mass of the electron = $m_e = 9.1095 \times 10^{-31}$ kg

= 0.000549 u = 0.511 MeV/C^2

Mass of the proton = $m_p = 1.6726 \times 10^{-27}$ kg

= 1.007276 u = 938.28 MeV/C^2

Mass of the neutron = $m_n = 1.6750 \times 10^{-27}$ kg

= 1.008665 u = 939.573 MeV/C^2 .

NUCLEAR FORCES :

The attractive force which holds the nucleons together in the nucleus is called nuclear force.

Properties of nuclear forces :

1) ***Nuclear forces are strongest forces in nature.*** Nuclear forces are about 10^{38} times as strong as gravitational forces. The relative strengths of the gravitational, Coulomb's and nuclear forces are

$$F_g : F_e : F_n = 1 : 10^{36} : 10^{38}$$

2) ***Nuclear forces are short range forces .***

3) ***Nuclear forces are basically strong attractive forces, but contain a small component of repulsive forces.***

4) ***Nuclear forces are saturated forces.***
5) ***Nuclear forces are charge independent.***
6) ***Nuclear forces are spin-dependent.***
7) ***Nuclear forces are exchange forces.***
8) ***Nuclear forces are non-central forces.***

MASS DEFECT, BINDING ENEGRY, EINSTEIN'S MASS ENERGY RELATION

➢ When matter is completely annihilated, energy released is $E = mc^2$

➢ The energy equivalent to 1 amu is 931.5 M eV $= 1.4925 \times 10^{-10} J$.

MASS DEFECT: Atomic mass is always less than the sum of the masses of constituent particles. The difference between the total mass of the nucleons and mass of the nucleus of an atom gives mass defect.

$$\Delta m = \left[\left[ZM_p + (A - Z)M_n\right] - M_{nucleus}\right]$$

Z = Atomic number; M_p = Mass of proton
M_n = Mass of neutron; A=Mass number
$M_{nucleus}$ = Mass of nucleus.

BINDING ENERGY: The energy required to bring the nucleons from infinity to form the nucleus is called binding energy or it is the energy required to split a nuclens into nucleons.

It is energy equivalent of mass defect BE= $[\Delta m]C^2$

NOTE: BE = mass defect x 931.5 MeV if mass is expressed in a.m.u.

B.E. per nucleon = Binding fraction

$$\frac{\text{Binding Energy}}{\text{Mass Number}} = \frac{\Delta m \times 931 \text{ Mev}}{A}$$

Averge Binding energy or Binding energy fraction: It is the Binding energy per nucleon (or) the average energy needed to separate a nuclei in to its individual nucleons.

➢ Binding energy is not a measure of stability of a nucleus.

PACKING FRACTION OF A NUCLEUS :

Packing fraction : It is defined as the mass defect per nucleon. Packing fraction $= \frac{\Delta m}{A} = \frac{M - A}{A}$

If the packing fraction is negative then the nucleus is more stable.
If the packing fraction is positive then the nucleus is unstable.

Packing fraction is zero for ${}_6C^{12}$

➢ Packing fraction measures the stability of a nucleus. Smaller the value of packing fraction, large is the stability of the nucleus.

VARIATION OF B.E. PER NUCLEON WITH MASS NUMBER

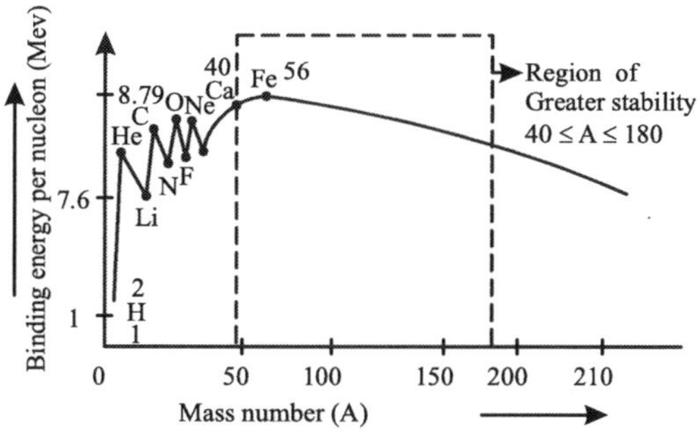

The main features of binding energy curve shown in figure are :

(1) The minimum value of binding energy per nucleon is in the case of deuteron (1.11MeV).

(2) The maximum value of $\frac{BE}{A}$ is 8.7McV for the nucleus $_{28}Fe^{56}$ (iron) which is the most stable.

(3) Binding energy is high in the range 28<A<138. The binding energy of these nuclei is very close to 8.7 MeV.

(4) Further increase in the mass number, binding energy per nucleon decreases and consequently for the heavy nuclei like uranium it is 7.6 MeV.

(5) In the region of smaller mass numbers, the binding energy per nucleon curve shows the characteristic minima and maxima. Minima are associated with nuclei containing an odd number of protons and neutrons such as $_3^6Li,_5^{10}B,_7^{14}N$ and the maxima are associated with nuclei having an even number of protons and neutrons such as $_2^4He,_6^{12}C,_8^{16}O$.

(6) Nuclei with A> 220 are distinctly unstable. That means from A> 220 single heavy nucleus breaks into two nearly equal nuclei with mass number A< 150 and so which are most stable. This process takes at right of the BE curve as shown in figure . This process explains the nuclear fission.

(7) Light nuclei such as hydrogen combine to form heavy nucleus to form helium for greater stability. This process takes at left of the BE curve as shown in figure. This process explains the nuclear fusion.

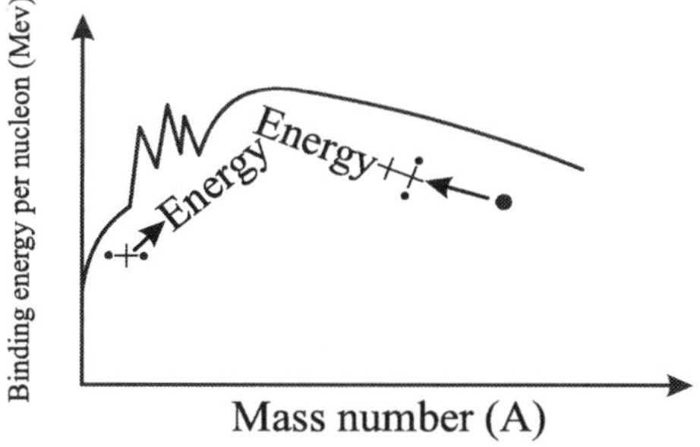

Note : Iron $\left(_{28}Fe^{56}\right)$ whose binding energy per nucleon stands maximum at 8.7 MeV is most stable and will undergo neither fission nor fusion.

Exo-ergic Reaction : ***The reaction in which energy will be released is called exo-ergic Reaction.*** $A+B \rightarrow C+D+Q$

Here A and B are called Reactants

C and D are called Products

Q is the amount of energy released

In an Exo - ergic Reaction

Mass of reactants > Mass of products

$\Delta m = M_R - M_P = (M_A + M_B) - (M_C + M_D)$

Energy Released $Q = \Delta m \times C^2$ joule [Δm is in kg] $= \Delta m \times 931.5$ MeV (Δ m is in amu)

If Binding energies are given then for Exo-ergic reactions.

(B.E) Products > (B.E) Reactants

Energy released $Q = (B.E)_P - (B.E)_R = [(B.E)_C + (B.E)_D] - [(B.E)_A + (B.E)_B]$

Endo-ergic Reaction : ***The reaction in which energy will be absorbed is called Endo-ergic Reaction.*** $A + B \rightarrow C + D - Q$

Here A and B are called Reactants

C and D are called Products

Q is the amount of energy absorbed

In an Endo - ergic Reaction

mass of reactants < Mass of products

$$\Delta m = M_P - M_R = (M_C + M_D) - (M_A + M_B)$$

Energy absorbed $Q = \Delta m \times C^2$ joule [Δm is in kg]

$\Rightarrow \Delta m \times 931.5$ MeV (Δ m is in amu)

If Binding energies are given then for

Endo-ergic reaction.

(B.E) Products < (B.E) Reactants

Energy absorbed $Q = (B.E)_R - (B.E)_P = [(B.E)_A + (B.E)_B] - [(B.E)_C + (B.E)_D]$

Note: A nuclear reaction can occur only if certain conservation laws are followed. These are :

1. Conservation of mass number A.
2. Conservation of charge.
3. Conservation of energy, linear momentum and angular momentum.

W.E-3 : ***Find the binding energy of $^{56}_{26}Fe$. Atomic mass of Fe is 55.9349u and that of Hydrogen is 1.00783u and mass of neutron is 1.00876u***

Sol. Mass of the hydrogen atom $m_H = 1.00783$u; Mass of neutron $m_n = 1.00867$ u; Atomic number of iron $Z = 26$; mass number of iron $A = 56$; Mass of iron atom $M_a = 55.9349$u

Mass defect $\Delta m = [Zm_H + (A-Z)m_n] - M_a$

$= [26 \times 1.00783 + (56-20)1.00867] - 55.93493$

u = 0.5287 u.

$\therefore$ Binding energy $= (\Delta m)c^2 = (0.52878)$

$c^2 = (0.52878)(931.5\text{MeV}) = 492.55$ MeV

W.E-4 : ***Find the energy required to split $^{16}_{8}O$ nucleus into four α - particles. The mass of an α -particle is 4.002603u and that of oxygen is 15.994915u.***

Sol. Mass of a-particle = 4.002603 u

Mass of oxygen = 15.994915u

B.E= [Mass of 4 particles-Mass of oxygen] x 931.5MeV

B.E = [4 x 4.002603 - 15.994915] x 931.5 MeV

= (16.010412-15.994915) x 931.5 MeV = 0.015497 x 931.5 ; B.E = 14.43 MeV

W.E-5 : ***Calculate the binding energy per nucleon of $^{40}_{20}Ca$. Given that mass of $^{40}_{20}Ca$ nucleus = 39.962589 u, mass of proton = 1.007825 u. mass of Neutron = 1.008665 u and 1 u is equivalent to 931 MeV.***

Sol. A=40, Z = 20, A - Z = 20

$\Delta m = \{Zm_p + (A-Z)m_n\} - M_n$

= { (20 x 1.007825+(20x1.008665)} - 39.962589

= 40.329800 - 39.962589 ; $\Delta m = 0.367211$

Binding energy per nucleon =

$$\frac{\Delta m \times 931}{A} = \frac{0.367211 \times 931}{40} = 8.547 MeV.$$

W.E-6 : ***The binding energies per nucleon for deuterium and helium are 1.1 MeV and 7.0 MeV respectively. What energy in joules will be liberated when 2 deuterons take part in the reaction.***

Sol. $^2_1H + ^2_1H \rightarrow ^4_2He + Q$

Binding energy per nucleon of helium (^4_2He) =7 MeV

Binding energy = 4 x7 = 28 MeV

Binding energy per nucleon of deuterium (^2_1H)=1.1MeV

Binding energy = 2 x 1.1 = 2.2 MeV

Energy liberated (Q) = (28 -(2.2)2] = 23.6 Mev.

i.e. Q=23.6x10^6x1.6x10^{-19}; $Q = 37.76 \times 10^{-13} J$

W.E-7: ***The kinetic energy of α-particles emitted in the decay of $_{88}Ra^{226}$ into $_{86}Rn^{222}$ is measured to be 4.78 MeV. What is the total disintegration energy or the 'Q-value of this process'?***

Sol. The standard relation between the kinetic energy of the α-particle (KE_α) and the Q-value (or total disintegration energy) is

$$KE_a = \left(\frac{A-4}{A}\right).Q \qquad Q = \left(\frac{A}{A-4}\right).KE_a$$

$$= \left(\frac{226}{226-4}\right) \times 4.78\text{MeV} \quad = \frac{226}{222} \times 4.78\text{MeV}$$

$Q = 4.865\text{MeV} \approx 4.87\text{MeV}$

W.E-8: ***A nucleus X-initially at rest, undergoes alpha-decay, according to the equation.***

$$^A_{92}X \rightarrow ^{228}_{z} Y + \alpha$$

The α-particle in the above process is found to move in a circular track of radius $1.1 \times 10^2 m$ ***in a uniform magnetic field of*** $3.0 \times 10^3 T$.

The energy (in MeV) released during the process and binding energy of the parent nucleus X, respectively.

Given: $m_y = 228.03\,amu$; $m_\alpha = 4.003\,amu$, $m\left(^1_0n\right) = 1.009\,amu$; $m\left(^1_0H\right) = 1.008\,amu$,

$1\,amu = 1.66 \times 10^{-27}$; $kg \equiv 931.5\,MeV/c^2$

Sol: The given equation is ${}^{A}_{92}X \rightarrow {}^{228}_{z}Y + {}^{4}_{2}He$

$A = 228 + 4 = 232\ ;\ 92 = z + 2 \qquad \therefore z = 90$

$$\frac{m_\alpha v_\alpha^2}{r} = qv_\alpha B\ ;\ v_\alpha = \sqrt{\frac{rqB}{m_\alpha}}$$

$$= \sqrt{\frac{1.1\times10^2\times2\times1.6\times10^{-19}\times3\times10^3}{4.003\times1.66\times10^{-27}}}$$

$$= 4.0\times10^6\, m/s$$

From conservation of linear momentum, $m_\alpha v_\alpha = m_y v_y$

$$v_y = \frac{m_\alpha v_\alpha}{m_y} = \frac{(4.003)(4.0\times10^6)}{(228.03)} = 7.0\times10^4\, m/s$$

There fore, energy released during the process $= \frac{1}{2}\left[m_\alpha v_\alpha^2 + m_y v_y^2\right] = \frac{(1.66\times10^{-27})}{(2\times1.6\times10^{-13})}$

$$\left[(4.003)(4.0\times10^6)^2 + (228.03)(7.0\times10^4)^2\right] MeV$$

$$= 0.34\, MeV = \frac{0.34}{931.5} amu = 0.000365 amu$$

Therefore, mass of ${}^{232}_{92}X = m_y + m_\alpha + 0.000365 = 232.033365\ u$

Mass defect $\quad \Delta m = 92(1.008) + (232 - 92)$

$(1.009) - 232.033365$.

$\therefore$ Binding energy $= 1.962635\times931.5\ MeV = 1828.2\, MeV$

NATURAL RADIO ACTIVITY :

Spontaneous decay of naturally occurring unstable nuclei by emission of certain sub particles (like α, β, and γ radiation) is called natural radio activity.

The emission of these rays takes place because of the instability of the nucleus. In the process of emitting these rays a nucleus tries to attain the stability.

In general natural radioactivity takes place in heavy nuclei beyond lead in the periodic table. There are also naturally radioactive light nuclei, such as potassium isotope ${}_{19}K^{40}$, the carbon isotope ${}_{6}C^{14}$ and the rubidium isotope ${}_{37}Rb^{87}$.

Regarding radioactivity.

(i) It is completely uneffected by the physical and chemical conditions to which the nucleus is subjected i.e we cannot change the radio activity by applying high temperature, high pressure and strong electric field etc.

(ii) The nucleus can disintegrate immediately (or) it may take infinite time.

(iii) The energy liberated during the radioactive decay comes from individual nuclei.

MODES OF DECAY :

The radioactive nucleus before decay is called a parent nucleus, the nucleus resulting from its decay by particles (Radiation) emission is called daughter nuclei.

This daughter nuclei may be stable (or) unstable.

$$\underset{\text{Parent}}{X} \longrightarrow \underset{\text{Daughter}}{Y} + \underset{\text{Radiation}}{R} + \underset{\text{Energy}}{Q}$$

Here R may be either α particle (or) β particle (or) γ radiation. Q is the energy of the emitted particles (or radiation).

α **-DECAY :** ***When a nucleus disintegrates by radiating*** α ***-rays, it is said to undergo*** α ***-decay.*** An α - particle is a helium nucleus. Thus a nucleus emitting an α particle losses two protons and two neutrons, as a result its atomic number Z decreases by 2, the mass number A decreases by 4 and the neutron number N decreases by 2.

$${}_{Z}X^{A} \rightarrow {}_{Z-2}Y^{A-4} + {}_{2}He^{4} + Q$$

Ex: ${}_{88}Ra^{226} \rightarrow {}_{86}Rn^{222} + {}_{2}He^{4} + 4.87MeV$

Both electric charge and nucleon number are conserved in the process of α **decay.**

Application : When a stationary Radio active nucleus x decays into another nucleus y by emitting an α-particle. $x \rightarrow y + \alpha$ particle $+Q$

Applying LCLM if α particle moves forward with a momentum 'P' then daughter nucleus y recoils with same momentum 'P' so that total momentum of the system is zero. Hence $\boxed{P_y = P_\alpha}$

The energy released 'Q' is in the form of K.E of daughter nucleus 'y' and 'α' particle.

$$\boxed{Q = KE_y + KE_\alpha}$$

Ratio of kinetic energies $\boxed{\dfrac{KE_y}{KE_\alpha} = \dfrac{M_\alpha}{M_y}}$

$\left(\because KE = \dfrac{P^2}{2m}\right.$ and $KE \ \alpha \ \dfrac{1}{m}$ when 'P' is same)

$$1 + \frac{KE_y}{KE_\alpha} = 1 + \frac{M_\alpha}{M_y}; \quad \frac{KE_\alpha + KE_y}{KE_\alpha} = \frac{M_y + M_\alpha}{M_y}$$

$$\boxed{KE_\alpha = Q\left(\frac{M_y}{M_\alpha + M_y}\right)}; \quad \boxed{KE_y = Q\left(\frac{M_\alpha}{M_\alpha + M_y}\right)}$$

Notice that KE_α ***is very close to (but smallerthan) Q.***

β **-DECAY : When a nucleus disintegrates by radiating** $\beta -$ **rays, it is said to undergo** $\beta -$ **decay.**

i) β particles are nothing but electrons. Hence when a nucleus emits a β particle, the atomic number (Z) increases by 1 unit, but the mass number does not change.

The general form of $\beta -$ decay can be written as

$${}_{Z}X^{A} \rightarrow {}_{Z+1}Y^{A} + {}_{-1}e^{0}.$$

Ex: ${}_{90}Th^{234} \rightarrow {}_{91}Pa^{234} + {}_{-1}e^{0}$

Both electric charge and nucleon number are conserved in β decay also.

γ -DECAY: When a nucleus disintegrates by radiating $\gamma -$ rays, it is said to undergo $\gamma -$ decay.

Gamma rays are nothing but electromagnetic radiations of short wavelengths (not exceeding 10^{-10}m.)

The emission of $\gamma -$ rays from the nucleus does not alter either atomic number Z or mass number A. It just results in the change of the energy state of a nucleus.

When a parent nucleus emits an α or a β particle, the daughter nucleus may be formed in one of excited states. Such a nucleus will eventually comes to the ground state. In this process $\gamma -$ radiation will be emitted.

$${}_{Z}X^{A} \rightarrow {}_{Z}X^{A} + \gamma \; Photon(s)$$

Example: ${}_{38}{}^{*}Sr^{87} \rightarrow {}_{38}Sr^{87} + \gamma$.

${}_{38}Sr$ is isomer of ${}_{38}Sr$.

Note : When a Radio active nucleus emits an α - particle followed by two β - particles, its isotope is formed.

$${}_{Z}X^{A} \xrightarrow{\alpha} {}_{Z-2}Y^{A-4} \xrightarrow{2\beta^{-}} {}_{Z}X^{A-4}$$

Note : When a Radio active nucleus emits a β - particle its isobar is formed.

$${}_{Z}X^{A} \xrightarrow{\beta^{-}} {}_{Z+1}Y^{A}$$

Note: When a Radio active nucleus emits a γ - particle its isomer is formed

$${}_{Z}X^{A} \xrightarrow{\gamma} {}_{Z}X^{A}$$

Deflection of Radioactive radiations in electric and magnetic fields :

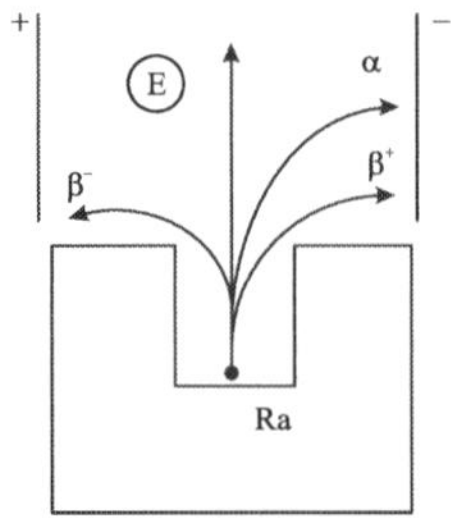

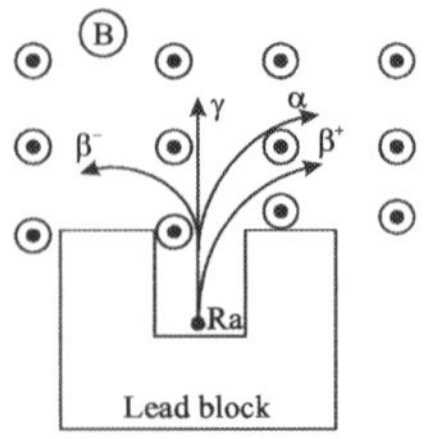

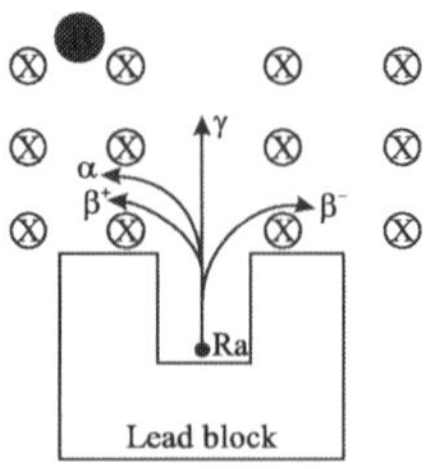

W.E-9: ***The nucleus $^{23}_{10}Ne$ decays by β^- emission. Write down the β-decay equation and determine the maximum kinetic energy of the electrons emitted. Given that :***
$m(^{23}_{10}Ne) = 22.994466u$; $m(^{23}_{11}Na) = 22.989770\ u$

Sol. $^{23}_{10}Ne \rightarrow ^{23}_{11}Na + \bar{e} + \bar{v} + Q$

For β^- - decay, Q = [M(x) - M(y)]C² = [22.994466-22.989770]931.5

= 0.004696 x 931.5 = 4.37 MeV

W.E-10: ***Calculate the binding energy of an α -particle. Given that mass of proton = 1.0073u, mass of neutron = 1.0087u. and mass of α -particle = 4.0015u.***

Sol. $m_P = 1.0073u$, $m_N = 1.0087u$, M = 4.0015u

N = A - Z = 4 - 2 = 2 $\quad \left(\because {}_2He^4 = {}_ZX^A\right)$

B.E = Δm x 931.5 MeV $= \left\{\left[Zm_p + (A-Z)m_n\right] - M\right\} \times 931.5$

$\left[\left[(2\times1.0073)+(2\times1.0087)-4.0015\right]\right]\times 931.5MeV$

= 0.0305 $\times$ 931.5 MeV ; B.E = 28.4 MeV

W.E-11: ***How many α and β - particles are emitted when uranium nucleus $({}_{92}U^{238})$ decay to ${}_{82}Pb^{214}$?***

Sol. Let n be the number of α - particles and m be the number of β - particles emitted.

${}_{92}U^{238} \rightarrow {}_{82}Pb^{214} + n_2He^4 + m\ {}_{-1}e^0.$

As mass is conserved, 238 = 214 + 4n + m (0)

= 214 + 4n ; 4n = 24; n = 6

As charge is conserved , 92 = 82 + 2n + m (-1)

10 = 2(6) - m $(\because n = 6)$; m = 2.

$\therefore 6\alpha$ - particles and 2 β - particles are emitted

RADIOACTIVE DECAY LAW:

Based on their experimental observations and analysis of certain radioactive materials Rutherford and Soddy formulated a theory of radioactive decay. According to them

After decay of a nucleus the new product (daughter) of nucleus has totally different physical as well as chemical properties.

The rate of radioactive decay (or) the number of nuclei decaying per unit time at any instant is directly proportional to the number of nuclei (N) present at that instant and is independent of the external physical conditions like temperature, pressure etc.

Let 'N' be the number of radioactive atoms present at a time 't' and N_0 is the initial number of radio active nuclei. Let dN atoms disintegrate in time 'dt'. According to the law of radioactive decay

$$\left(\frac{dN}{dt}\right) \propto N ; \left(\frac{dN}{dt}\right) = -\lambda N \ldots\ldots. (1)$$

The proportionality constant λ is called decay constant (or) disintegration constant. The negative sign indicates that as time increases N decreases.

From eqn (1) $\frac{dN}{N} = -\lambda\, dt$ ....(2)

Integrating eq (2) on both sides $\int \frac{dN}{N} = -\lambda \int dt$

$\log_e N = -\lambda t + C$ .....(3)

Here C is the constant of integration

At t = O, N= N_0 Substituting in eqn (3),

we get, $\log_e N_0 = C$

$\therefore \log_e N = -\lambda t + \log_e N_0$;

$\therefore \log_e N - \log_e N_0 = -\lambda t$

$\therefore \log_e \left(\frac{N}{N_0}\right) = -\lambda t ; \frac{N}{N_0} = e^{-\lambda t} ; \boxed{N = N_0 e^{-\lambda t}}$ ... (4)

This shows that the number of radioactive nuclei decreases exponentially with time.

Above equation is known as the decay law (or) the law of radio-active decay. It is an exponential law.

taking logarithm on both sides for the above equation. $\log_e N = \log_e N_0 - \lambda t$; $\lambda t = \log_e \frac{N_0}{N}$

$\therefore \boxed{t = \frac{1}{\lambda} \ln\left(\frac{N_0}{N}\right)}$

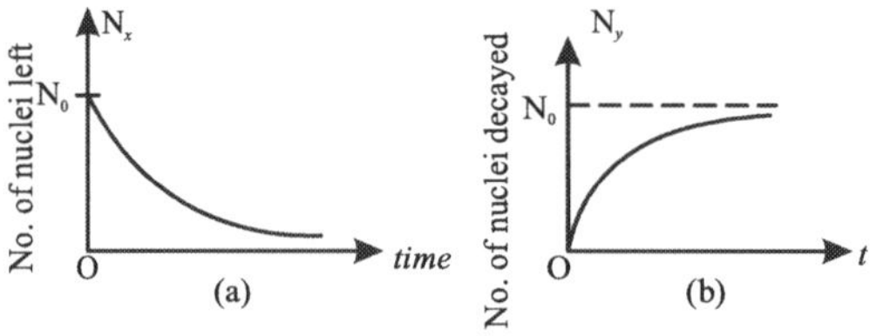

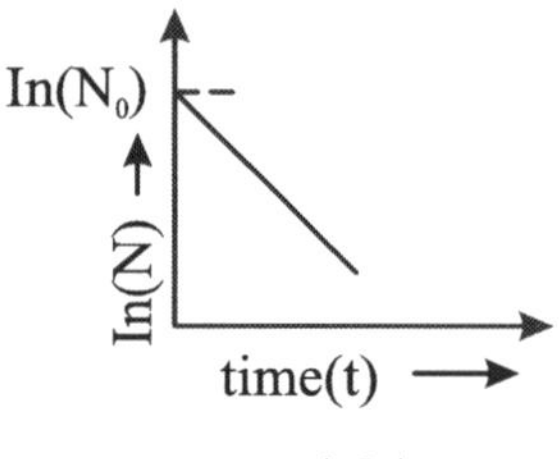

(C)

ACTIVITY (R) :

The number of decays per unit time (or) decay rate is called activity (R)

$|R| = \left|\frac{dN}{dt}\right| = \frac{d}{dt}\left(N_0 e^{-\lambda t}\right)$ (or) $R = \lambda N = \lambda N_0 e^{-\lambda t}$ (or)

$R = R_0 e^{-\lambda t}$, where $R_0 = \lambda N_0$ is the decay rate at $t = 0$, called initial activity.

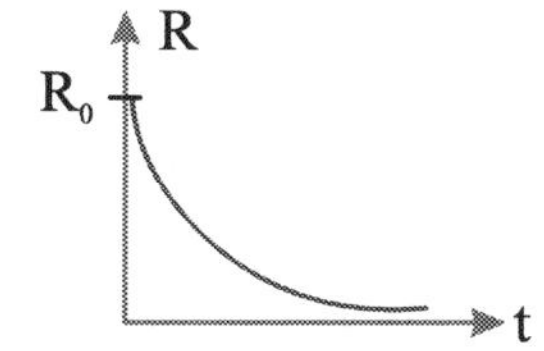

If a nucleus can decay simultaneously by n processes, which have activities R_1, R_2, ... and R_n. Then the resultant activity $R = R_1 + R_2 + ... + R_n$. **If nucleus decays simultaneously more than one process is called parallel decay.**

The S.I unit of activity is **Becquerel (Bq)** and other units are **curie (Ci) and Rutherford (Rd).**

1 Bq = 1 decay per second,

1 Rd = 10^6 decays per second.

1 Ci = 3.7×10^{10} decays per second.

Note : Curie is approximately equal to the activity of one gram of pure radium.

DECAY CONSTANT (λ) : It gives the ability of a nucleus to decay. ***The decay constant λ for a given radio active sample is defined as the reciprocal of the time during which the number of nuclei decreases to $\frac{1}{e}$ times their original value.***

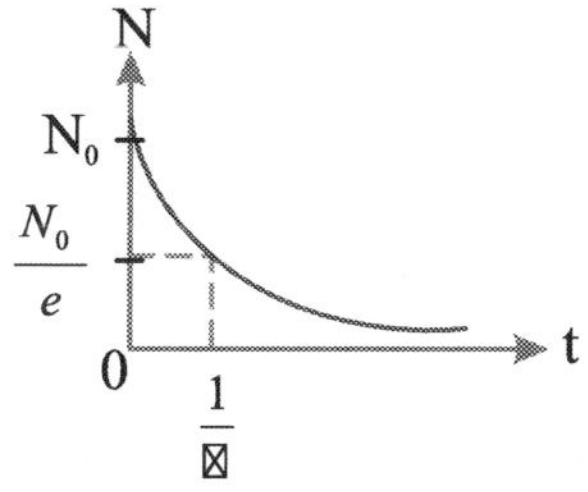

1) **Larger value of λ corresponding to decay in smaller time and vice versa.**
2) $\lambda = 0$ for stable nuclei.
3) Decay constant is the characteristic of the sample taken and does not vary with time.
4) If a nucleus can decay simultaneously by more than one process (say n), which have decay constants $\lambda_1, \lambda_2 ....and\ \lambda_n$, then the effective decay constant is $\lambda = \lambda_1 + \lambda_2 + ... + \lambda_n$. This is called parallel decay.

HALF LIFE (T) : As the name suggests, the half life of a radioactive sample is defined as ***"The time interval during which the activity of a radio active sample falls to half of its value, (or) The time interval during which the number of radio active nuclei of a sample disintegrate to half of its original number of nuclei"*** Half lives vary from isotope to isotope. While T may be as small as 10^{-16} s, its largest value may be as big as 10^9 years.

Eg: Half-life of uranium $\left({}^{238}_{92}U\right)$ is 4.47 x 10^9 years

half-life of krypton $\left({}^{89}_{36}Kr\right)$ is 3.16 minutes.

Relation between decay constant (λ) and half life period (T).

From Law of Radioactive decay $\frac{N}{N_0} = e^{-\lambda t}$

when $N=\frac{N_0}{2}$, $t=T$ $\therefore \frac{1}{2}=e^{-\lambda T}$ or $2=e^{\lambda T}$

taking logirthms on both sides $\ln 2=\lambda T$

(or) $\log_e 2=\lambda T$ $\therefore T=\frac{2.303\log_{10}2}{\lambda}=\frac{0.693}{\lambda}$

$$\therefore \boxed{T=\frac{\ln 2}{\lambda}=\frac{2.303\log 2}{\lambda}=\frac{0.693}{\lambda}}$$

The above relation establishes that the half - life (T) depends upon the decay constant λ of the radioactive substance. The value of λ is different for different radioactive substances.

Note :

i) Half life is the characteristic property of the sample and T cannot be changed by any known method.

ii) At any given instant whatever be the amount of the undecayed sample, it will be reduced to exactly half its value after a time equal to the half life of the sample.

iii) In parallel decay $\lambda=\lambda_1+\lambda_2+........\lambda_n$ hence $\frac{1}{T}=\frac{1}{T_1}+\frac{1}{T_2}+........\frac{1}{T_n}$, where T is the equivalent half-life and $T_1, T_2 ....T_n$ are the half-lives in individual decay.

Application :

In a radioactive sample the number of nuclides undecayed after n-half lives (i.e., $t=nT$) is

$$t=nT=\frac{1}{\lambda}\ln\left(\frac{N_0}{N}\right) \text{ or } \frac{n(\ln 2)}{\lambda}=\frac{1}{\lambda}\ln\left(\frac{N_0}{N}\right)$$

$$\text{or } 2^n=\frac{N_0}{N}; \text{ or } \boxed{N=N_0\left(\frac{1}{2}\right)^n}$$

Note: The number of nuclei remain in the sample after **half of half life period (t=1/2T) is** given by

$$N=N_0\left(\frac{1}{2}\right)^n \text{ here } n=\frac{1}{2} \text{ then } N=N_0\left(\frac{1}{2}\right)^{\frac{1}{2}}$$

$$\therefore \boxed{N=\frac{N_0}{\sqrt{2}}} \text{ taking } N_0=100, N=50\sqrt{2}=70.7$$

70.7% of nuclei remain and 29.3% of nuclei decayed.

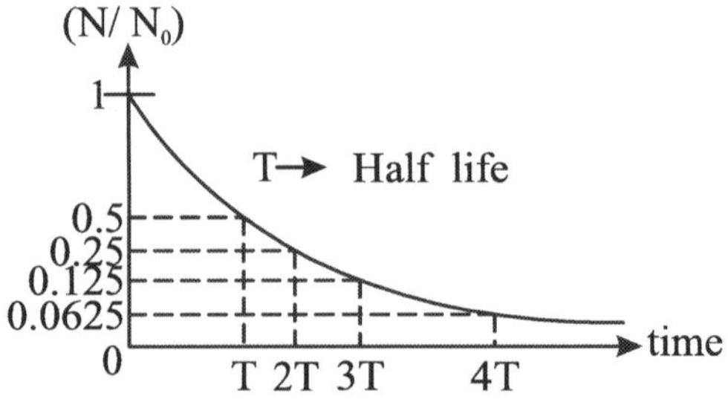

AVERAGE LIFE (OR) MEAN LIFE :

The phenomenon of radioactivity is random because we just can't predict which of the atoms in a given sample will decay first and when. Hence radioactivity process totally depends on chance. In decay process some of the atoms of the given sample may have very short life span, and others may not decay even after a very large span of time. So to determine the ability of the nucleus to decay it would be useful to calculate the average life. Hence average life is defined as ***the total life time of all the nuclei divided by the total number of original nuclei.***

i.e $\tau = \frac{\sum \text{life span of individual nucleus}}{\text{Total number of original nuclei}} = \frac{\sum t}{N_0}$

Let N_0 be the radio active nuclei that are present at $t = 0$ in the radioactive sample.

The number of nuclei which decay between t and $(t + dt)$ is dN i.e the life time of these nuclei is 't'.

The total life time of these dN nuclei is (t dN)

$\therefore$ The total life time of all the nuclei present initially in the sample $= \int_{t=0}^{t=\infty} t\,dN$ $[\because N = 0 \text{ at infinity}]$

Average life time $\tau = \frac{\int t\,dN}{N_0}$ But $\frac{-dN}{dt} = \lambda N$

$dN = -\lambda N dt = -\lambda N_0 e^{-\lambda t} dt \quad (\because N = N_0 e^{-\lambda t})$

$\tau = \int_0^{\infty} t \frac{\lambda N_0 e^{-\lambda t}}{N_0} dt \; ; \; \boxed{\tau = \frac{1}{\lambda}}$

The mean life (or) average life of a radio active sample is reciprocal to decay constant.

We know that $N = N_0 e^{-\lambda t}$; When $t = \tau$, $\quad N = N_0 e^{-\frac{1}{\tau} \times \tau} = \frac{N_0}{e} = 0.37 N_0 = 37\% \text{ of } N_0$

Hence average life period of a radio active sample can also be defined as ***"The time interval during which 63% of sample decays or sample reduces to 37% of its original amount"***.

RELATION BETWEEN HALF LIFE PERIOD AND AVERAGE LIFE PERIOD

We know that $T = \frac{0.693}{\lambda}$ & $\tau = \frac{1}{\lambda}$

Hence $T = 0.693\tau$ (or) $\boxed{\tau = \frac{T}{0.693} = 1.443T}$

From the above equation it is clear that average life period is 44.3% greater than half life period.

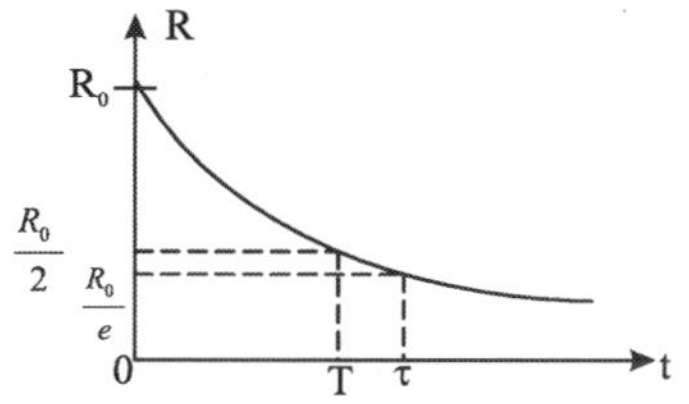

Determination of decay constant (λ) and half life period (T) of a radioactive sample graphically

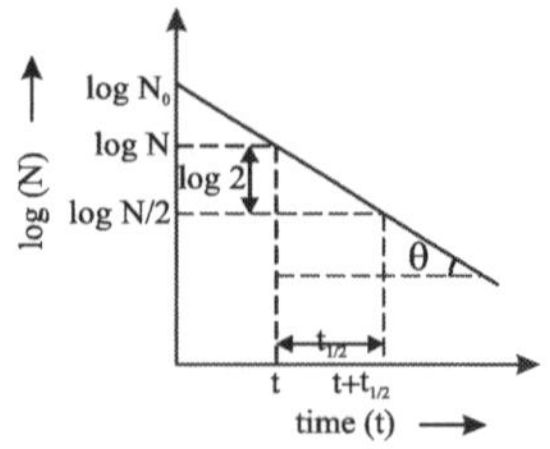

If N_0 and N be the number of atoms present undecayed initially and after a time t, then

We know that $N=N_0\, e^{-\lambda t}$ taking log on both sides

$$\log_e N = \log_e N_0 - \lambda t \Rightarrow \log N = \log N_0 - \frac{\lambda t}{2.303}$$

$$\log N = \left(\frac{-\lambda}{2.303}\right)t + \log N_0$$

Slope of the graph $m = -\tan\theta = \frac{-\lambda}{2.303}$

$\Rightarrow \boxed{\lambda = 2.303 \tan\theta}$

Half life period $T = \frac{2.303 \log 2}{\lambda}$

$T = \frac{2.303 \log 2}{2.303 \tan\theta}$ $\quad\therefore \boxed{T = (\log 2)\cot\theta}$

Note : In radioactive sample decay

1) The probability survival of nucleus after time $P_s = \frac{N}{N_0} = e^{-\lambda t}$.

2) The probability of nucleus to disintegrate in time t is $P_d = 1 - P_s = 1 - e^{-\lambda t}$.

W.E-12: ***A radioactive sample has an activity of 5.13 x10⁷ Ci. Express its activity in 'becqueral' and 'rutherford'.***

Sol. Since 1 Ci = 3.7×10^{10} decays per second,

activity = 5.13×10^7 Ci

$= 5.13 \times 10^7 \times 3.7 \times 10^{10}$ Bq $= 1.9 \times 10^{18}$ Bq

Since , 1×10^6 decay per second = 1Rd

Activity $= 1.9 \times 10^{18}$ Bq $= \frac{1.9\times10^{18}}{1\times10^{6}} Rd = 1.9 \times 10^{12}$ Rd.

W.E-13: ***A radioactive substance has 6.0×10^{18} active nuclei initially. What time is required for the active nuclei of the same substance to become 1.0×10^{18} if its half-life is 40 s.***

Sol. The number of active nuclei at any instant of time t,

$$\frac{N_0}{N} = e^{\lambda t}\ ; \quad \log_e\left(\frac{N_0}{N}\right) = \lambda t$$

$$\therefore t = \frac{\log_e\left(\frac{N_0}{N}\right)}{\lambda} = \frac{2.303\log_{10}\left(\frac{N_0}{N}\right)}{\lambda}$$

In this problem, the initial number of active nuclei,

$N_0 = 6.0 \times 10^{18}$; $N=1.0\times10^{18}, T=40s$,

$$\lambda=\frac{0.693}{T}=\frac{0.693}{40}=1.733\times10^{-2}\,s^{-1}.$$

$$t=\frac{2.303\log_{10}\left(\frac{6.0\times10^{18}}{1.0\times10^{18}}\right)}{1.733\times10^{-2}}=\frac{2.303\log_{10}(6)}{1.733\times10^{-2}}=\frac{2.303\times0.7782}{1.733\times10^{-2}}=103.4s.$$

W.E-14: A radioactive sample can decay by two different processes. The half-life for the first process is T_1 and that for the second process is T_2. Find the effective half-life T of the radioactive sample.

Sol. Let N be the total number of atoms of the radioactive sample initially. Let $\frac{dN_1}{dt}$ and $\frac{dN_2}{dt}$ be the initial rates of disintegrations of the radioactive sample by the two processes respectively. Then

$$\frac{dN_1}{dt}=\lambda_1 N \text{ and } \frac{dN_2}{dt}=\lambda_2 N$$

Where λ_1 and λ_2 are the decay constants for the first and second processes respectively.

The initial rate of disintegrations of the radioactive sample by both the processes

$$=\frac{dN_1}{dt}+\frac{dN_2}{dt}=\lambda_1 N+\lambda_2 N=(\lambda_1+\lambda_2)N.$$

If λ is the effective decay constant of the radioactive sample, its initial rate of disintegration.

$$\frac{dN}{dt}=\lambda N$$

But $\frac{dN}{dt}=\frac{dN_1}{dt}+\frac{dN_2}{dt}$

$$\lambda N=(\lambda_1+\lambda_2)N$$

$$\lambda=\lambda_1+\lambda_2$$

$$\frac{0.693}{T_1}+\frac{0.693}{T_2}=\frac{0.693}{T}$$

$$\frac{1}{T}=\frac{1}{T_1}+\frac{1}{T_2};T=\frac{T_1T_2}{T_1+T_2}.$$

W.E-15: *Plutonium decays with a half life of 24,000 years. If plutonium is stored for 72,000 years, what fraction of it remains?*

A. $T_{1/2}$ = 24,000 years

Duration of time (t) = 72,000 years

$$\text{Number of half lifs (n)}=\frac{t}{T_{1/2}}=\frac{72000}{24000}=3$$

$$\therefore\ 1g\xrightarrow{1}\frac{1}{2}g\xrightarrow{2}\frac{1}{4}g\xrightarrow{3}\frac{1}{8}g$$

$\therefore$ Fraction of plutonium remains $= \frac{1}{8} g$

W.E-16: ***A certain substance decays to 1/32 of its initial activity in 25 days. Calculate its half -life.***

Sol. $1g \xrightarrow{1} \frac{1}{2}g \xrightarrow{2} \frac{1}{4}g \xrightarrow{3} \frac{1}{8}g \xrightarrow{4} \frac{1}{16}g \xrightarrow{5} \frac{1}{32}g \therefore n = 5$

$$(n) = \frac{t}{t_{1/2}} \Rightarrow t_{1/2} = \frac{t}{n} = \frac{25}{5};\ t_{1/2} = 5 \text{ days}$$

W.E-17: ***The half -life period of a radioactive substance is 20 days. What is the time taken for 7/8th of its original mass to disintegrate?***

Sol. Let the initial mass be one unit.

Mass reamaining $= 1 - \frac{7}{8} = \frac{1}{8}$

A mass of 1 unit becomes $\frac{1}{2}$ unit in 1 half life

$\frac{1}{2}$ unit becomes $\frac{1}{4}$ unit in 2nd half life

$\frac{1}{4}$ unit becomes $\frac{1}{8}$ unit in 3rd half life

$\therefore$ Time taken = 3 half lifes = 3 x 20 = 60 days

W.E-18: ***How many disintegrations per second will Occur in one gram of $^{238}_{92}U$, if its half-life against α -decay is 1.42 x 10^{17}s?***

Sol. Given Half-life period (T) $= \frac{0.693}{\lambda} =$

1.42×10^{17} s

$$\lambda = \frac{0.693}{1.42 \times 10^{17}} = 4.88 \times 10^{-18}$$

Avagadro number (N) = 6.023×10^{23} atoms

n = Number of atoms present in 1 g of $^{238}_{92}U = \frac{N}{A}$

$$= \frac{0.623 \times 10^{23}}{238} = 25.30 \times 10^{20}$$

Number of disintegrations $= \frac{dN}{dt} = \lambda n$

$= 4.88 \text{ x} 10^{-18} \text{ x } 25.30 \text{ x } 10^{20} = 1.2346 \text{ x } 10^4$ disintegrates/sec

W.E-19: ***One gram of radium is reduced by 2 milligram in 5 years by α -decay. Calculate the half-life of radium.***

Sol: Initial mass = 1 g, t = 5 years

Reduced mass = 2mg = 2 x 10^{-3} g = $\frac{2}{1000}$ g

Remaining mass = 1 - $\frac{2}{1000} = \frac{998}{1000}$

$\frac{N}{N_0} = \frac{998}{1000}$ ($\because$ *Mass* $\propto$ *Number of atoms*)

$\frac{N}{N_0} = e^{-\lambda t}$; $\frac{998}{1000} = e^{-\lambda t}$

$\frac{1000}{998} = e^{\lambda t} = e^{5\lambda} \Rightarrow \log_e\left(\frac{1000}{998}\right) = 5\lambda$

2.303 (3.0000 -2.9991) = 5 λ

$\lambda = \frac{2.303 \times 1 \times 0.0009}{5}$

$(T_{1/2}) = \frac{0.693}{\lambda} = \frac{0.693 \times 5}{2.303 \times 0.0009} = 1671.7$ years

W.E-20: ***The half-life of a radioactive substance is 5000 years . In how many years, its activity will decay to 0.2 times of its initial value? Given $\log_{10}5 = 0.6990$.***

Sol. T = 5000 years,

$\frac{N}{N_0} = 0.2 = \frac{2}{10} = \frac{1}{5}$

$\lambda = \frac{0.693}{T} = \frac{0.693}{5000}$

$\frac{N}{N_0} = e^{-\lambda t}$

$\frac{1}{5} = \frac{1}{e^{\lambda t}} \Rightarrow 5 = e^{\lambda t}$

$\log_e^5 = \lambda t$

$2.303 \times 0.6990 = \lambda t$

$t = \frac{2.303 \times 0.6990 \times 5000}{0.693}$

t = 11614.6years = 1.1615×10^4 years

W.E-21: ***Obtain the amount of $^{60}_{27}$Co necessary to provide a radioactive source of 8.0 mCi strength. The half -life of $^{60}_{27}$Co is 5.3 years.***

Sol. Half - life of $^{60}{}_{27}Co$ = 5.3 years

= 5.3x365x24 x 60 x 60 S = 5.3 x 3.15 x 10^7 s

Now `x` gm of $^{60}{}_{27}Co$ contains $\frac{10^{-3}x}{60}$ k - mole $= \frac{x\times10^{-3}}{60}\times6.025\times10^{26}$

$N = 0.1004x \times 10^{23} = 1.004\, x \times 10^{22}$ atoms

Required strength of $^{60}{}_{27}Co$ = 8 m Ci

$= 8 \times 10^{-3} \times 3.7 \times 10^{10}$ dis/s

We know that, decay rate (R) = λN

$$N = \frac{R}{\lambda} = \frac{8\times3.7\times10^{7}}{0.693/T_{1/2}}$$

$$= \frac{29.6\times10^{7}}{0.693} \times 5.3 \times 3.15 \times 10^{7} = 713.0909 \times 10^{14}$$

$1.004x \times 10^{22} = 713.0909 \times 10^{14}$

$x = 710.2499 \times 10^{-8}$ kg $= 7.1 \times 10^{-6}$ g

ARTIFICIAL TRANSMUTATION OF ELEMENTS

The conversion of one element into another by artificial means is called artificial transmutation of the element. Rutherford performed number of experiments in which the atoms of different stable elements, such as nitrogen, aluminium, phosphorus, etc, were bombarded by high speed α – particles from natural radioactive substances. Finally in 1919, he discovered the phenomenon of artificial transmutation.

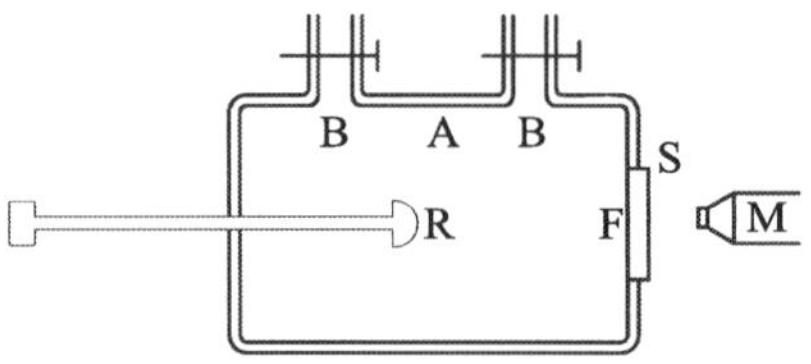

The apparatus used by Rutherford is as shown in Fig.

i. It consists of a chamber A provided with an adjustable rod, carrying a radio - active substance R (Radium C).

ii. The side of the glass tube facing 'R' is covered by metal plate with a central hole which is closed by a thin silver foil 'F'.

iii. A screen 'S', coated with a fluorescent material like zinc sulphide is arranged infront of the silver foil and the scintillations produced on it can be observed through the microscope 'M'.

iv. The side tubes B, B were used to fill various gases in the chamber.

v. The source of α - particles, Ra was placed on a small disc at R. Its distance from F was adjustable.

vi. The radio-active substance emits α -particles whose range in air was found to be about 7cm.

vii. When the glass tube is filled with nitrogen gas, scintillations are observed, even when 'R' is at a distance of 40cm from the foil.

viii. These particles producing scintillations can not be α -particles as they can not have such a long range.

ix. Rutherford concluded that nitrogen nucleus hit by an α ($_2He^4$)-particle transmutes into oxygen nucleus along with a proton ($_1H^1$).

x. The nuclear reaction causing artificial transmutation can be represented as $_7N^{14} + {}_2He^4 \rightarrow {}_8O^{17} + {}_1H^1$
Thus an atom of nitrogen is transformed into an isotope of oxygen. This process is called transmutation of elements.

* High energy α - particles were used in the discovery of artificial transmutation and neutron because α - particles produce intense ionisation of the medium through which they pass and can be stopped after travelling a few mm in air.

Significance :

i) It leads to the discovery of proton and neutron.

ii) It helps to produce radio isotopes.

iii) It helps to produce transuranic elements.

RADIO ISOTOPES AND THEIR USES:

Radio isotopes have very short half lives and hence used for various purposes.

1) Medical applications :

a) $_{13}Al^{27} +_0 n^1 \rightarrow_{11} Na^{24*} +_2 He^4$

Radio – sodium is used to find out how a given medicine is circulated in the body. It is also used to find out circulatory disorders in blood vessels.

b) $_{15}P^{31} +_1 H^2 \rightarrow_{15} P^{32*} +_1 H^1$

Radio – phosphorus is used in the treatment of skin diseases. It is also used for the treatment of blood disorders.

c) $_{53}I^{127} +_0 n^1 \rightarrow_{53} I^{128*} + \gamma - \text{rays}$

Radio – iodine is used in the treatment of thyroid glands. Radio - iodine (I^{131}) is used for diagnosis and treatment of brain tumor and for the study of pumping condition of heart.

d) $_{27}Co^{59} +_0 n^1 \rightarrow_{27} Co^{60*} + \gamma - \text{rays}$ Radio – cobalt is used in the detection and treatment of cancer

e) Radio-iron is used to detect anemia and treat anemia.

2) In Geology:

a) Radio carbon(C^{14}) is used to determine the age of fossils by radio - carbon dating

b) Radio isotopes are used to determine the age of rocks by the ratio of U^{238} to Pb^{206}

3) In industry :

a) Radio – isotopes are used to find the wear and tear of machine parts

b) Radio isotopes are used to detect flaws in metal structures

c) Radio isotopes are used for treatment of alloys such as quenching , annealing and hardening.

d) Radio isotopes are used in the selection of appropriate lubricants.

4) In research: Radio - isotopes are used in the study of nuclear disintegrations of elements.

5) In food preservation: By exposing vegetables and other food stuffs to radiations from radio - active isotopes, their shelf life can be increased.

6) In agriculture:

a) Radio phosphorus (P^{32}) is used to study the uptake of phosphorus by plants using.

b) Radio sulphur (S^{34}) is used to study the transport of minerals in plants .

c) Radio zinc is used to develop new species of plants by causing genetic mutation.

d) Irradiation by γ - radiations of seeds to improve yields.

7) In Chemistry :

a) Radio oxygen (O^{18}) is used to study the mechanisms of photosynthesis and hydrolysis of ester

b) Radio isotopes are used in the chemical analysis of solubility of sparingly soluble salts such as $PbSO_4$ and AgCl and determination of trace amounts of elements in industrial raw materials and products.

NEUTRON

- It is electrically neutral and its mass is slightly greater than that of proton. It was discovered by chadwick
- Bothe – Becker equation:

$$_4Be^9 + _2He^4 \rightarrow \left[_6C^{13}\right] \rightarrow _6C^{12} + _0n^1$$

- Neutron is unstable outside the nucleus.

$_0n^1 \rightarrow _1H^1 + _{-1}e^0 + \bar{v}$ (anti neutrino)

- It has high penetrating power and low ionizing power.
- Slow moving neutrons are called thermal neutrons. Fast moving neutrons convert into thermal neutrons when they pass through a substance called moderator.
- Thermal neutrons have an average energy of nearly 0.025 eV. Fast moving neutrons have an average energy of 2 MeV.

NUCLEAR FISSION

- Nuclear Fission is a nuclear reaction in which a heavy atomic nucleus like U^{235} splits into two approximately equal parts, emitting neutrons and liberating large amount of energy.
- Bohr and Wheeler proposed liquid drop model to explain this fission process.
- Nucleus of U^{235} undergoes fission when it is struck by slow neutrons. This fission is not due to the impact of neutron.
- Energy of about 200 MeV is released during one fission reaction of $_{92}U^{235}$. The most probable nuclear fission reaction is

$$_{92}U^{235} + _0n^1 \rightarrow _{56}Ba^{141} + _{36}Kr^{92} + 3\,_0n^1 + \text{energy}$$

There is no guarantee that U^{235} always breaks into Barium and Krypton.

- On an average, in the fission of U^{235}, 2.5 neutrons are emitted per fission when fission occurs due to slow neutrons. U^{235} undergoes fission with fast neutrons also. But this probability is minimum.
- Fission fragments are unstable and emit neutrons some time after fission reaction which are called "delayed neutrons"
- 99% of neutrons emitted during fission process are prompt.
- Delayed neutrons play an important role in chain reaction

CHAIN REACTION:

- If the mass of fissionable material exceeds a critical value, chain reaction or self propagating fission reaction takes place.
- The rate of reaction increases in geometric progression during uncontrolled chain reaction.
- Chain-reaction : The process of continuation of nuclear fission which when once started continues spontaneously without the supply of additional neutrons from outside is defined as chain reaction.

Reproduction factor (K): "It is the ratio of number of neutrons in any particular generation to the number of neutrons in the preceeding generation.

Case(i): $K<1$; Chain reaction is not maintained.(sub-critical state)

Case (ii): $K=1$: Chain reaction is maintained at steady rate. (critical state). In the state electricity is produced in the reactors at steady rate

Case (iii): $K>1$: Chain reaction becomes self sustained and lead to atomic explosion(super critical state)

- Uncontrolled chain reaction takes place in atom bomb.

NUCLEAR REACTOR OR ATOMIC PILE:

- Nuclear reactor is a device in which nuclear fission is produced by controlled self sustaining chain reaction. And is used for the production of nuclear power (energy).
- The essential parts of a nuclear reactor are (i) the fuel, (ii) moderator, (iii) control rods, (iv) coolant, (v) radiation shields.
- **THE FUEL:** The common fuels used are uranium $\left(U^{238}\right)$, enriched uranium (U^{235}) and plutonium $\left(Pu^{236}\right)$ and Th^{232}.

MODERATOR:

- The function of a moderator is to slow down the fast moving neutrons to increase the rate of fission.
- The commonly used moderators in the order of efficiency are (i) Heavy water (ii) graphite, (iii) Berillium and Berillium Oxide
- Heavy water is a best moderator
- A good moderator should have

  (1) low atomic mass

  (2) poor absorption of neutrons

  (3) good scattering property.

  (4) The size of moderator atom should be nearly of same size as that of the size of a prompt neutron.

CONTROL RODS:

- The function of a control rod is to absorb (capture) the neutrons.
- Cadmium, Boron and steel rods are used as control rods in a nuclear reactor.
- Cadmium rods are best control rods
- They regulate the net rate of neutron production and hence they control the intensity of fission process.

COOLANT:

- The function of a coolant is to keep the reactor temperature at a low value so that there may not be any danger of heat damage to the reactor.
- Air and CO_2 are used as gaseous coolants. Water, Organic liquids, Helium, Liquid Sodium are used as liquid coolants. Liquid sodium is best coolant.
- **Protective shield:** The process of preventing radioactive effect around nuclear reactor is called Protective Shield.
- During the working of a nuclear reactor dangerous radiations such as high energy neutrons, gamma rays and thermal radiations are produced. To protect the persons working there, the reactor is thoroughly shielded with concrete wall of several feet thick and lined with metals like lead.

POWER OF A NUCLEAR REACTOR

In the nuclear reactor, large amount of heat will be generated in the core. These reactors have elaborate cooling systems that use water. This water absorbs the heat and produces steam. This steam in turn is used to run the steam turbines which ultimately generate electric power. Such reactors are called power reactors.

The power generated by a nuclear reactor is $\boxed{P=\frac{nE}{t}}$ here $\frac{n}{t}$ be the number of fissions per second and E be the energy released in each fission

$E = 200\text{ MeV} = 200\times10^{6}\times1.6\times10^{-19}\text{ J}$

$= 3.2 \times 10^{-11}\text{J}$

Note : Number of fissions per sec in a reactor of power 1 W is given by $\frac{n}{t}=\frac{P}{E}$

$=\frac{1}{3.2\times10^{-11}}$ = 3.125 x 10^{10} fissions per sec

Note : If only x% of energy released in fission is converted into electrical energy then out put power of reactor is $P=\frac{x}{100}\left(\frac{nE}{t}\right)$

Note: If 'x' gm of fuel with mass number 'A' completely undergo nuclear fission in time t sec in a reactor then its power is given by

Number of moles in x gm of fuel $=\frac{x}{A}$

Number of atoms (nuclei) present in x gm of fuel $n=\left(\frac{x}{A}\right)N_A$. Where N_A is Avogadro number

$\therefore$ power $P=\frac{nE}{t}\Rightarrow \boxed{P=\frac{xN_AE}{At}}$

Uses of Nuclear Reactors:

1) To generate electric power.

2) To produce nuclear fuel plutonium -239 and other radioactive materials which have a wide variety of applications in the fields of medicine, industry and research.

USES OF ATOMIC ENERGY :

1. **Generation of electric power** : The coolant in a nuclear reactor absorbs the heat generated as a result of the chain reaction and it releases the heat to the water which is converted into high pressure steam. This steam is used to drive turbine and operate the electric generator.

2. **Production of radio isotopes** : A small amount of the pure element is placed in an aluminium container and the container is placed in the reactor for a few days. The element absorbs neutrons and the element becomes radioactive isotope.

   $_{53}I^{127}+_0n^1\rightarrow_{53}I^{128*}+\gamma$.

   Radioiodine obtained in this way can be used to treat the thyroid gland. These radio isotopes have a number of applications in the field of medicine , agriculture, industry and basic research.

3. **Source of neutrons :** A large number of neutrons are produced in a reactor. They are used in research . The effect of neutrons on biological tissues is studied. A new branch of physics called Neutron Physics has come up.

4. Atomic energy is used to create artificial lakes, to divert the course of a river , to make tunnels for laying new railway tracks etc.

5. Atomic energy is used for driving automobiles, submarines and war - planes.

6. Atomic energy is used in war - fare for creating destructive atom bombs and hydrogen bombs.

NUCLEAR FUSION:

➢ The phenomenon in which two lighter nuclei combine to form a heavier nucleus of mass less than the total mass of the combining nuclei is called nuclear fusion. This mass defect appears as energy.

- At temperatures of about 10^7K, light nuclei combine to give heavier nuclei. Hence, fusion reactions are called thermo nuclear reactions.
- Nuclear fusion takes place in the sun and other stars.
- Energy produced in a single fission of $_{92}U^{235}$ is larger than that in a single fusion of Hydrogen into Helium.
- But fusion produces more energy than fission per nucleon.
- In fission, 0.09% of mass is converted into energy. In fusion 0.66% of mass is converted into energy.
- Hydrogen bomb is a fission – fusion bomb.

STELLAR AND SOLAR ENERGY:

Stellar and solar energy is due to fusion.

The cycles that occur are.

Proton - Proton Cycle & Carbon - Nitrogen Cycle

PROTON - PROTON CYCLE:

The Thermonuclear reactions involved are:

$$2(_1H^1)+2(_1H^1) \rightarrow 2(_1H^2) + 2(_{+1}e^0) + Q_1$$

$$2(_1H^2)+2(_1H^1) \rightarrow 2(_2He^3)+Q_2$$

$$_2He^3 + _2He^3 \rightarrow (_2He^4) + 2(_1H^2) + Q_3$$

On adding up these reactions, we obtain.

$$4(_1H^1) \rightarrow _2He^4 + 2\ _{+1}e^0 + Q$$

Where $Q=Q_1+Q_2+Q_3$ is the total energy evolved in the fusion of 4 hydrogen nuclei (protons) to form Helium nucleus. The value of Q as calculated from mass defect comes out to be 26.7MeV

CARBON - NITROGEN CYCLE:

Proposed by bethe. It consists of following reactions.

$$_6C^{12} +_1 H^1 \rightarrow_7 N^{13} + Q_1 \quad ; \quad _7N^{13} \rightarrow _6C^{13} + (_1e^0) + Q_2$$

$$_6C^{13} +_1 H^1 \rightarrow_7 N^{14} + Q_3; \quad _7N^{14} +_1 H^1 \rightarrow_8 O^{15} + Q_4$$

$$_8O^{15} \rightarrow _7N^{15} + (_{+1}e^0) + Q_5;$$

$$_7N^{15} +_1 H^1 \rightarrow_6 C^{12} +_2 He^4 + Q_6$$

On adding all these 6 equations, we get

$$4\ _1H^1 \rightarrow _2He^4 + 2(_{+1}e^0) + Q$$

Where $Q=Q_1+Q_2+Q_3+Q_4+Q_5+Q_6$ The value of Q as calculated from mass defect is 26.7 Mev.

- In the sun both proton - proton cycle and carbon- nitrogen cycles occur with equal probabilities. In stars, whose interior temperatures are less than that of the sun, proton -proton cycle dominates the energy generation. Again in stars, whose interior temperatures are more than that of the sun, the energy generation is mainly due to carbon- nitrogen cycle.
- The core temperature of heavier stars may be larger than that of the sun and much larger nuclei may be formed.

NOTE: Due to enormous energy released in Sun and Stars the atmosphere of them will be in ionised state which is called Plasma (Which contains fast moving neutrons and electrons). Nuclear fusion can not be controlled.

Nuclear Fission	Nuclear Fusion
1) Neutrons are required for it	1) Protons are required for it
2) It is possible at normal pressure and temperature	2) It is possible at high pressure and temperature
3) Energy released per nucleon ≅ 0.9MeV	3) Energy released per nucleon ≅ 6MeV
4) % of mass getting converted into energy =0.1%	4) % of mass getting converted into energy = 0.7%
5) Fissionable materials are expensive.	5) Fusion materials are cheap
6) Harmful reactions are produced.	6) Harmful reactions are not produced.

W.E-22: ***An explosion of atomic bomb releases an energy of 7.6 x 10^{13} J. If 200 MeV energy is released on fission of one ^{235}U atom calculate (i) the number of uranium atoms undergoing fission. (ii) the mass of uranium used in the atom bomb***

Sol: E=7.6x10^{3}J; Energy released per fission = 200 MeV

$= 200 \times 10^6 \times 1.6 \times 10^{-19} = 3.2 \times 10^{-11}$ J

$$\text{Number of uranium atoms (n)} = \frac{\text{Total energy}}{\text{Energy per fission}}$$

$$n = \frac{7.6 \times 10^{13}}{3.2 \times 10^{-11}} = 2.375 \times 10^{24} \text{ atoms}$$

Avagadro number (N) = 6.023×10^{23} atoms

Mass of uranium =

$$\frac{n \times 235}{N} = \frac{2.375 \times 10^{24} \times 235}{6.023 \times 10^{23}} = 92.66\text{g}$$

W.E-23: ***Calculate the energy released by fission from 2 g of $^{235}_{92}U$ in kWh. Given that the energy released per fission is 200 MeV.***

Sol. Mass of uranium = 2g

Energy released per fission = 200 MeV

$= 200 \times 10^6 \times 1.6 \times 10^{-19} = 3.2 \times 10^{-11}$ J

Number of atoms in 2 gram of uranium is

$$n = \frac{2 \times 6.023 \times 10^{23}}{235} = 5.125 \times 10^{21} \text{ atoms}$$

Total energy released = No. of atoms x energ released per fission

$= 5.125 \times 10^{21} \times 3.2 \times 10^{-11} = 16.4 \times 10^{10}$ J

$$\therefore \text{Energy in Kwh} = \frac{16.4 \times 10^{10}}{36 \times 10^5} \text{Kwh}$$

$= 0.455 \times 10^5$ Kwh $= 4.55 \times 10^4$ Kwh

W.E-24: ***200 Mev energy is released when one nucleus of ^{235}U undergoes fission. Find the number of fissions per second required for producing a power of 1 megawatt.***

Sol. Energy released = 200MeV

$= 200 \times 10^6 \times 1.6 \times 10^{-19} = 3.2 \times 10^{-11}J$

P = 1 mega watt = 10^6 watts.

$$\text{No.of fissions per second (n)} = \frac{\text{Total energy}}{\text{Energy per fission}}$$

$$n = \frac{10^6}{3.2 \times 10^{-11}} = 3.125 \times 10^{16} \text{ Fissions}$$

W.E-25: ***How much ^{235}U is consumed in a day in an atomic power house operating at 400 MW, provided the whole of mass ^{235}U is converted into energy?***

Sol. Power = 400 MW = 400×10^6 W;

time = 1 day = 86, 400 s.

Energy produced , E = power × time = $400 \times 10^6 \times 86{,}400 = 3.456 \times 10^{13}$ J.

As the whole of mass is converted into energy , by Einstein's mass -energy relation.

$E = Mc^2$

$$\frac{E}{c^2} = \frac{3.456 \times 10^{13}}{(3 \times 10^8)^2} = 3.84 \times 10^{-4}\ kg = 0.384\ g.$$

W.E-26: ***How long can an electric lamp of 100W be kept glowing by fusion of 2.0 kg of deuterium ? Take the fusion reaction as $^2_1H + ^2_1H \rightarrow ^3_2He + n + 3.27$ MeV***

Sol. $_1H^2 +_1 H^2 \rightarrow^3_2 He + n + 3.27MeV$

No. of atoms in 2 kg of $_1H^2$ = $2/2 \times 6.023 \times 10^{26} = 6.023 \times 10^{26}$ atoms

In the above reaction two deuterium nuclei are combined

Power (p) = w x rate of fusion.

$$= 3.27 \text{ MeV x } \frac{Number\ of\ atoms}{Time\ expended}$$

$$100 = 3.27 \times 10^6 \times 1.6 \times 10^{-19} \times \frac{6.023 \times 10^{26}}{2x}$$

$$\therefore x = \frac{3.27 \times 1.6 \times 6.023 \times 10^{+11}}{2} = 15.756 \times 10^{11} S$$

$$= \frac{15.756 \times 10^{11}}{365 \times 24 \times 60 \times 60} = \frac{15.756 \times 10^{11}}{3.15 \times 10^7} = 5 \times 10^4 \text{ years}$$

W.E-27: ***Suppose India had a target of producing by 2020 AD, 200,000 MW of electric power, ten percent of which was to be obtained from nuclear power plants. Suppose we are given that, on an avedrage, the efficiency of utilization (i.e conversion to electric energy) of thermal energy produced in a reactor was 25%. How much amount of fissionable uranium would our country need per year by 2020 ? Take the heat energy per fission of ^{235}U to be about 200 MeV.***

Sol. Required power from nuclear plants

$= 10\%$ of 2,00,000 Mw $= 2 \times 10^{10}$w

Required electric energy from nuclear plants in one year $= 2 \times 10^{10} \times 365 \times 24 \times 60 \times 60$

$= 2 \times 10^{10} \times 3.15 \times 10^{7} = 6.30 \times 10^{7}$ J

Available electric energy per fission = 25% of 200 MeV = 50 MeV $= 8 \times 10^{-12}$J

Req. no. of fissions per year $= \frac{6.30\times10^{17}}{8\times10^{-12}} = 0.7875\times10^{29}$

Req. no. of moles of $U^{238} = \frac{0.7875\times10^{29}}{6.023\times10^{23}} = 0.1307\times10^{6}$

Required mass of $U^{238} = 0.1307 \times 235 \times 10^{6}$ g

$= 30.71 \times 10^{6}$ gm $= 30.71 \times 10^{6} \times 10^{-3}$ kg

$= 0.03071 \times 10^{6}$ kg $= 3.071 \times 10^{4}$ kg

W.E-28: Calculate the energy released by the fission 1 g of ^{235}U in joule, given that the energy released per fission is 200 MeV.
(Avogadro's number = 6.023×10^{23})

Sol. The number of atoms in 1 g of ^{235}U

$$= \frac{Avogadro's\ number}{Mass\ number} = \frac{6.023\times10^{23}}{235} = 2.563\times10^{21}$$

Energy released per fission = 200 MeV

$$= 200\times10^{6}\times1.6\times10^{-19} = 3.2\times10^{-11}J.$$

Energy released by 1 g of ^{235}U

= Number of atoms $\times$ energy released per fission

$$= 2.563\times10^{21}\times3.2\times10^{-11}J = 8.202\times10^{10}J$$

W.E-29: ***In the process of nuclear fission of 1 gram uranium, the mass lost is 0.92 milligram. The efficiency of power house run by it is 10%. To obtain 400 megawatt power from the power house, how much uranium will be required per hour? ($c = 3 \times 10^{8}\ ms^{-1}$)***

Sol. Power to be obtained from power house = 400 mega watt\ Energy obtained per hour = 400 meagwatt x 1hour = $(400 \times 10^{6}$ watt$) \times 3600$ second $= 144 \times 10^{10}$ joule

Here only 10% of input is utilised. In order to obtain 144×10^{10} joule of useful energy, the output energy from the power house $\frac{10E}{100} = 144 \times 10^{10}$J

$E = 144 \times 10^{11}$ joule

Let, this energy is obtained from a mass–loss of Δm kg.

Then $(\Delta m)c^2 = 144\times10^{11}$joule

$$\Delta m = \frac{144\times10^{11}}{(3\times10^{8})^2} = 16\times10^{-5}kg \quad = 0.16\ g$$

Since 0.92 milli gram ($= 0.92 \times 10^{-3}$ g) mass is lost in 1 g uranium, hence for a mass loss of 0.16g, the uranium required is $= \frac{1\times0.16}{0.92\times10^{-3}} = 174\ g$

Thus to run the power house, 174 gm uranium is required per hour.

➢ **PAIR AND PRODUCTION AND PAIR ANNIHILATION :** When an energetic γ - photon falls on a heavy nucleus, it is absorbed by the nucleus and a pair of electron and positon is produced. This phenomenon is called as *pair production* and can be represented by the following equation:

$$\underset{(\gamma-photon)}{h\nu} = \underset{(Positron)}{{}_{+1}\beta^0} + \underset{(electron)}{{}_{-1}\beta^0}$$

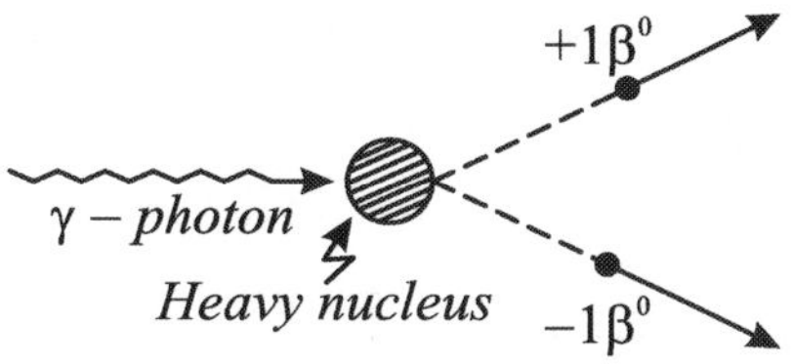

The rest mass energy of electron or positron is:

$$E_0 = m_0c^2 \qquad = (9.1\times10^{-31})\times(3\times10^8)^2$$

$$= 8.2\times10^{-14} J \simeq 0.51 MeV.$$

Hence for pair production, the minimum energy of γ -photon must be $2\times0.51 = 1.02MeV$. If the energy of γ -photon is less than this, there may be Compton's effect. If energy of γ -photon is greater than E_0, then extra energy will become kinetic energy of the particles. If E is the energy of γ -photon,

then kineric energy of each particle will be, $K_{electron} = K_{positron} = \dfrac{E-2E_0}{2}$

The inverse process of pair production is called *pair annihilation*. According to it when electron and a positron come close to each other, annihilate each other and produces minimum two γ -photons. Thus

$$\underset{(Positron)}{{}_{+1}\beta^0} + \underset{(electron)}{{}_{-1}\beta^0} = \underset{(\gamma-photon)}{2hf}$$

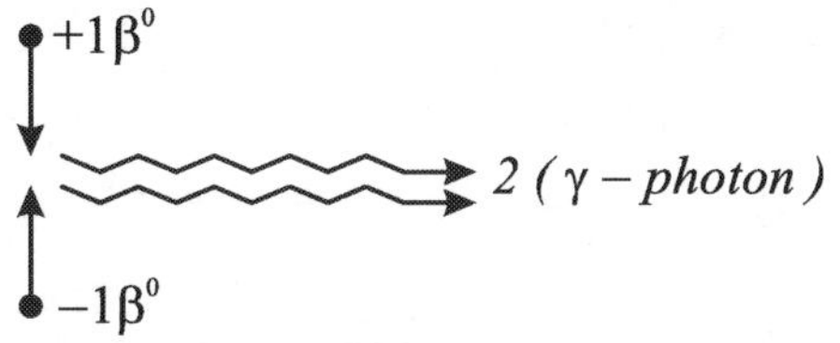

ADDITIONAL INFORMATION

ELEMENTARY PARTICLES :

We have realized, so far that there are only four fundamental constituents of matter. We can describe various physical processes involving atoms, molecules and nuclei in terms of electrons, protons, neutrons and photons. The first three are the building blocks of atoms and hence matter. The fourth one (i.e photon) is the quantized energy which is exchanged whenever electronic or nucleonic transition is involved.

Subsequently many more elementary particles and antiparticles have been discovered, using giant and modern accelerating machines.

The particles which are not constituted by any other particles are called Elementary particles. A brief discussion of important fundamental particles is as follows.

i) **Electron :** It was discovered in 1897 by Thomson. Its charge is –e and mass is 9.1 X 10^{-31} kg. Its symbol is e^- (or $_{-1}\beta^0$). It is a stable particle having spin $= \frac{1}{2}$

ii) **Proton :** It was discovered in 1919 by Rutherford in artificial nuclear disintegration. It has a positive charge +e and its mass is 1836 times (1.673 X 10^{-27}kg) the mass of electron. In free state, the proton is a stable particle. Its symbol is P^+. It is also written as $_1H^1$. It is a stable particle having spin = 1/2.

iii) **Neutron :** It was discovered in 1932 by Chadwick. Electrically it is a neutral particle. Its mass is 1839 times (1.675 X 10^{-27} kg) the mass of electron. In free state the neutron is unstable. Inside the nucleus the neutron is stable. Its symbol is n (or) $_0n^1$.

iv) **Positron :** It was discovered by Anderson in 1932. It is the antiparticle of electron, i.e., its charge is +e and its mass is equal to that of electron. Its symbol is e^+ (or $_{+1}\beta^0$)

v) **Antiproton :** It is the antiparticle of proton. It was discovered in 1955. Its charge is –e and its mass is equal to that of proton. Its symbol is P^-.

vi) **Antineutron :** It was discovered in 1956. It has no charge and its mass is equal to the mass of neutron. **The only difference between neutron and antineutron is that their magnetic momenta will be equal in magnitude and opposite in direction**. The symbol for antineutron is $\bar{n}$.

vii) **Neutrino and antineutrino :** The existence of these particles was predicted in 1930 by Pauli while explaining the emission of β - particles from radioactive nuclei, but these particles were actually observed experimentally in 1956. Their rest mass and charge are both zero but they have energy and momentum. These are mutually antiparticles of each other. They have the symbol ν and $\bar{\nu}$

viii) **Pi - Mesons :** The existence of pi - mesons was predicted by Yukawa in 1935, but they were actually discovered in 1947 in cosmic rays. Nuclear forces are explained by the exchange of pi-mesons between the nucleons. pi - mesons are of three types, positive π - mesons (π^+), negative pi-mesons (π^-) and neutral π - mesons (π^0). Charge on $\pi^\pm$ is $\pm$e. Whereas mass of $\pi^\pm$ is 274 times the mass of electron. π^0 has mass nearly 264 times the electronic mass. These are unstable having half life 10^{-8} sec and spin = 0

ix) **Mu-Mesons :** These were discovered in 1936 by Anderson and Neddermeyer. These are found in abundance in the cosmic rays at the ground level. There are two types of mu-mesons. Positive mu-meson (μ^+) and negative mu-meason (μ^-). There is no neutral mu-meson. Both the mu-mesons have the same rest mass 207 times the rest mass of the electron. These are unstable having half life 10^{-6} sec and spin = 1/2.

x) **Photon :** These are bundles of electromagnetic energy and travel with the speed of light. Energy and momentum of a photon of frequency ν are $h\nu$ and $\frac{h\nu}{c}$ respectively. They posses no charge. no mass and spin = 1 and are stable.

xi) **Gravitons :** Hypothetical particles that carry gravitational energy are called Gravitons. They possess no mass, no charge and spin = 2 as proposed by Dirac.

Antiparticles : An antiparticle is a form of matter that has the same mass as the particle but carries an opposite charge and / or a magnetic moment that is oriented in an opposite direction relative to the spin.

Name of the particle	Antiparticle
Electron (e^{-1})	Positron (e^{+1})
Proton (P^+)	Anti proton (P^-)
Neutron (n)	Anti Neutron ($\bar{n}$)
Neutrino (ν)	Anti Neutrino ($\bar{\nu}$)
Positive Pi-Meson (π^+)	Negative Pi-Meson (π^-)
Positive Mu-Meson (μ^+)	Negative Mu-Meson (μ^-)

Note: A few electrically neutral particles, like the photon and neutral π meson are their own antiparticles. A collision between a particle and an antiparticle results in annihilation of matter.

Classification of particles based on spin

1) **Bosons :** These particles have spin in the integral multiples of unity
2) **Fermions :** These particles have spin in the integral multiples of 1/2.

Classification of particles based on rest mass

1) **Photons :** Particles with zero rest mass
2) **Leptons :** Lighter particles
3) **Mesons :** Particles with intermediate mass
4) **Baryons :** Heavier particles

Classification of particles based on interaction

1) **Photons:** representing electromagnetic interactions.
2) **Leptons:** representing weak interactions.
3) **Hadrons:** representing strong interactions.
4) **Gravitons:** representing gravitational interactions.

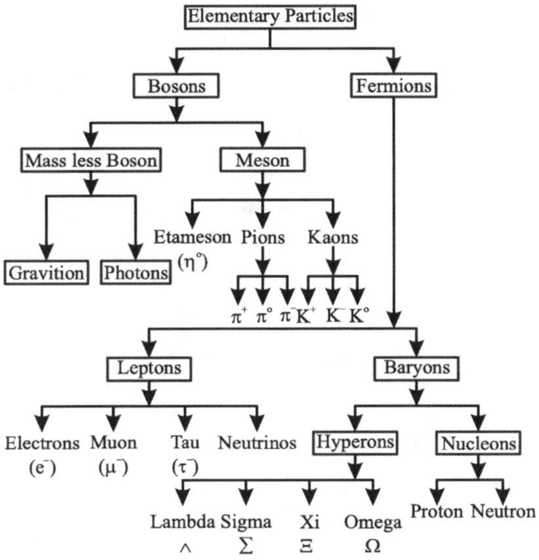

THE QUARK MODEL:

According to this model, each baryon and meson particle is made up of quarks. Quarks are supposed to be the ultimate building blocks of these particles and hence matter. They are point particles with no internal structure. Unlike any other charged particles, quarks are supposed to have fractional charges and spin 1/2.

Name	Quarks Symbol	Quarks Charge	Antiquarks Symbol	Antiquarks Charge
Up	u	$+\frac{2}{3}e$	$\overline{u}$	$-\frac{2}{3}e$
Down	d	$-\frac{1}{3}e$	$\overline{d}$	$+\frac{1}{3}e$
Strange	s	$-\frac{1}{3}e$	$\overline{s}$	$+\frac{1}{3}e$
Charmed	c	$+\frac{2}{3}e$	$\overline{c}$	$-\frac{2}{3}e$
Top	t	$+\frac{2}{3}e$	$\overline{t}$	$-\frac{2}{3}e$
Bottom	b	$-\frac{1}{3}e$	$\overline{b}$	$+\frac{1}{3}e$

These quarks combine to form some well known Mesons and Baryons. A meson consists of a quark and an antiquark whereas a baryon contains three quarks.

Particle	Quarks	charge	spin
proton	(u, u, d)	e	$\frac{1}{2}\uparrow$
neutron	(u d d)	0	$\frac{1}{2}\downarrow$
pi-meson π^+	$(u\overline{d})$	e	0

Photons and leptons are considered to be elementary particles and they are not composed of quarks.

Note : Isolated quark doesn't exist.

W.E-30: ***An electron-positron pair is produced when a γ -ray photon of energy 2.36MeV passes close to a heavy nucleus. Find the kinetic energy carried by each particle produced, as well as the total energy with each.***

Sol. The reaction is represented by

$\gamma \to ({}_{-1}e^0) + ({}_{+1}e^0)$, so that

$E = m_0C^2 + K.E_{electron} + m_0C^2 + K.E_{positron}$ $2.36\text{MeV} = 2m_0.C^2 + \text{K.E}_{(\text{electron})} + \text{K.E}_{(\text{positron})}$

$= 1.02\text{ MeV} + K.E_{(e^-)} + K.E_{(e^+)}$

$\text{K.E. of } (e^-) = K.E_{(e^+)} = \frac{1}{2}(2.36 - 1.02)MeV,$

(K.E. carried each) = 0.67 MeV (motional energy)

Total energy shared by each particle is obviously $m_0C^2 + K.E = 0.51MeV + 0.67MeV = 1.18\text{MeV}$.

W.E-31: ***A gamma ray photon of energy 1896 MeV annihilates to produce a proton-antiproton pair. If the rest mass of each of the particles involved be 1.007276 a.m.u approximately, find how much K.E these will carry?***

Sol. Working on the same lines as an electron-positron pair production, we notice that the reaction.

$\gamma \to$ proton + antiproton, has the energy balance

$E = m_{0\,(proton)} C^2 + K.E_{(proton)} +$

$m_{0\,(antiproton)} C^2 + K.E_{(antiproton)}$

But m_0C^2 = energy equivalent of 1.007276 a.m.u $\approx$ 938 MeV. [$\because$ 1.007276 x 931 $\approx$ 938 MeV]

Thus K.E of each particle

$$= \frac{1}{2}[1896 \text{ MeV} - 2 \times 938 \text{ MeV}] = 10\text{MeV}.$$

W.E-32: Obtain the maximum kinetic energy of β-particles, and the radiation frequencies of γ decays in the decay scheme shown in Fig. 14.6. You are given that

m(^{198}Au) = 197.968233 u; m(^{198}Hg) = 197.966760 u

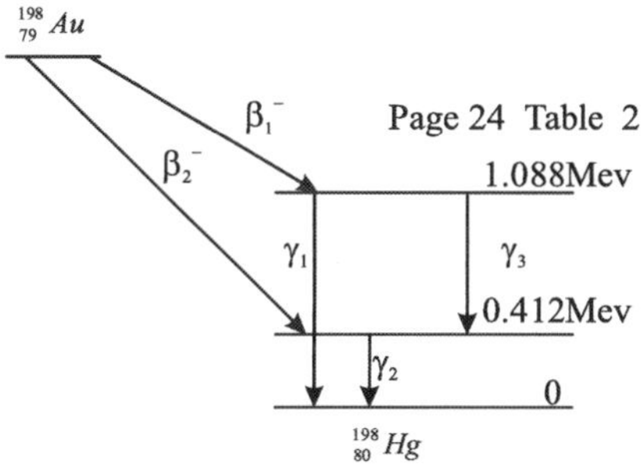

Sol. γ-rays are electro magnetic radiations having energy $E = h\upsilon \Rightarrow \upsilon = \frac{E}{h}$ where h = plank`s constant = 6.625 x 10^{-34} J.S

1) Frequencies of γ_1, γ_2 and γ_3 are calculated as follows

$$\gamma_1 = \frac{\Delta E}{h} = \frac{(1.088-0)MeV}{6.625\times10^{-34}J.s} = \frac{1.088\times10^6\times1.6\times10^{-19}}{6.625\times10^{-34}}$$

= 0.2627 x 10^{21} = 2.627 x 10^{20} Hz

$$\gamma_2 = \frac{\Delta E}{h} = \frac{(0.412-0)MeV}{6.625\times10^{-34}J.s} = \frac{0.412\times1.6\times10^{-19}\times10^6}{6.625\times10^{-34}}$$

= 0.0995 x10^{21} = 9.95 x 10^{19}Hz

$$\gamma_3 = \frac{\Delta E}{h} = \frac{(1.088-0.412)\times10^6\times1.6\times10^{-19}}{6.625\times10^{-34}}$$

= 0.1632 x 10^{21} = 1.632 x 10^{20} Hz

2) Now maximum K.E of β_1^- = [M($^{198}{}_{79}$Au)

$$M(^{198}{}_{80}Hg) - \frac{1.088}{931.5}]\ c^2$$

($\because$ 1 amu = 931.5 MeV $\Rightarrow$ 1MeV = $\frac{1}{931.5}U$)

$= [197.968233-197.966760-0.001168]931.5\text{ MeV}$

$= 0.000305 \times 931.5 = 0.284\text{ MeV}$

Maximum K.E of $\beta_2^- = [M({}^{198}{}_{79}Au)-M({}^{198}{}_{80}Hg)-\frac{0.412}{931.5}]c^2$

$= [197.968233-197.966760-0.000442]931.5$

$= 0.001031 \times 931.5 = 0.9603\text{ MeV}$

CONCEPTUAL

COMPOSITION OF NUCLEUS, ISOTOPES, ISOBARS, ISOTONES SIZE OF THE NUCLEUS

1. **The particle A is converted to C via following reactions then**

   $A \rightarrow B + {}_2He^4$ $\quad\quad$ $B \rightarrow C + 2\,{}_{-1}e^0$

   1) A and C are isobars $\quad$ 2) A and C are isotopes
   3) A and B are isobars $\quad$ 4) A and B are isotopes

2. **The particles which can be added to the nucleus of an atom without changing its chemical properties are**

   1) electrons $\quad$ 2) protons $\quad$ 3) neutrons $\quad$ 4) positron

3. **The radius of the nuclues is proportional to, (if A is the atomic mass number)**

   1) A $\quad$ 2) A^3 $\quad$ 3) $A^{1/3}$ $\quad$ 4) $A^{2/3}$

4. **The radius of a nucleus mainly depends on**

   1) Proton number $\quad$ 2) Electron Number $\quad$ 3) Mass number $\quad$ 4) Neutron number

5. **The nuclei ${}_6C^{13}$ and ${}_7N^{14}$ can be described as**

   1) isotones $\quad$ 2) isobars $\quad$ 3) isomers $\quad$ 4) isotopes

6. **The graph of ℓn (R/R_0) versus ℓn A (R =radius) of nucleus. A is atomic number is**

   1) Straight line $\quad$ 2) Parabola $\quad$ 3) Ellipse $\quad$ 4) Circle

7. **The nucleus of ${}_{56}Ba^{141}$ contains**

   1) 85 protons, 56 neutrons $\quad$ 2) 55 protons ,86 neutrons
   3) 56 protons,85 neutrons $\quad$ 4) 86 protons,55neutrons.

8. **The nuclear size is measured in units of**

   1) Angstrom $\quad$ 2) Fermi $\quad$ 3) Bar $\quad$ 4) Light-year

9. **Nuclides which have the same mass number are called**

   1) Isotopes $\quad$ 2) Isobars $\quad$ 3) Isotones $\quad$ 4) Isomars

10. **Observe the following statements regarding isotones**

    i) ${}^{39}K_{19}$ and ${}^{40}Ca_{20}$ are isotones

    ii) Nuclides having different atomic numbers (z) and mass number (A) but same number of neutrons (n) are called Isotones

    iii) ${}^{19}F_9$ and ${}^{23}Na_{11}$ are isotones

    The correct answer is

    1) i,ii and iii are correct $\quad$ 2) only i and ii are correct
    3) only i and iii are correct $\quad$ 4) only ii and iii are correct

11. A and B are isotopes. B and C are isobars. All three are radioactive. Which one of the following is true.

1) A, B and C must belong to the same element
2) A, B and C may belong to the same element
3) It is possible that A will change to B through a radioactive-decay process
4) It is possible that B will change to C through a radioactive-decay process

MASS DEFECT, BINDING ENEGRY, PACKING FRACTION AND MASS ENERGY RELATION

12. M, M_n & M_p denotes the masses of a nucleus of $_ZX^A$ a neutron, and a proton respectively. If the nucleus is separated in to its individual protons and neutrons then

1) $M=(A-Z)M_n+ZM_p$
2) $M=ZM_n+(A-Z)M_p$
3) $M>(A-Z)M_n+ZM_p$
4) $M<(A-Z)M_n+ZM_p$

13. The difference between the mass of a nucleus and the combined mass of its nucleons is

1) zero
2) positive
3) negative
4) zero, positive or negative

14. The mass number of a nucleus is

1) Always less than atomic number
2) Always more than atomic number
3) Equal to atomic number
4) Sometimes more or equal to atomic number.

15. If M is atomic weight, A is mass number then (M-A)/A represent

1) Mass defect
2) Packing fraction
3) Binding Energy
4) Chain Reaction

16. The difference between mass of the nucleus and total mass of its constituents is called

1) Packing fraction
2) Mass defect
3) Binding energy
4) Binding energy per neucleon

17. The parameter used to measure the stability of the nucleus is

1) Average binding energy
2) No of protons
3) No of neutrons
4) No of electrons

18. When the number of nucleons in a nuclues increases the binding energy per nucleon

1) Increase continuously with mass number
2) Decreases continuously with mass number
3) Remains constant with mass number
4) First increases and then decreases with increase in mass number

19. Maximum value of binding energy per nucleon is

1) 8MeV
2) 8.8MeV
3) 7.6MeV
4) 1.1MeV

20. The binding energy per nucleon is maximum at A=56 and its value is around_Mev/ Nucleon

1) 8.4
2) 8.7
3) 9
4) 7.8

21. Average binding energy per nucleon over a wide range is

1) 8MeV
2) 8.8MeV
3) 7.6MeV
4) 1.1MeV

22. The wrong statement about binding energy is

1) It is the sum of the rest mass energies of nucleons minus the rest mass energy of the nucleus.
2) It is the energy released when the nucleons combine to form a nucleus.
3) It is the energy required to break a given nucleus into its constituent nucleons.
4) It is the sum of the kinetic energies of all the nucleons in the nucleus.

23. The binding energies of a deutron and an α-particle are 1.125, 7.2 MeV/nucleon respectively. The more stable of the two, is

1) deutron
2) α-particle
3) both
4) sometimes deutron and sometimes α-particle

24. Mass defect of an atom refers to

1) inaccurate measurement of mass of neutrons
2) mass annihilated to produce energy to bind the nucleons
3) packing fraction
4) difference in the number of neutrons and protons in the nucleus

25. The stability of a nucleus can be measured by

1) Average binding energy
2) Packing fraction
3) Ratio of number of neutrons and protons
4) All the above.

26. In a nuclear reaction some mass converts into energy. In this reaction total B.E of reactants when compared with that of products is

1) always greater
2) always les
3) either greater or less
4) always equal

RADIOACTIVITY, α, β, γ RAYS, HALF LIFE, MEAN LIFE, DECAY CONSTANT

27. The age of pottery is determined by archeologists using a radioisotope of

1) carbon
2) cobalt
3) iodine
4) phosphorus

28. During an artificial transmutation the nucleus emits

1) β-particles
2) α-particles
3)always neutrons
4) may emit protons or neutrons

29. When two deuterium nuclei fuse together to form a tritium nucleus, we get a

1) neutron
2) deuteron
3) alpha particle
4) proton

30. Identify the correct statement/statements

a) Radiation causes genetic mutation
b) Restriction in blood circulation can be detected using radio-iodine
c) Hydrocarbon plastics are used as moderators in a nuclear reactor
d) The damage caused due to α-radiation is small due to its small penetrating power

1) a,b,c
2) a,c,d
3) b,c,d
4) a,b,d

31. Identify the correct ascending order of α, β and γ with reference to their ionizing power

I) α-ray **II) γ-ray** **III) β-ray**

1) II,III,I
2) I,III,II
3) II,I,III
4) I,II,III

32. Two identical nuclei A and B of the same radioactive element undergo β^- decay. A emits a β^- particle and changes to A'. B emits a β^- particle and then a γ–photon immediately afterwards, and changes to B'.

1) A' and B' have the same atomic number and mass number
2) A' and B' have the same atomic number and different mass numbers
3) A' and B' have different atomic numbers but the same mass number
4) A' and B' are isotopes

33. Identify the correct ascending order of α, β and γ with reference to their penetrating power

I) β-ray **II) α-ray** **III) γ-ray**

1) II,III,I
2) II,I,III
3) I,II,III
4) III,I,II

34. If a beam consisting of α, β and γ radiation is passed through an electric field perpendicular to the beam, the deflections suffered by the components, in decreasing order, are

1) α,β,γ 2) α,γ,β 3) β,α,γ 4) β,γ,α

35. Decrease in atomic number is observed during

A) α-emission B) β-emission C) Positron emission D) electron capture

1) B is correct 2) A and B are correct

3) A,C and D are correct 4) Only C

36. When $_{15}P^{30}$ decays to become $_{14}Si^{30}$,the particle released is

1) electron 2) α-particle 3) neutron 4) positron

37. During β^-- decay, a neutron inside nucleus converts into proton, electron and x. Then the particle x is

1) π^+- meson 2) neutrons 3) anti-neutrino 4) π^--meson

38. If a nucleus emits a gamma-ray, its atomic and mass number____ but there will be ___in the energy of the nucleus.Select suitable pair

1) Remain same, increase 2) Remain same, decrease

3) Decrease, increase 4) increase, decrease

39. In the following nuclear reaction $_{13}Al^{27}+_2He^4 \rightarrow {}_{15}P^{30}$+X,X will be

1) Proton 2) Electron 3) Neutron 4) α-particle

40. In nuclear reaction $_4Be^9+_2He^4$-->$_6C^{12}$+X,X will be

1) Proton 2) Neutron 3) β-particle 4) α-particle

41. In nuclear reaction $_2He^4+_zX^A \rightarrow {}_{z+2}Y^{A+3}+R$ R denotes

1) electron 2) positron 3) proton 4) neutron

42. A positron is emitted by radioactive nucleus of proton no 90. The product nucleus will have proton number

1) 91 2) 90 3) 89 4) 88

43. $_{13}Al^{27}+\alpha$-particle $\rightarrow$ neutron +'X' then 'X' is

1) $_{15}p^{31}$ 2) $_{14}Si^{30}$ 3) $_{15}p^{30}$ 4) $_{15}Si^{30}$

44. The penentrating power of beta particle compared to alpha particle is

1) Less 2) More 3) Equal 4) Can be more or less

45. In a nuclear reactor , heavy water is used as a

1) Controlling material 2) Moderator 3) Fuel 4) Heat exchanger

46. The units of radioactivity is

1) Fermi 2) Farad 3) Curie 4) Hertz

47. The half-life of a radioactive isotope is 3 hours. The value of its disintegration constant is

1) 0.3 hour^{-1} 2) 0.693 hour^{-1} 3) 0.231 hour^{-1} 4) 0.231 min^{-1}

48. $_{92}U^{238} \rightarrow {}_{82}Pb^{206} + 8{}_2^4He$. The number of β particles released in this reaction is

1) 6 2) 3 3) 1 4) 10

49. The radioactivity in any nucleus is measured in

1) Curie 2) Rutherford 3) Both 1 &2 4) Newton

50. **The α-particles are**
1) high energy electrons
2) positively charged hydrogen ions
3) high energy α-radiation
4) doubly positively charged helium nuclei

51. **α-particles carries**
1) Mass 1 2) Mass 2 3) Mass 3 4) Mass 4

52. **At a specific instant, the emission of radio active compound is deflected in a magnetic field. The compound can emit**
i) electrons ii) Positrons iii) He^{+2} iv) neutrons
1) i,ii,iii 2) i,ii,iii,iv 3) iv 4) ii,iii

53. **The atomic number (A) and mass number (M) of the nuclide formed where three alpha (α) and two (β) particles are emitted from $^{238}_{92}U$**
1) $A=87, M=233$ 2) $A=86, M=226$ 3) A =88, M = 227 4) A = 88, M = 226

54. **Element $_ZM^A$ emits one α (alpha) particle followed by two β (beta) particles. Among the following the daughter element is**
1) $_{Z-2}M^{A-4}$ 2) $_{Z-2}M^{A}$ 3) $_ZM^{A-4}$ 4) $_{Z+2}M^{A-4}$

55. **Particles which can be added to the nucleus of an atom without changing its chemical properties are called.**
1) Neutrons 2) Electrons 3) Protons 4) Alpha Particles

56. **An Electric field can deflect**
1) α - particles 2) X - rays 3) Neutrons 4) γ - rays

57. **On the bombardment of Boron with neutron, α - particle is emitted and product nucleus formed is ....**
1) $_6C^{12}$ 2) $_2Li^6$ 3) $_3Li^8$ 4) $_4Be^9$

58. **The one has maximum radioactive activity**
1) Uranium 2) Plutonium 3) Radium 4) Thorium

59. **Among the following one is not emitted by a radioactive substance**
1) Electrons 2) β - rays 3) Positron 4) Protons

60. $^{228}_{90}Th \rightarrow ^{212}_{83}Bi + \alpha + \beta$. **the number of α and β particals given out during the process are**
1) $4\alpha, 7\beta$ 2) $4\alpha, 1\beta$ 3) 4α 4) 7β

61. **The reciprocal of radioactive decay constant is called**
1) Half life period
2) Whole life period
3) Average life period
4) Avagadro number

62. **The missing particle in the reaction** $^1_1p \longrightarrow \quad + \; ^0_{-1}e$
1) deuteron 2) proton 3) neutron 4) β - particle

63. **In the Radioactive transformation**
$R \xrightarrow{\alpha} X \xrightarrow{\beta} Y \xrightarrow{\beta} Z$; **the neuclii R and Z are**
1) Isotopes 2) Isobars 3) Isomers 4) Isotones

64. **In the reaction ${}_{15}P^{30} \rightarrow {}_{14}Si^{30}$, The change requires the emission of**

1) α - particle 2) β -particle 3) neutron 4) positron

65. **When a radioactive substance is subjected to a vacuum, the rate of disintegration per second**

1) increases considerably 2) is not affected
3) increases only if the products are gases 4) suffers a slight decrease

66. **A radioactive nuclide can decay simultaneously by two different processes which have decay constants λ_1 and λ_2. The effective decay constant of the nuclide is λ.**

1) $\lambda = \lambda_1 + \lambda_2$ 2) $\lambda = \frac{1}{2}(\lambda_1 + \lambda_2)$ 3) $\frac{1}{\lambda} = \frac{1}{\lambda_1} + \frac{1}{\lambda_2}$ 4) $\lambda = \sqrt{\lambda_1 \lambda_2}$

67. **A sample of radioactive material is used to provide desired doses of radiation for medical purposes. The total time for which the sample can be used will depend**

1) only on the number of times radiation is drawn from it
2) only on the intensity of doses drawn from it
3) on both (a) and (b) 4) neither on (a) nor on (b)

68. **A fraction f_1 of a radioactive sample decays in one mean life, and a fraction f_2 decays in one half-life.**

1) $f_1 > f_2$ 2) $f_1 < f_2$ 3) $f_1 = f_2$
4) May be (a), (b) or (c) depending on the values of the mean life and half-life.

NUCLEAR FORCES

69. **The short range attractive nuclear forces that are responsible for the binding of nucleons in a nucleus are supposed to be caused by the role played by the particles called**

1) Positron 2) m-Meson 3) K-Meson 4) π - Meson

70. **The strong interaction exists in**

1) Gravitational forces 2) Electrostatic force of attraction
3) Nuclear forces 4) Magnetic force on a moving charge

71. **Nuclear forces are**

1) Non-central forces 2) saturated 3) Spin dependent 4) All the above

72. **Identify the correct statement/statements**

a) At greater distances nuclear forces are negligible
b) Nuclear forces are non central forces
c) Nuclear forces are weakest in nature
d) Nuclear forces are charge dependent forces

1) a, b 2) b, c 3) c, d 4) a, d

73. **Which of the following is not correct about nuclear forces?**

1) They are short range attractive forces 2) They are independent of charge
3) They change to repulsion at very close distance 4) They obey inverse square law

74. **Among the following, short range, charge independent and spin dependent forces are**

1) Gravitational forces 2) Nuclear forces
3) Electromagnetic forces 4) Weak forces

75. **Let F_{pp},F_{pn} and F_{nn} denote the magnitudes of the nuclear force by a proton on a proton ,by a proton on a neutron and by a neutron on a neutron respectively when the separation is less than one fermi, then (2008 E)**

1) $F_{pp}>F_{pn}=F_{nn}$ 2) $F_{pp}=F_{pn}=F_{nn}$ 3) $F_{pp}>F_{pn}>F_{nn}$ 4) $F_{pp}<F_{pn}=F_{nn}$

76. **Two protons are kept at a seperation of 10nm.Let F_n be the nuclear force and F_e be electrostatic force between them. Then**

1) $F_e=F_n$ 2) $F_e>>F_n$ 3) $F_e<<F_n$ 4) $F_n= 3F_e$

77. **Two nucleons are at a separation of $1x10^{-15}$m. The net force between them is F_1 if both are neutrons, F_2 if both are protons and F_3 if one is a proton and other is a neutron. In such a case**

1) $F_2>F_1>F_3$ 2) $F_1=F_2>F_3$ 3) $F_1=F_2=F_3$ 4) $F_1=F_3>F_2$

78. **Two protons attract each other when**

1) the distance between them is 10^{-10} m
2) the distance between them is 10^{-1} m
3) the distance between them is 10^{-15} m
4) the distance between them is 10^{-6} m

79. **Among gravitational,electrostatic and nuclear forces,the two attractive forces between two neutrons are**

1) Electrostatic and nuclear
2) Electrostatic and gravitational
3) Gravitational and nuclear
4) Electrostatic

80. **Among the following interactions one is of least significant in nuclear physics is**

1) nuclear interaction
2) gravitational interaction
3) electrostatic interaction
4) electromagnetic interaction

81. **The origin of nuclear force between nucleons is due to the exchange of**

1) Mesons 2) Photons 3) Positrons 4) Electrons

DISCOVERY OF NEUTRON

82. **Which of the following particles is most unstable?**

1) Neutron 2) Proton 3) Electron 4) α particle

83. **A free neutron decays spontaneously into:**

1) a proton, an electron and an anti-neutrino
2) a proton, an electron and a neutrino
3) a proton and electron
4) a proton, an electron, a neutrino and an anti-neutrino

84. **Neutron was discovered by the experiment of**

1) Artificial transmutation of $\left({}_4Be^9\right)$ by α - particles

2) Artificial transmutation of $\left({}_7N^{11}\right)$ by α -particles

3) Rutherfored scattering of alpha particles by heavy nuclei
4) Bequerel with radio activity

85. **The average life of an isolated neutron is**

1) 1500 s 2) 1000 s 3) 1200 s 4) 3 minutes

86. **Thermal neutrons energy is**

1) < 1 ev 2) > 1 ev 3) = 2 Mev 4)= 4 Mev

87. **A nucleus with an excess of neutrons may decay with the emission of**

1) a neutron 2) a proton 3) an electron 4) a positron

88. **The most penetrating atom smashing particle is**

1) neutron 2) proton 3) alpha particle 4) deuteron

89. Which of the followig is formed by decay of a free neutron?
1) A number of electrons
2) Two Protons
3) A proton and an electron
4) An α - particle

90. In neutron discovery experiment Beryllium target is bombarded by
1) Protons
2) Alpha particles
3) Neutrons
4) Deutrons

91. Slow neutrons are sometimes refer to as thermal neutrons because
1) they are sort of heat radiations
2) they are in thermal equilibrium
3) they are capable of generating heat
4) their energies are of same order as that of molecular energies at ambient temperatures.

92. Thermal neutrons are
1) Prompt neutrons
2) Slow neutrons
3) Neutrons which are in the nucleus
4) Neutrons from the sun

93. In neutron discovery experiment, Be is bombarded with
1) Proton
2) Deutrons
3) α -particle
4) β -particle

94. The process of producing a new stable nucleus from the other stable nucleus is called
1) Nuclear reaction
2) Artificial transmutation
3) Nuclear fusion
4) Nuclear fission

NUCLEAR FISSION

95. At least how many thermal neutrons should be available to start a fission reaction
1) 2
2) 3
3) 1
4) 4

96. Which of the following changes in the artificial transmutation of elements?
1) number of neutrons
2) number of electrons
3) atomic weight
4) nucleus

97. During the fission process of Uranium,the amount of energy liberated per fission is nearly
1) 100MeV
2) 200MeV
3) 150MeV
4) 300MeV

98. The number of neutrons that are released on an average during the fission of U^{235}nucleus is
1) 3
2) 1
3) 2.5
4) 5

99. Nuclear fisssion can be explained by
1) Optical model of the nucleus
2) Shell model of nucleus
3) Collective model of the nucleus
4) Liquid drop model of the nucleus

100. Percentage of mass lost during the fission of $_{92}U^{235}$ approximately is
1) 0.01%
2) 0.1%
2) 0.7%
4) 0.9%

101. Most of energy released in the fission is carried by
1) neutrons
2) fission fragments
3) neutrons and fragments carry equally
4) positrons

102. Regarding Prompt neutrons
1) They are highly energetic
2) They constitute 99%
3) Cannot initiate chain reaction
4) 1,2,3 are correct

103. Nuclear reactions obey the law of conservation of
1) Mass and energy
2) Charge
3) Momentum
4) All the above

104. The critical mass of a fissionable material is

1) 0.1kg equivalent
2) The minimum mass needed for chain reaction
3) The rest mass equivalent to 1020 joule
4) 0.5kg

105. For fast chain reaction, the size of U^{235} block, as compared to its critical size, must be

1) greater
2) smaller
3) same
4) anything.

106. The critical mass of fissionable uranium-235 can be reduced by

1) adding impurities
2)heating material
3) surrounding it by a neutron-reflecting material
4) surrounding it by a neutron-absorbing material

107. Nuclear energy is released in fission since binding energy per nucleon is

1) smaller for fission fragments than for parent nucleus
2) the same for fission fragments and parent nucleus
3) larger for fission fragments than for parent nucleus
4) sometimes larger and sometimes smaller

108. In a critical chain reaction

1) energy is released at increasing rate
2) energy is released at steady rate
3) energy is released at decreasing rate
4) energy is not released

109. Among the following one is wrong

1) The energy of thermal neutrons is about 25 meV
2) In a nuclear reactor, when neutrons multiplication factor, K = 1 then the reaction is said to be critical
3) ${}_{92}U^{235}$ undergoes fission by bombardment of high energy neutron
4) On average 2.5 neutrons are emitted per fission of ${}_{92}U^{235}$

110. When 1gm of U^{235} is completely annihilated energy liberated is E_1 and when 1 gm of U^{235} completely undergoes fission the energy liberated is E_2, then

1) $E_1 > E_2$
2) $E_1 = E_2$
3) $E_1 < E_2$
4) $E_1 \leq E_2$

111. In the process of fission, the binding energy per nucleon

1) Increases
2) Decreases
3) Remains unchanged
4) Increases for mass number A < 56 nuclei but decreases for mass number A> 56

112. Assertion (A) : Fragments produced in the fission of U^{235} are radioactive.

Reason (R) : The fragments have abnormally high proton to neutron ratio

1) Both A and R are true and R is correct explanation of A
2) Both A and R are true and R is not correct explanation of A
3) A is true but R is false
4) A is false but R is true

113. The products of the fission of U^{235} by thermal neutron are

1) Ba^{141} and Kr^{92} and 3 neutron always
2) Xe^{140}, Sr^{94} and 2 ${}_0n^1$ always
3) can be different in each fission
4) should have same mass number

114. Consider the following statements A and B. Identify the correct in the given answer.

A) p-n; p-p; n-n forces between nucleons are not equal and charge dependent.

B) In nuclear reactor the fission reaction will be in accelerating state if the value of neutron reproduction factor k>1.

1) Both A and B are correct
2) Both A and B are wrong
3) A is wrong B is correct
4) A is correct B is wrong

115. The process of fission is responsible for the release of energy in

1) The hydrogen bomb
2) The atom bomb
3) The sun
4) The star

116. The working principle in atom bomb is

1) under-critical chain reaction
2) Critical chain reaction
3) super-critical chain reaction
4) All the above

117. Heavy water is

1) Water at 4^0C 2)Watercontaining various salts
3) Compound of heavy oxygen and hydrogen 4) Compound of oxygen and deuterium.

118. Nuclear reactor is surrounded by concrete walls to

1) Strengthen the construction 2) Control the chain reaction
3) form a protective shield 4) as moderator

119. The operation of a nuclear reactor is said to be critical, if the multiplication factor (K) has a value

1) 1 2) 1.5 3) 2.1 4) 2.5

120. Cadmium and Boron rods are used in a nuclear reactor to

1) Slow down the neutrons 2) Absorb excess number of thermal neutrons
3) speed up neutrons 4) absorb fast neutrons

121. The coolant in the nuclear reactor is

1) Liquid sodium 2) cadmium 3) Deuterium 4) Liquid hydrogen

122. Substance used to slow down the fast neutrons released during nuclear fission is called.

1) Fuel 2) Moderator 3) Controlling rods 4) Reflecting rods

123. If the neutron reproduction factor K is

a) greater than 1 the fission reaction is accelerated b) less than 1 the fission reaction retards
c) equal to 1 the fission reaction is at steady state

1)only a,b are true 2)only b,c are true 3)only a,c are true 4)only a,b,c true

124. Unstable fission fragments decay by emitting neutrons and electrons, neutrons so emitted are called

1) prompt nuetrons 2) delayed neutrons 3) stray neutrons 4) sustained neutrons

125. Chain reaction can be initiated by

1) prompt neutrons 2) delayed neutrons 3) slowed prompt neutrons 4) 2 or 3

126. The man-made element which was made in the nuclear reactor is

1) polonium 2) plutonium 3) thorium 4) uranium

127. In a fast breeder reactor, the main charm is that the nuclear ash is that it is

1) more fissile than parent fuel 2) not dangerous as a potential pollutant
3) easily disposed off 4) stable in terms of further decay

128. If "x" gm of a nuclear fuel of mass number "A" undergoes fission inside a reactor then the number of fissions will be (N-Avagadro number)

1) NA/x 2) NAx 3) Nx/A 4) Ax/N

129. The reactor which produces power due to fission by fast neutron and at the same time regenerates more fissionable material than it consumes is

1) Thermal reactor 2) Breeder reactor 3) Both the above 4) Neither 1 & 2

130. The reactor in which the number of fissionable nuclides produced are more than the used is called

1) breeder reactor 2) Pressurised reactor
3) Heterogeneous reactor 4) Homogenerous reactor

131. A good moderator should

1) be a gas 2) have appetite for neutrons
3) be lighter in mass number 4) heavier in mass number

132. Who designed the atomic reactor?

1) Wilson 2) Fermi 3) Rutherford 4) Teller

133. From the following that are conserved in nuclear reactions are

1) mass number and energy
2) mass number and charge number
3) charge number and mass
4) mass number, charge number and energy

134. (A) Fission is a thermonuclear process
(B) Fusion is a thermonuclear process
(C) Fusion is exothermic
(D) Fission is exothermic

Choose the correct answer

1) A and B are correct
2) B and C are correct
3) A and C are correct
4) B,C and D are correct

135. Consider the following statements A and B. Identify the correct choice in the given answer.

(A) p-p,p-n,n-n forces between nucleons are not equal and charge dependent

(B) In nuclear reactor the fission reaction will be in accelerating state if the value of neutron reproduction factor k >1

1) Both A and B are correct
2) Both A and B are wrong
3) A is wrong and B is correct
4) A is correct and B is wrong.

136. A chain reaction in fission of Uranium is possible because

1) Large amount of energy is released
2) Two intermediate size nuclear fragments are formed
3) More than one neutron is given out in each fission
4) Fragments in fission are radioactive

137. A slow neutron can cause fission in

1) U^{238} 2) U^{235} 3) Pb^{206} 4) Sr^{90}

138. Heavy water is used as moderator in a nuclear reactor. The function of the moderator is

1) To slow down the neutrons to thermal energies
2) To control the energy released in the reactor
3) To cool the reactor faster
4) To absorb neutrons and stop chain reaction

139. To control fission process of the reactor , the following material is used_______

1) Graphite 2) Cadmium 3) Gold 4) Uranium

140. Nuclear fission is caused by

1) fast protons 2) fast neutrons 3) Slow protons 4) slow neutrons

141. The liquid drop model of nucleus was proposed by

1) Bohr,Wheeler 2) Fermi 3) Rutherford 4) Chadwick

142. In nuclear reactor, Cadmium rods are used as

1) Fuel 2) Moderator 3) Control Rods 4) None

143. The material used to slow neutrons in a reactor is called

1) Controlrod 2) Moderator 3) Fuel 4) Heat exchanger

NUCLEAR FUSION

144. Atomic mass of the most useful material for fusion reaction is

1) 1 2) 4 3) 235 4) 292.

145. Average *K.E* of thermal neutron is of the order of (in KeV)

1) 3.0 2) 0.03 3) 0.3 4) 0.003

146. Inside the sun

1) Four nuclei of hydrogen combine to form two nuclei of helium
2) Four nuclei of hydrogen combine to form four nuclei of helium
3) Four nuclei of hydrogen combine to form one nucleus of helium
4) Four nuclei of hydrogen is transformed into one nucleus of helium

147. As the age of star increases

1) Helium quantity increases
2) Helium quantity decreases
3) Helium quantity does not change
4) Helium, Hydrogen both quantities increases

148. In the carbon cycle of nuclear fusion carbon acts like a

1) Moderator 2) Activator 3) Catalyst 4) Controller

149. In a fusion process a proton and neutron combine to give a deuterium nucleus.If m_o and m_p be the mass of neutron and proton respectively the mass of deuterium nucleus is

1) equal to m_o+m_p 2) more than m_o+m_p 3) less than m_o+m_p
4) can be less than or more than(m_o+m_p)

150. The binding energies of the atoms of elements P and Q are E_P and E_Q respectively. Three atoms of elements Q fuse to form one atom of element P. In this process the energy released is e. The correct relation between E_P, E_Q and e will be

1) $E_Q=3E_P+e$ 2) $E_Q=3E_P-e$ 3) $E_P=3E_Q+e$ 4) $E_P=3E_Q-e$

151. The $\frac{B.E.}{A}$ for deutron and an α - particle are X_1 and X_2 respectively. The energy released in the fusion of deuterium into α -particle is

1) $4(X_2-X_1)$ 2) $2(X_2-X_1)$ 3) $4(X_2+X_1)$ 4) $\frac{X_2-X_1}{4}$

152. If Q_1 and Q_2 are the energies released in the fusion of hydrogen in Carbon - nitrogen cycle and proton - proton cycle respectively then

1) $Q_1 > Q_2$ 2) $Q_1 = Q_2$ 3) $Q_1 < Q_2$ 4) $Q_1 > Q_2$

153. Fusion reaction is initiated with the help of

1) low temperature 2) high temperature 3) neutrons 4) any particle

154. In an exo-ergic reaction the binding energies of reactants and products are E_1, E_2 respectively then

1) $E_1<E_2$ 2) $E_1=E_2$ 3) $E_1>E_2$ 4) $E_1 \geq E_2$

155. In an endo-ergic reaction the binding energies of reactants and products are E_1, E_2 respectively

1) $E_1<E_2$ 2) $E_1=E_2$ 3) $E_1>E_2$ 4) $E_1 \geq E_2$

156. Among the following reactions which is impossible

1) ${}_2He^4+{}_4Be^9= {}_0n^1+{}_6C^{12}$
2) ${}_2He^4+{}_7N^{14}= {}_1H^1+{}_8O^{17}$
3) $4({}_1H^1) ={}_2He^4+2 ({}_{-1}e^0)$
4) ${}_3Li^7+{}_1H^1 = {}_4Be^8$

157. If the nuclei of masses X and Y are fused together to form a nucleus of mass m and some energy is released, then

1) X+Y=m 2) X+Y<m 3) X+Y> m 4) X-Y=m

158. Fusion reactions take place at about

1) 3×10^2 K 2) 3×10^3 K 3) 3×10^4 K 4) 3×10^6 K

159. The percentage of mass lost during nuclear fusion is

1) 0.1% 2) 0.4% 3) 0.5% 4) 0.65%

160. Nuclear fusion requires high temperature because

1) All nuclear reactions absorb heat
2) The particles can not come together unless they are moving rapidly
3) The binding energy must be supplied from an external source
4) The mass defect must be supplied

161. Among the following true option is

1) Energy released per nucleon is same in both fission and fusion reactions
2) Energy released per nucleon is more in fission than in fusion reaction
3) Energy released per nucleon is less in fission than in fusion reaction
4) No energy is released in fusion reaction

162. Fusion reaction take place at high temperature because

1) atoms are ionized at high temperature 2) molecules break up at high temperatures
3) nuclei break up at high temperature
4) kinetic energy is high enough to overcome repulsion between nuclei

163. In the carbon cycle from which stars hotter than the sun obtain their energy the $_6C^{12}$ isotope

1) splits up into three alpha particles 2) fuses with another $_6C^{12}$ nucleus to form $_{12}Mg^{24}$
3) is completely converted into energy 4) is regenerated at the end of the cycle

164. Source of solar energy can be said to be due to natural fusion in which hydrogen gets converted into helium with carbon serving as a natural catalyst. This carbon cycle was proposed by

1) Bethe 2) Yukawa 3) Fermi 4) Soddy

165. In Carbon-Nitrogen fusion cycle , protons are fused to form a helium nucleus, positrons and release some energy.The number of protons fused and the number of positrons released in this process respectively are

1) 4,4 2) 4,2 3) 2,4 4) 4,6

166. Nuclear fission and fusion can be explained on the basis of ____

1) Einstein theory of relativity 2) Einstein specific heat equation
3) Einstein mass-energy relation 4) Einstein photo electric equation

167. Energy in the sun is due to

1) Fossil fuels 2) Radioactivity 3) Fission 4) Fusion

168. The overall process of carbon nitrogen fusion cycle results in the fusion of 4 protons to yield helium nucleus and ---

1) positron 2) two electrons 3) two positrons 4) An electron.

169. The nucleus finally formed in fusion of protons in proton-proton cycle is that of

1) Heavy hydrogen 2) Carbon 3) Helium 4) Lithium

170. $4\,_1H^1 \rightarrow _2He^4 + 2e^+ + 26MeV$ **. The above reaction represents ....**

1) Fusion 2) Fission 3) b-decay 4) g-decay

171. The source of stellar energy is __process

1) Nuclear fission 2) Nuclear fusion
3) Nuclear fission&fusion 4) Nuclear decay

172. Fusion reaction takes place at very high temperature because.

1) atoms are ionised at high temperatures.
2) molecules breakup at high temperature.
3) nuclei break up at high temperature.
4) kinetic energy is high enough to overcome repulsion between nuclei

173. In the carbon cycle, from which stars hotter than the sun obtain their energy the $_6C^{12}$ isotope

1) splits into three alpha particles
2) fuse with another $_6C^{12}$ nucleus to form $_{12}Mg^{24}$
3) is completely converted into energy.
4) is regenerated at the end of the cycle.

PAIR PRODUCTION & PAIR ANNIHILATION

174. The phenomenon of pair production is

1) The production of an electron and a positron from γ radiations
2) Ejection of an electron from a metal surface when exposed to ultraviolet light
3) Ejection of an electron from a nucleus
4) Ionization of a neutral atom

175. When the particle and its antiparticle unite, the result is

1) a heavier particle
2) two or more smaller particles
3) photons
4) partly matter and partly photons.

176. Particles and their antiparticles have

1) The same masses but opposite spins
2) The same masses but opposite magnetic moment
3) The same masses and same magnetic moment
4) Opposite spins and same magnetic moment

177. To produce pair production, the minimum energy of γ -ray should be (in MeV)

1) 0.15
2) 1
3) 1.02
4) 1.5

178. The rest mass energy of electron or positron is (in MeV)

1) 0.51
2) 1
3) 1.02
4) 1.5

179. A positron and an electron come close together to give a neutral one called

1) Electronium
2) Positronium
3) γ -photon
4) β -particle

180. Positronium is converted into

1) 2 Photons each of energy 0.51MeV
2) 1 Photon of energy 1.02 MeV
3) 2 Photons each of energy 1.02MeV
4) One Photon of energy 0.51MeV

181. In pair annihilation two γ -ray photons are produced it is due to

1) Law of conservation of energy
2) Law of conservation of mass
3) Law of conservation of momentum
4) Law of conservation of angular momentum

182. In pair annihilation the least number of γ - ray photons produced is

1)2
2)3
3)4
4)1

183. The number of protons, electrons and neutrons in the nucleus of $_{13}Al^{27}$ is

1) 13, 13, 14
2) 13, 0, 14
3) 14,14, 13
4) 14, 0, 13

184. $^{39}_{19}K$ and $^{40}_{20}Ca$ are

1) Isotopes
2) Isobars
3) Isotones
4) Isodiaphers

185. K^{40}, Ar^{40}, Ca^{40} are

1) Isobars
2) Isotopes
3) Isotones
4) Isogonals

186. Of the following atoms $_{6}C^{14}, _{7}N^{13}, _{88}Ra^{236}, _{7}N^{14}, _{8}O^{16}$ and $_{86}Rn^{232}$ a pair of isobars is :

1) $_{6}C^{11}, _{7}N^{13}$ 2) $_{7}N^{13}, _{7}N^{14}$ 3) $_{6}C^{14}, _{7}N^{14}$ 4) $_{6}C^{14}, _{8}O^{16}$

187. Of the following a pair of isotones is

1) $_{6}C^{11}, _{7}N^{13}$ 2) $_{7}N^{13}, _{7}N^{14}$ 3) $_{6}C^{14}, _{7}N^{14}$ 4) $_{6}C^{14}, _{8}O^{16}$

188. Of the following a pair of isotopes is

1) $_{6}C^{11}, _{7}N^{13}$ 2) $_{7}N^{13}, _{7}N^{14}$ 3) $_{6}C^{14}, _{7}N^{14}$ 4) $_{6}C^{14}, _{8}O^{16}$

189. Of the following a pair of isodiaphers is

1) $_{88}Ra^{236}, _{86}Ra^{232}$ 2) $_{7}N^{13}, _{7}N^{14}$ 3) $_{6}C^{14}, _{7}N^{14}$ 4) $_{6}C^{14}, _{8}O^{16}$

ASSERTION & REASONS

These Questions consist of two statements each printed as Assretion and Reason. While answering these questions you are required to choose any one of the following four responses

1) A and R are true and R is the correct explanation of A.
2) A and R are true and R is not the correct explanation of A.
3) A is true, R is false. 4) A is false, R is true.

190. (A): Free Neutron decays into proton, electron and antinuetrino
(R): Neutron is unstable outide the nucleus

191. (A): Nuclear forces arise from strong coulombic interactions between protons and neutrons
(R): Nulear forces are independent of charge of the nucleons

MATCHING TYPE

192. Match list I with list II

List I	List II
a) Red dwarfs cycle	e) Carbon nitrogen
b) stars having higher temperatures	f) Isotope of carbon
c) Blood circulation problems	g) proton - proton cycle
d) Radio carbon dating	h) radio sodium

1) a- e, b - g, c - h , d - f 2) a - g, b - e, c - f, d- h
3) a - g, b - e, c - h, d - f 4) a - f, b - e, c - g , d - h

193. Matching

List – I	List – II
A) K < 1	**e) Critical state**
B) K = 1	**f) Sub critical state**
C) K > 1	**g) Super critical**

1) $A \to f, B \to e, C \to g$ 2) $A \to e, B \to f, C \to g$
3) $A \to g, B \to e, C \to f$ 4) $A \to e, B \to g\ C \to f$

194. Matching.

List - I	List - II
A) Radio Iodine	e) Lukamia
B) Radio sodium	f) age of ancient objects
C) Radio phosphorous	g) Restriction in blood circulation
D) Radio carbon	h) Functioning of thyroid gland

1) $A \to h, B \to g, C \to e, D \to f$ 2) $A \to g, B \to h, C \to f, D \to e$
3) $A \to h, B \to e, C \to g, D \to f$ 4) $A \to h, B \to g, C \to f, D \to e$

195. Match the following.

List – I	**List – II**
A) Moderator	e) Absorbs heat
B) Control rods	f) Prevent neutrons exposed outside
C) Radiation shielding	g) Absorb neutrons
D) Coolant	h) Slow down neutrons

1) A – h, B – g, C – f, D – e
2) A – g, B – h, C – f, D – e
3) A – h, B – g, C – e, D – f
4) A – h, B – f, C – e, D – g

196. This question contains Statement -1 and Statement -2. Of the four choice given after the statements, choose the one that best describes the two statements.

Statement - 1: Energy is released when heavy nuclei undergo fission or light nuclei undergo fusion. and

Statement - II : For heavy nuclei,binding energy per nucleon increases with increasing Z while for light nuclei it decrease with increasing Z.

1) Statement - 1 is false,Statement -2 is true

2) Statement - 1is true, Statement -2 is true ; Statement -2 is correct explanation for Statement -1.

3) Statement - 1 is true, Statement -2 is true ; Statement -2 is ***not*** a correct explanation for statement -1

4) Statement -1 is true, Statement -2 is False.

CONCEPTUAL - KEY

1) 2	2) 3	3) 3	4) 3	5) 1	6) 1	7) 3	8) 2	9) 2	10) 2
11) 4	12) 4	13) 3	14) 4	15) 2	16) 2	17) 1	18) 4	19) 2	20) 2
21) 1	22) 4	23) 2	24) 2	25) 4	26) 4	27) 1	28) 4	29) 4	30) 2
31) 1	32) 1	33) 2	34) 3	35) 3	36) 4	37) 3	38) 2	39) 3	40) 2
41) 4	42) 3	43) 3	44) 2	45) 2	46) 3	47) 3	48) 1	49) 3	50) 4
51) 4	52) 1	53) 4	54) 3	55) 1	56) 1	57) 4	58) 3	59) 4	60) 2
61) 3	62) 3	63) 1	64) 4	65) 2	66) 1	67) 4	68) 1	69) 4	70) 3
71) 4	72) 1	73) 4	74) 2	75) 2	76) 3	77) 4	78) 3	79) 3	80) 2
81) 1	82) 1	83) 1	84) 1	85) 2	86) 1	87) 3	88) 1	89) 3	90) 2
91) 4	92) 2	93) 3	94) 2	95) 3	96) 4	97) 2	98) 3	99) 4	100) 2
101) 2	102) 4	103) 4	104) 2	105)1	106) 3	107) 3	108) 2	109) 3	110) 1
111) 1	112) 3	113) 3	114) 3	115) 2	116) 3	117) 4	118) 3	119) 1	120) 2
121) 1	122) 2	123) 4	124) 2	125) 4	126)2	127) 1	128) 3	129) 2	130) 1
131) 3	132) 2	133)4	134) 4	135) 3	136) 3	137) 2	138) 1	139) 2	140)4
141) 1	142) 3	143) 2	144) 1	145) 2	146) 3	147)1	148) 3	149) 3	150) 3
151) 1	152) 2	153) 2	154)1	155) 3	156) 3	157) 3	158) 4	159) 4	160) 2
161)3	162) 4	163) 4	164) 1	165) 2	166) 3	167) 4	168)3	169) 3	170) 1
171) 2	172) 4	173) 4	174) 1	175)3	176) 2	177) 3	178) 1	179) 2	180) 1
181) 3	182)1	183) 2	184) 3	185) 1	186) 3	187) 4	188) 2	189)1	190) 1
191) 4	192) 3	193) 1	194) 1	195) 1	196)4				

LEVEL - I -1

SIZE OF THE NUCLEUS

1. **The density of a nucleus in which mass of each nucleon is 1.67×10^{-27}kg and $R_0 = 1.4\times10^{-15} m$ is**

1) $1.453\times10^{17} kg/m^3$ 2) $1.453\times10^{16} kg/m^3$
3) $1.453\times10^{21} kg/m^3$ 4) $1.453\times10^{10} kg/m^3$

2. **r_1 and r_2 are the radii of atomic nuclei of mass numbers 64 and 27 respectively. The ratio (r_1/r_2) is**

1) 64 / 27 2) 27 / 64 3) 4 / 3 4) 1

3. **The mass number of a nucleus is 216. The size of an atom without changing its chemical properties are called**

1) $7.2\times10^{-13} cm$ 2) $7.2\times10^{-11} cm$ 3) $7.2\times10^{-10} cm$ 4) $3.6\times10^{-11} cm$

MASS DEFECT, BINDING ENERGY, PACKING FRACTION AND MASS ENERGY RELATION

4. **Energy released as mass of 2 amu is converted into energy is**

1) 1.5 x 10^{-10} J 2) 3 x 10^{-10} J 3) 1863 J 4) 931.5 Mev

5. **A 1 MeV positron encounters a 1 MeV electron travelling in opposite direction. The total energy released is (in MeV)**

1) 2 2) 3.02 3) 1.02 4) 2.04

6. **The binding energies of the nuclei A and B are E_a and E_b respectively. Three nuclei of the element B fuse to give one nucleus of element 'A' and an energy 'Q' is released. Then E_a, E_b, Q are related**

1) $E_a - 3E_b = Q$ 2) $3E_b - E_a = Q$ 3) $E_a + 3E_b = Q$ 4) $E_b + 3E_a = Q$

7. **The binding energies per nucleon for deuterium and helium are 1.1 MeV and 7.0 MeV respectively. The energy in joules will be liberated when 10^6 deuterons take part in the reaction**

1) $18.88\times10^{-3} J$ 2) $18.88\times10^{-5} J$ 3) $18.88\times10^{-7} J$ 4) $18.88\times10^{-10} J$

8. **1 Kg of iron (specific heat 120 Cal $kg^{-1}C^{-1}$) is heated by 1000°C. The increase in its mass is**

1) Zero 2) 5.6 x 10^{-8} Kg 3) 5.6 x 10^{-16} Kg 4) 5.6 x 10^{-12} Kg

9. **In nuclear fission, 0.1% mass is converted into energy. The energy released in the fission of 1Kgmass is**

1) 2.5 x 10^5 KWH 2) 2.5 x 10^7 KWH 3) 2.5 x 10^9 KWH 4) 2.5 x 10^{-7} KWH

α-DECAY, β – DECAY, γ – DECAY

10. **After the emission of one α-particle followed by two β-particles from $^{238}_{92}U$, the number of neutrons in the newly formed nucleus is**

1) 140 2) 142 3) 144 4) 146

11. **A radioactive nucleus undergoes a series of decays according to the sequence $A \xrightarrow{\beta} A_1 \xrightarrow{\alpha} A_2 \xrightarrow{\alpha} A_3$. If the mass number and atomic number of A_3 are 172 and 69 respectively, then the mass number and atomic number of A is**

1) 56, 23 2) 180, 72 3) 120, 52 4) 84, 38

12. The number of α and β particles emitted in the conversion of $_{90}Th^{232}$ to $_{82}Pb^{208}$ are

1) 6, 4 2) 4, 6 3) 8, 6 4) 6, 8

RADIOACTIVITY, HALF LIFE, MEAN LIFE, DECAY CONSTANT

13. The decay constant of a radio active element, which disintegrates to 10 gms from 20 gms in 10 minutes is

1) $0.693\,min^{-1}$ 2) $6.93\,min^{-1}$ 3) $0.693\,sec^{-1}$ 4) $0.0693\,min^{-1}$

14. Half life period of radium is 1600 years. 2 gm of radium undergoes decay and gets reduced to 0.125 gms in

1) 3200 years 2) 25600 years 3) 8000 years 4) 6400 years

15. After a certain lapse of time, the fraction of radioactive polonium undecayed is found to be 12.5% of initial quantity. The duration of this time lapse is (if the half-life of polonium is 138 days)

1) 414days 2) 407 days 3) 421 days 4) 410 days

16. Two radioactive substances X and Y initially contain an equal number of atoms. Their half-lives are 1 hour and 2 hours respectively. Then the ratio of their rates of disintegration after two hours is

1) 1:1 2) 2:1 3) 1:2 4) 2:3

17. 1 g of a radio active substance disintegrates at the rate of 3.7×10^{10} disintegrations per second. The atomic mass of the substance is226. Calculate its mean life.

1) $1.2\times10^{5}s$ 2) $1.39\times10^{11}s$ 3) $2.1\times10^{5}s$ 4) 7.194×10^{10}S

NUCLEAR FISSION-NUCLEAR REACTOR

18. No. of uranium 235 nuclei required to undergo fission to give 9×10^{13} joule of energy is

1) 2.8125×10^{24} 2) 28.125×10^{24} 3) 281.25×10^{24} 4) 28215×10^{24}

19. The energy supplied by a power plant is 40 million kilowatt hour. It is supplied by annihilation of matter, the mass that is annihilated is.

1) 1.6 gm 2) 1.6 kg 3) 1.6 mg 4)1.6 amu.

NUCLEAR FUSION

20. The amount of energy released in the fusion of two $_1H^2$ to form a $_2He^4$ nucleus will be

[Binding energy per nucleon of $_1H^2$ = 1.1 Mev Binding energy per nucleon of $_2He^4$ =7 Mev]

1) 8.1 MeV 2) 5.9 MeV 3) 23.6 MeV 4) 2 MeV

PAIR PRODUCTION & PAIR ANNIHILATION

21. The minimum amount of energy released in annhilation of electron-positron is

1) 1.02MeV 2) 0.58MeV 3)185MeV 4) 200MeV

LEVEL - I-1 - KEY

1) 1 2) 3 3) 1 4) 2 5) 2 6) 1 7) 3 8) 4 9) 2 10) 2
11) 2 12) 1 13) 4 14) 4 15) 1 16) 3 17) 4 18) 1 19) 1 20) 3
21) 1

LEVEL - I -1 - HINTS

1. $d = \dfrac{Am}{\frac{4}{3}\pi R^3} = \dfrac{3Am}{4\pi\left(R_0 A^{\frac{1}{3}}\right)^3} = \dfrac{3Am}{4\pi R_0^{\,3} A} = \dfrac{3m}{4\pi R_0^{\,3}}$

2. As $\dfrac{r_1}{r_2} = \left(\dfrac{A_1}{A_2}\right)^{1/3} = \left(\dfrac{64}{27}\right)^{1/3} = \dfrac{4}{3}$

3. $R = R_0 A^{1/3} = 1.2\times10^{-13}\times(216)^{1/3} = 7.2\times10^{-13}$ cm

4. $E = MC^2$

5. Total energy=rest mass energy+KE

6. $3E_b \rightarrow E_a + Q$

7. $E = 4\left(\dfrac{B.E.}{A}\ of\ He - \dfrac{B.E.}{A}\ of\ Deuterium\right)$

8. $\Delta mc^2 = J(MS\Delta t)$

9. $E = \dfrac{0.1}{100}mc^2; in\ kwh = \dfrac{E}{36\times10^5}$

10. ${}_{92}U^{238} \rightarrow {}_{92}X^{234} + {}_2He^4 + 2{}_{-1}e^0$

    no of neutrons=234-92=142

11. $A \xrightarrow{\beta} A_1 \xrightarrow{\alpha} A_2 \xrightarrow{\alpha} A_3$

    ${}_{72}X^{180} \xrightarrow{{}_{-1}e^0} {}_{73}X^{180} \xrightarrow{{}_2He^4} {}_{71}X^{176}$

    ${}_{71}X^{176} \xrightarrow{{}_2He^4} {}_{69}X^{172}$

12. ${}_{90}Th^{232} \rightarrow {}_{82}Pb^{208} + 6{}_2He^4 + 4{}_{-1}e^0$

13. $\lambda = 0.693/T_{\frac{1}{2}}$

14. $N = N_0 e^{-\lambda t}$

15. $\dfrac{dN}{dt} = \lambda N$, here $N =$ avagadrono /238,

    $\lambda = 0.693/T_{\frac{1}{2}}$

16. $A = \lambda N$

17. $\dfrac{N}{N_0} = \dfrac{1}{2^n}$

18. $200MeV \rightarrow 1$ fission

    $9\times10^{13}J \rightarrow ?$

    Number of fissions $= \dfrac{9\times10^{13}}{200\times10^6\times1.6\times10^{-19}}$

19. $E = 40\times10^6\times36\times10^5 J$; $E = mc^2$

$40\times10^6\times36\times10^5 = m\times9\times10^{16}$

20. ${}_1H^2 + {}_1H^2 \rightarrow {}_2He^4 + Q$

E = B.E. of products - B.E. of reactants

= 4(7) - 4(1.1) = 23.6 MeV

21. 0.51 MeV + 0.51 MeV = 1.02 MeV

LEVEL - I -2

SIZE OF THE NUCLEUS

1. Assume that the nuclear mass is of the order of 10^{-27}kg and the nuclear radius is of the order of 10^{-15}m. The nuclear density is of the order of

1) 10^2Kg/m^3 2) 10^{10}kg/m^3 3) 10^{17}Kg/m^3 4) 10^{31} kg/m^3

2. Highly energetic electrons are bombarded on a target of an element containing 30 neutrons. The ratio of radii of nucleus

1) 25 2) 26 3) 56 4) 30

MASS DEFECT, BINDING ENEGRY, PACKING FRACTION AND MASS ENERGY RELATION

3. Sun radiates energy at the rate of 3.6x10^{26}J/ s. The rate of decrease in mass of sun is (Kgs^{-1}).

1) 12 x 10^{10} 2) 1.3 x 10^{20} 3) 4 x 10^9 4) 3.6 x 10^{36}

4. A slow neutron strikes a nucleus of ${}^{235}_{92}U$ splitting it into lighter nuclei of ${}^{141}_{56}Ba$ and ${}^{92}_{36}Kr$ along with three neutrons. The energy released in this reaction is (The masses of uranium, barium and krypton of this reaction are 235.043933, 140.917700 and 91.895400 u respectively. The mass of a neutron is 1.008665u)

1) 740.69 MeV 2) 156.9 MeV 3) 186.9 MeV 4) 209.8 MeV

5. The energy required to separate the typical middle mass nucleus ${}^{120}_{50}Sn$ into its constituent nucleons ($Mass\ of\ {}^{120}_{50}Sn = 119.902199u$; mass of proton=1.007825 u and mass of neutron = 1.008665 u)

1) 951 MeV 2) 805 MeV 3) 1021MeV 4)1212 MeV

6. The mass defect in a nucleus is 3.5 amu. Then the binding energy of the nucleus is

1) 32.58MeV 2) 325.85 MeV 3) 3260.25MeV 4) 3.258 MeV

7. Consider the nuclear reaction:

$X^{200} \rightarrow A^{110} + B^{90} + \text{energy}$ **The binding energy per nucleon for X^{200}, A^{110} and B^{90} is respectively 7.4MeV, 8.2MeV and 8.2Mev. The energy released is**

1) 200MeV 2) 160MeV 3) 110MeV 4) 90MeV

α-DECAY, β – DECAY, γ – DECAY

8. The isotope ${}^{238}_{92}U$ decays successively to form ${}^{234}_{90}Th$, ${}^{234}_{91}Pa$, ${}^{234}_{92}U$, ${}^{230}_{90}Th$ and ${}^{226}_{88}Ra$, The radiations emitted in these five steps are

1) $\alpha,\alpha,\alpha,\beta,\beta$ 2) $\alpha,\alpha,\beta,\beta,\alpha$ 3) $\alpha,\beta,\beta,\alpha,\alpha$ 4) $\beta,\beta,\alpha,\alpha,\alpha$

9. **The nuclide which disintegrates by emitting a β- particle to form $^{14}_{7}N$ contains**

1) 8 neutrons 2) 10 neutrons 3) 7 neutrons 4) 6 neutrons.

10. **A nucleus X initially at rest, undergoes alpha decay according to the equation $^{232}_{z}X \rightarrow ^{A}_{90}Y + \alpha$ What fraction of the total energy released in the decay will be the kinetic energy of the alpha particle**

1) $\frac{90}{92}$ 2) $\frac{228}{232}$ 3) $\sqrt{\frac{228}{232}}$ 4) $\frac{1}{2}$

RADIOACTIVITY, HALF LIFE, MEAN LIFE, DECAY CONSTANT

11. **A radio active sample contains 600 radio active atoms. Its half life period is 30 minutes. The no. of radio active atoms remaining, if the decay occurs for 90 minutes is**

1) 300 2) 200 3) 400 4) 75

12. **Radio active carbon - 14, in a wood sample decays with a half life of 5700 years. The fraction of the radio active carbon - 14, that remains after a decay period of 17,100 years is**

1) 1/4 2) 3/4 3) 1/8 4) 7/8

13. **The half-life of ^{238}U for $\alpha-$ decay is 4.5×10^{9} years. The number of disintegrations per second occur in 1 g of ^{238}U is (Avogadro's number= $6.023\times10^{23}mol^{-1}$)**

1) $1.532\times10^{4}s^{-1}$ 2) $1.325\times10^{4}s^{-1}$ 3) $1.412\times10^{4}s^{-1}$ 4) $1.235\times10^{4}s^{-1}$

14. **A certain substance decays to 1/32 of its initial activity in 25 days. The half-life is**

1) 1 day 2) 3 days 3) 5 days 4) 7 days

NUCLEAR FISSION-NUCLEAR REACTOR

15. **The energy released by the fission of 1 g of ^{235}U in joule, given that the energy released per fission is 200 MeV. (Avogadro's number = 6.023×10^{23})**

1) 8.202×10^{12} 2) 8.202×10^{8} 3) 8.202×10^{10} 4) 8.202×10^{14}

NUCLEAR FUSION

16. **The ratio of the amounts of energy released as a result of the fussion of 1 kg hydrogen (E_1) and fission of 1 kg of $_{92}U^{236}$ (E_2) will be**

1) 1.28 2) 3.28 3) 5.28 4) 7.28

LEVEL - I-2 - KEY

1) 3 2) 2 3) 3 4) 4 5) 3 6) 3 7) 2 8) 3 9) 1 10) 2
11) 4 12) 3 13) 4 14) 3 15) 3 16) 4

LEVEL - I -2 - HINTS

1. $d = \frac{3M_n}{4\pi R^3}$ $R = radius$

2. As $R = R_0A^{1/3} \Rightarrow \frac{R_1}{R_2} = \left(\frac{A_1}{A_2}\right)^{1/3} \Rightarrow \frac{R}{R_{He}} = \left(\frac{A}{4}\right)^{1/3}$

$$\Rightarrow (14)^{1/3} = \left(\frac{A}{4}\right)^{1/3} \Rightarrow A = 56 \text{ (or) } Z = A - N$$

$$Z = 56 - 30 = 26$$

3. $\frac{dE}{dt} = \frac{dm}{dt}c^2$

4. $BE = \Delta mc^2 = \Delta m \times 931.5 MeV$

5. $\Delta E = \Delta m\ 931.5 MeV$

6. Binding energy= $\Delta m \times 931.5 MeV$

7. Energy released=BE of products-BE of reactants

9. ${}_7N^{14} \rightarrow {}_6X^{14} + {}_{-1}e^0$

10. $\frac{\text{K.E}\alpha}{\text{K.E}} = \frac{\text{A}-4}{\text{A}}(\text{Where A} = 232)$

11. $\frac{N}{N_0} = \left(\frac{1}{2}\right)^n, n = \frac{t}{T_{\frac{1}{2}}}$

12. $n = \frac{t}{T_{\frac{1}{2}}}; \frac{N}{N_0} = \left(\frac{1}{2}\right)^n$

13. $N = \frac{N_0}{2^n}$

14. $\frac{dN}{dt} = \lambda N$ here N = avagadrono / 226, $\lambda = 1/\tau$

15. E = (avagadrono./235)X-200x10^6x1.602x10^{-19}J

16. Energy relesed for nucleon in fussion = 7.28 (energy released per nucleon in fission) $E_1 = 7.28E_2$

LEVEL - II -1

SIZE OF THE NUCLEUS

1. A nucleus x ^{235} splits into two nuclei having the mass numbers in the ratio 2:1. The ratio of the radii of those two nuclei is

1) 2:1 2) 1:2 3) $2^{1/3}$:1 4) 1:$2^{1/3}$

2. A match box of 5 cm x 5cm x 1 cm dimensions is filled with nuclear matter. Its weight is in the order of

1) 10g 2) 10^8 g 3) 10^{12}g 4) 10^{15}g

MASS DEFECT, BINDING ENEGRY, PACKING FRACTION AND MASS ENERGY RELATION

3. If the speed of light were 2/3 of its present value, the energy released in a given atomic explosion will be decreased by a fraction of

1) 2/3 2) 4/9 3)4/3 4) 5/9

4. **The binding energy per nucleon of C^{12} is 7.68 MeV and that of C^{13} is 7.47 Mev. The energy required to remove one neutron from C^{13} is**

1) 495 MeV 2) 49.5 MeV 3) 4.95 MeV 4) 0.495 MeV

5. **The binding energy per each nucleon in the neighbourhood of medium nuclei is 8.5 MeV and the binding energy per each nucleon is about 7.6 MeV in the neighbourhood of Uranium. The energy released in the fission of U^{236} is**

1) 212 eV 2) 212 MeV 3) 2.12 MeV 4) 0.9 MeV

α - DECAY, β – DECAY, γ – DECAY

6. **^{22}Ne nucleus, after absorbing energy, decays into two α - particals and an unknown nucleus. The unknown nucleus is**

1) Carbon 2) Nitrogen 3) Boron 4) oxygen

RADIOACTIVITY, HALF LIFE, MEAN LIFE, DECAY CONSTANT

7. **The mass of one curie of U^{234} is**

1) $3.7\times10^{10}\,g$ 2) $3.7\times10^{-10}\,g$ 3) $6.25\times10^{-34}\,g$ 4) $1.438\times10^{-11}\,g$

8. **A radio active isotope having a half life of 3 days was received after 9 days. It was found that there was only 4 gms of the isotope in the container. The initial weight of the isotope when packed was**

1) 8 g 2) 64 g 3) 48 g 4) 32 g

9. **Half life of a radio active element is 5 min. 10 sec. Time taken for 90% of it to disintegrate is nearly**

1) 100 min 2) 1000sec 3) 10^4 sec 4) 10^4 min

10. **The half life of $^{238}_{92}U$ undergoing $\alpha-$decay is 4.5×10^9 years. Its activity of 1 g sample is**

1) 1.23×10^4 Bq 2) 2.4×10^5 Bq 3) 1.82×10^6 Bq 4) 4.02×10^8 Bq

NUCLEAR FISSION-NUCLEAR REACTOR

11. **In a thermo nuclear reaction $10^{-3}\,Kg$ of hydrogen is converted into $0.99\times10^{-3}\,kg$ of helium. If the efficiency of the generator is 50%, the electrical energy generated in KWH is**

1) 10^5 2) 1.5×10^5 3) 1.25×10^5 4) 1.3×10^5

12. **A nuclear reactor generated power at 50% efficiency by fission of $_{92}U^{235}$ into equal fragment of $_{46}Pd^{116}$ with the emission of two γ -rays of 5.2 MeV each and three neutrons. The average B.E. per particle of U^{235} and Pd^{116} is 7.2 MeV and 8.2 MeV respectively. The amount of U^{235} consumed per hour to produce 1600Mw power is**

1) 128 *gm* 2) 1.4 *kg* 3) 140.5 *gm* 4) 281 *gm*

NUCLEAR FUSION

13. **In nuclear fusion,One gram hydrogen is converted into 0.993gm.If the efficiency of the generator be 5%,then the energy obtained in KWH is**

1) 8.75×10^3 2) 4.75×10^3 3) 5.75×10^3 4) 3.73×10^3

PAIR PRODUCTION & PAIR ANNIHILATION

14. A photon of energy 1.12 Mev splits into electron positron pair. The velocity of electron is (Neglect relativistic correction)

1) $3 \times 10^8 ms^{-1}$ 2) $1.33 \times 10^8 ms^{-1}$ 3) $6 \times 10^8 ms^{-1}$ 4) $9 \times 10^8 ms^{-1}$

LEVEL - II (C.W.) - KEY

1) 3	2) 4	3) 2	4) 3	5) 2	6) 1	7) 4	8) 4	9) 2	10) 1
11) 3	12) 1	13) 1	14) 2						

LEVEL - II (C.W.) - HINTS

1. $R \alpha A^{1/3}$

2. $D = \frac{M}{V}$

3. $E = mc^2$

4. Energy required $= 13 \times 7.47 - 12 \times 7.68 = 4.95$ MeV

5. $Q = BE_{product} - BE_{reactants}$

6. ${}^{22}_{10}Ne \to 2\,{}_2He^4 + {}_6c^{12}$

7. $N \to 234gm$

$$3.7\times10^{10} \to xgm, x = \frac{3.7\times10^{10} \times 234}{N}$$

8. $n = \frac{t}{t_{\frac{1}{2}}} = \frac{9}{3} = 3;\ W = \frac{W_0}{2^n}$

$W_0 = 2^n,\ W = 2^3(4) = 32gm$

9. $t = \frac{t_{\frac{1}{2}}}{\log 2} \log\left(\frac{N_0}{N}\right)$

10. Activity or rate of decay is $R = \lambda N$ Where $\lambda = \frac{0.693}{T_{1/2}}$ and $N =$ No.of atom is 1 gm of

$${}^{238}_{92}U = \frac{1\times6.025\times10^{23}}{238} = 25.3\times10^{20}\text{atoms};\ R = \lambda N$$

11. $\frac{E}{t} = \frac{mc^2}{t}$; effieiency$= \frac{\text{output}}{\text{input}}$

12. $\frac{50P}{100} = 1600MW; P = \frac{\left[BE_{product} - BE_{reactant}\right]}{t}$

13. Efficency$= \frac{output}{\Delta mc^2}$

14. $E_\gamma = \left(\frac{1}{2}mv^2\right)2 + 2E_0$

LEVEL - II -2

SIZE OF THE NUCLEUS

1. **A nucleus splits into two nuclear parts having radii in the ratio 1:2. Their velocities are in the ratio**

   1) 8:1 2) 6:1 3) 4:1 4) 2:1

MASS DEFECT, BINDING ENEGRY, PACKING FRACTION AND MASS ENERGY RELATION

2. **The atomic mass of $_7N^{15}$ is 15.000108 amu and that of $_8O^{16}$ is 15.994915 amu. The minimum energy required to remove the least tightly bound proton is (mass of proton is 1.007825 amu)**

   1) 0.01 3018 amu 2) 12.13 MeV 3) 13.018 meV 4) 12.13 eV

3. **Assume that a neutron breaks into a proton and electron. The energy released during this process is (Mass of neutron=1.6725×10^{-27} Kg =mass of proton=1.6725×10^{-27} Kg, mass of electron = 9×10^{-31} Kg.)**

   1) 0.73 MeV 2) 7.10 MeV 3) 6.30 MeV 4) 5.4 MeV

α - DECAY, β – DECAY, γ – DECAY

4. **A nucleus with mass number 220 initially at rest emits an α – particle. If the Q value of the reaction is 5.5MeV then the kinetic energy of the α – particle is**

   1) 4.4 MeV 2) 5.4 MeV 3) 5.6 MeV 4) 6.5 MeV

RADIOACTIVITY, HALF LIFE, MEAN LIFE, DECAY CONSTANT

5. **If the activity of ^{108}Ag is 3 micro curie, the number of atoms present in it are(λ=0.005 sec^{-1})**

   1) 2.2 x 10^7 2) 2.2 x 10^6 3) 2.2 x 10^5 4) 2.2 x 10^4

6. **The half life period of Pb^{210} is 22 years. If 2g of Pb^{210} is taken, then after 11 years the amount of Pb^{210} will be present is**

   1) 0.1414 g 2) 1.414 g 3) 2.828 g 4) 0.707 g

7. **$^{221}_{87}Ra$ undergoes radiorctive decay with a half life of 4 days. The probability that a Ra nucleus will disintegrate in 8 days is**

   1) 1 2) 1/2 3) 1/4 4) 3/4

NUCLEAR FISSION-NUCLEAR REACTOR

8. **When $_{92}U^{235}$ undergoes fission. About 0.1% of the original mass is conveted into energy. Then the amount of $_{92}U^{235}$ should undergo fission per day in a nuclear reactor so that it provides energy of 200 mega watt electric power is**

   1) $9.6\times10^{-2}\,kg$ 2) $4.8\times10^{-2}\,kg$ 3) $19.2\times10^{-2}\,kg$ 4) $1.2\times10^{-2}\,kg$

PAIR PRODUCTION & PAIR ANNIHILATION

9. **A γ -ray photon creates an electron, positron pair. The rest mass of electron is 0.5 MeV. KE of the electron - positron pair system is 0.78 MeV. Then the energy of γ -ray photon is (in MeV)**

   1) 1.78 2) 0.28 3) 1.28 4) 0.14

LEVEL-II -2- KEY

1) 1 2) 2 3) 1 4) 2 5) 1 6) 2 7) 4 8) 3 9) 1

LEVEL-II-2 - HINTS

1. $m_1v_1 = m_2v_2$; $\frac{m_1}{m_2} = \left(\frac{R_1}{R_2}\right)^3 = \frac{V_2}{V_1}$

2. $\Delta m = m_p + m_{nitrogen} - m_{oxygen}$

   energy required $= \Delta m \times 931.5$ Mev

3. $\Delta m = (m_p + m_e) - m_n$;Energy released= $\Delta m \times c^2$

4. $K.E_\alpha = \left(\frac{A-4}{A}\right)Q = \frac{216}{220} \times 5.5 = 5.4MeV$

5. $\lambda = \frac{\left(\frac{dN}{dt}\right)}{N}$

6. $t = \frac{t_{\frac{1}{2}}}{\log 2} \log\left(\frac{N_0}{N}\right)$

7. $n = \frac{t}{t_{1/2}} = \frac{8}{4} = 2$; $N_o = 2^n N$; $\frac{N_o}{N} = 2^2 = 4$

   Remaining atoms $= N = \frac{N_o}{4}$

   Dacayed atoms $= N_o - N = N_o - \frac{N_o}{4} = \frac{3N_o}{4}$

   $\therefore$ Probability of decay $= \frac{N_o - N}{N_o} = \frac{3}{4}$

8. $E=MC^2$

9. $E_\gamma = 2(E_K + E_0)$

LEVEL - III

1. 50% of a radio active substance decays in 5 hours. The time required for the 87.5% decay is

1) 10hours 2) 15hours 3) 12.5hours 4) 17.5hours

2. 4 grams of radioactive substance A left 1/2 gm after some time. 1 gram of another radioactive substance B left 1/4 gm in the same period. If half life of B is 2 hours, the half life of A is

1) 3/4 hours 2) 4/3 hours 3) 1/4 hours 4) 1/2 hours

3. One mole of a α emitter of half life equal to 2days was placed in a sealed tube for 4 days at S.T.P. volume of helium collected is

1) 22.4 lit 2) 16.8 lit 3) 11.2 lit 4) 5.6 lit

4. **3 rutherfords of a radio active isotope of half-life equal to 3 days was received after 12 days. Initial isotope packed was**

1) 48 rutherfords 2) 12 rutherfords 3) 25 rutherfords 4) 36 rutherfords

5. **The half life of a radio active substance is 6 hours. The amount of the substance undergone disintegration when 36 gms of it undergoes decay for 18 hours is**

1) 31.5 gm 2) 4.5 gm 3) 18 gm 4) 9 gm

6. **The radio active nuclides A and B have half lives t and 2t respectively. If we start an experiment with one mole of each of them, the mole ratio after time interval of 6t will be**

1) 1 : 2 2) 1 : 8 3) 1 : 6 4) 1 : 1

7. **20% of a radio active element disintegrates in 1hr. The percentage of the radio active element disintegrated in 2hrs will be**

1) 36% 2) 64% 3) 60% 4) 40%

8. **The C^{14} to C^{12} ratio in a certain piece of wood is 25% of that in atmosphere. The half life period of C^{14} is 5,580 years. The age of wood piece is**

1) 5,580 years 2) 2790 years 3) 1395 years 4) 11,160 years

9. **A radioactive sample can decay by two differentprocesses. The half-life for the first process is T_1 and that for the second process is T_2. The effective half-life T of the radioactive sample is**

1) $T = T_1 + T_2$ 2) $\frac{1}{T} = \frac{1}{T_1} + \frac{1}{T_2}$ 3) $T = \frac{T_1 + T_2}{T_1 T_2}$ 4) $T = \frac{T_1 - T_2}{T_1 T_2}$

10. **The age of the wood if only 1/16 part of original C^{14} is present in its piece is (T of C^{14} is 5,580 years)**

1) 5580 years 2) 11,160 years 3) 22320 years 4) 16740 years

11. **A piece of wood is found to have the $\frac{C^{14}}{C^{12}}$ ratio to be 0.5 times of that in a living plant The number of years back the plant died will be (T of C^{14} = 5,580 years)**

1) 2,790 years 2) 5,580 years 3) 11,160 years 4) 27,900 years

12. **A piece of wood collected from cro-Magnon caves gave 4 disintegrations / min. A freshly cut wood of the same weight gives 16 d.p.m. The cro-magnon man lived about (Half life of C^{14} is 5760 years. Assume the activity is due to C^{14} only)**

1) 5700 years ago 2) 2900 years ago 3) 11520 years ago 4) 1400 years ago

13. **The number of U^{238} nuclei in a rock sample equal to the number of Pb^{206} atoms. The half life of U^{238} is 4.5×10^9 years.The age of the rock is**

1) 4.5×10^9 y 2) 9×10^9 y 3) 13.5×10^9 y 4) 18×10^9 y

14. **Equal masses of two samples A and B of charcoal are burnt and the activity of resulting carbon-di-oxide from two samples is measured. The gas from sample A gives 10^4 counts per month and that from sample B gives 2.5×10^3 counts per month. The age difference of two sample is (half life of C^{14} is 5730 years)**

1) 5730y 2) 11460y 3) 17190y 4) 22920y

15. The half life of a radioactive substance is 20 minutes.The approximate time interval $(t_2 - t_1)$ between the time t_2, when $\frac{2}{3}$ of it has decayed and time t_1 and $\frac{1}{3}$ of it had decayed is:

1) 14 minutes 2) 20 minutes 3) 28 minutes 4) 7 minutes

16. A charged capactor of capacitance C is discharged through a resistance R. A radio active sample decays with an average life t. Find R interms of C and t in order that the ratio of the electrostatic energy stored in the capacitor to the activity of the radio active sample remains constant with time

1) $\frac{2t}{C}$ 2) $\frac{C}{2t}$ 3) 2t C 4) t C

17. Uranium-238 decays to thorium-234 with half-life $5\times10^9\ yr$ .The resulting nucleus is in the excited state and hence further emits $\gamma -$ rays to come to the ground state. It emits $20\gamma -$ rays per second. the emission rate will drop to $5\gamma -$ rays per second in

1) $1.25\times10^9\ yr$ 2) $10^{10}\ yr$ 3) $10^{-8}\ yr$ 4) $1.25\times10^{-9}\ s$

18. A sample of radioactive material has mass m, decay constant λ and molecular weight M. Avagadro constant $= N_A$. The initial activity of the sample is

1) λm 2) $\frac{\lambda m}{M}$ 3) $\frac{\lambda m N_A}{M}$ 4) $mN_A e^{\lambda}$

19. A sample of radioactive material has mass m, decay constant λ and molecular weight M. Avagadro constant $= N_A$. The activity of the sample after time t will be

1) $\left(\frac{mN_A}{M}\right)e^{-\lambda t}$ 2) $\left(\frac{mN_A\lambda}{M}\right)e^{-\lambda t}$ 3) $\left(\frac{mN_A}{M\lambda}\right)e^{-\lambda t}$ 4) $\frac{m}{\lambda}\left(1-e^{-\lambda t}\right)$

20. In moon rock sample the ratio of the number of stable argon-40 atoms present to the number of radioactive potassium-40 atoms is 7:1. Assume that all the argon atoms were produced by the decay of potassium atoms, with a half-life of $2.5\times10^9\ yr$. The age of the rock is

1) $2.5\times10^9\ yr$ 2) $5.0\times10^9\ yr$ 3) $7.5\times10^9\ yr$ 4) $10^{10}\ yr$

21. The half-life of a radioactive sample is T. If the activities of the sample at time t_1 adn t_2 ($t_1 < t_2$) are R_1 and R_2 respec tively, then the number of atoms disintegrated in time $t_2 - t_1$ is proportional to

1) $(R_1 - R_2)T$ 2) $(R_1 + R_2)T$ 3) $\frac{R_1R_2}{R_1+R_2}T$ 4) $\frac{R_1+R_2}{T}$

22. Considering a hypothetical annihilation of a stationary electron with a stationary positron, the wavelengthof resulting radiation? (with usual notations)

1) $\frac{h}{2m_0C}$ 2) $\frac{2h}{m_0C}$ 3) $\frac{h}{m_0C}$ 4) $\frac{h\sqrt{2}}{m_0C}$

23. A radioactive nucleus can decay by two different processes. The half life for the first process is 2t and that for the second process is t. The effective disintegration constant of nucleus is

1) $\frac{3}{2t\ln 2}$ 2) $\frac{3\ln 2}{2t}$ 3) $\frac{\ln 2}{3t}$ 4) $\frac{3\ln 2}{t}$

24. **A proton with kinetic energy, K strikes another proton at rest. If the collision is head - on, find the correct graph between K and the distance of closest approach, r.**

1) 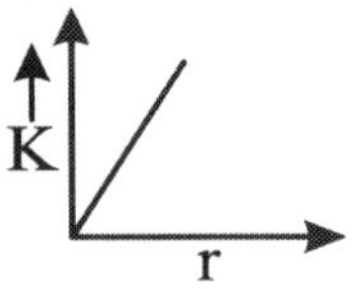

2) 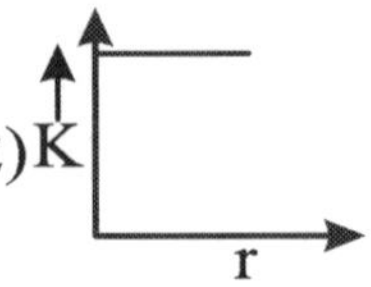

3) 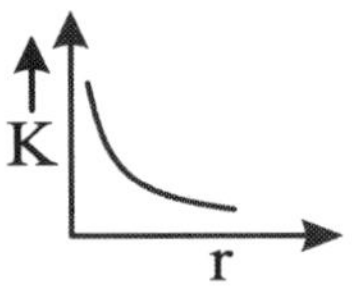

4) 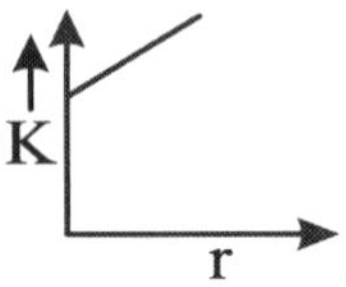

25. **The fraction of a radiactive sample will decay during half of its half-life period is**

1) $\frac{1}{\sqrt{2}}$ 2) $\frac{1}{\sqrt{2}-1}$ 3) $\frac{\sqrt{2}-1}{\sqrt{2}}$ 4) $\frac{1}{2}$

26. **A small quantity of a solution containing N^{24} radio - nuclide of half - life T and activity R_0 is injected into blood of a person. $1cm^3$ of sample of blood taken from the blood of the person shows activity R_1. If the total volume of the blood in the body of the person is V, find the timer after which sample is taken.**

1) $\frac{T}{\ln(2)}\left[1n\frac{R_0}{VR_1}\right]$ 2) $\frac{T}{\ln(2)}\left[1n\frac{VR_0}{R_1}\right]$

3) $\frac{T}{\ln(2)}\left[1n\frac{VR_1}{R_0}\right]$ 4) $\frac{T}{\ln(2)}\left[1n\frac{R_1}{VR_0}\right]$

27. **A nucleus with mass number 220 initially at rest emits an α - particle. If the Q value of the reaction is 5.5 MeV, calculate the kinetic energy of α -particle.**

1) 4.4 MeV 2) 5.4 MeV 3) 5.6MeV 4) 6.5 MeV

28. **Some amount of radioactive substance (half-life=10 days) is spread inside a room and consequently the level of radiation becomes 50times the permissible level for normal occupancy of the room. The room be safe for occupation after**

1) 20days 2) 34.8 days 3) 56.4 days 4) 62.9 days

29. **In the options given below, let E denote the rest mass energy of a nucleus and n aneutron. The correct option is**

1) $E\left({}^{236}_{92}U\right) > E\left({}^{137}_{53}I\right) + E\left({}^{97}_{39}Y\right) + 2E(n)$ 2) $E\left({}^{236}_{92}U\right) < E\left({}^{137}_{53}I\right) + E\left({}^{97}_{39}Y\right) + 2E(n)$

3) $E\left({}^{236}_{92}U\right) < E\left({}^{140}_{56}Ba\right) + E\left({}^{94}_{36}Kr\right) + 2E(n)$ 4) $E\left({}^{236}_{92}U\right) = E\left({}^{140}_{56}Ba\right) + E\left({}^{94}_{36}Kr\right) + 2E(n)$

30. **Four different radioactive elements are kept in separate containers. In the begining the container A has 200 g-atom with half-life of 2 days , B has 20 g-atom with half-life of 20 days, C has 2g-atom with half-life 200 days and D has 100g-atoms with half-life of 10 days. In the begining the maximum activity exhibited by the container is**

1) A 2) B 3) C 4) D

31. Binding energy per nucleon versas mass number curve for nuclei is shown in the fig. W,X,Y and Z are four nuclei indicated on the curve. The process that would release energy is

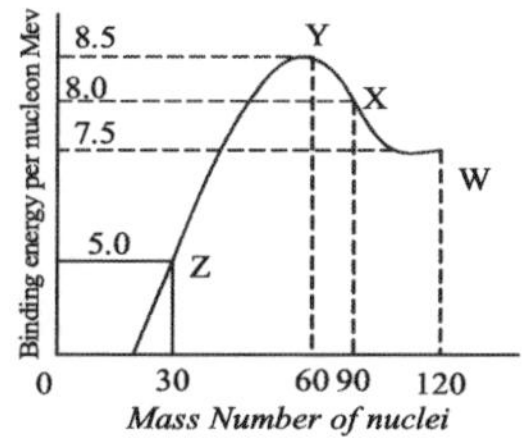

1) $Y \rightarrow 2Z$ 2) $W \rightarrow X + Z$ 3) $X \rightarrow Y + Z$ 4) $W \rightarrow 2Y$

32. When ${}_3Li^7$ nuclei are bombarded by protons, and the resultant nuclei are ${}_4Be^8$, the emitted particles will be.

1) alpha particles 2) beta particles 3) gamma photons 4) neutrons

33. A sample of uranium is a mixture of three isotopes ${}_{92}U^{234}$, ${}_{92}U^{235}$ and ${}_{92}U^{238}$ present in the ratio 0.006%, 0.71% and 99.284% respectively. The half lives of then isotopes are 2.5×10^5 years, 7.1×10^8 years and 4.5×10^9 years respectively. The contribution to activity (in %) of each isotope in the sample respectively

1) $51.41\%, 2.13\%, 46.46\%$ 2) $51.41\%, 46.46\%, 2.13\%$

3) $2.13\%, 51.41\%, 46.46\%$ 4) $46.46\%, 2.13\%, 51.41\%$

34. The table that follows shows some measurements of the decay rate of a sample of ${}^{128}I$, a radio nuclide often used medically as a tracer to measure the rate at which iodine is absorbed by the thyroid gland.

Time(min)	A(counts/s)	Time(min)	A (counts)/s
4	392.2	132	10.9
36	161.4	164	4.56
68	65.5	196	1.86
10	26.8	218	1.00

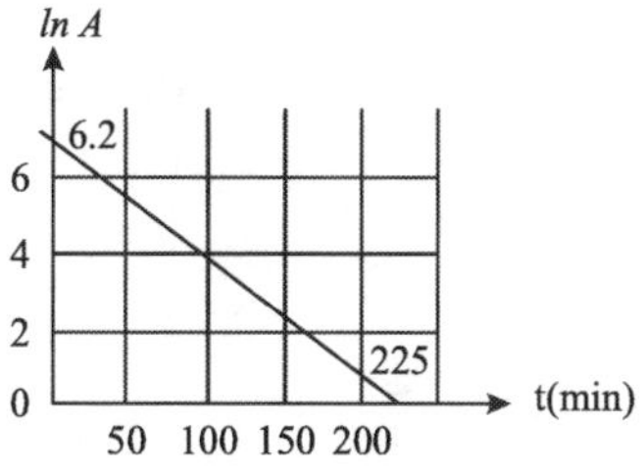

The half life $t_{1/2}$ for this radio nuclide.

1) 25 min 2) 50 min 3) 2.5 min 4) 5 min

35. **The fraction *f* of radioactive element decayed change with respect to time (*t*). The curve representing the correct variation is**

1) 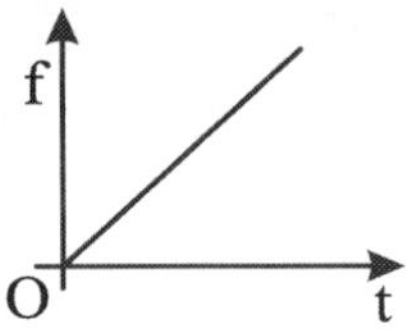

2) 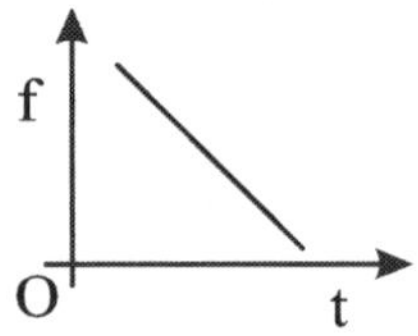

3) 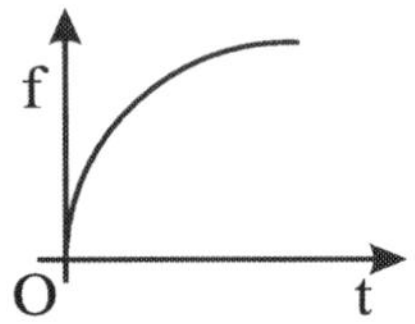

4) 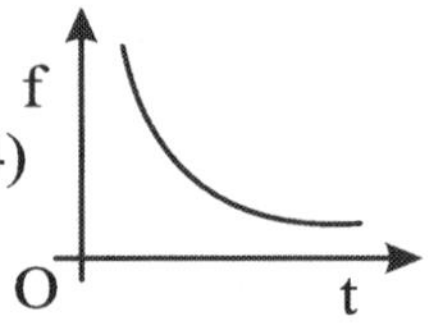

36. **The rate of decay (R) of nuclei in a radiocative sample is plotted against time (*t*). Which of the following best represents the resulting curve?**

1) 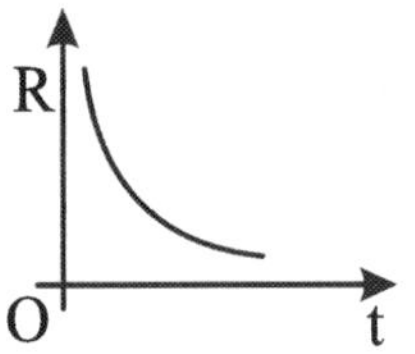

2) 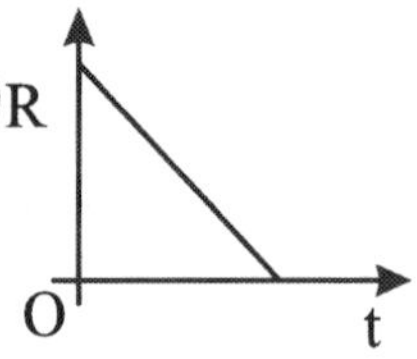

3) 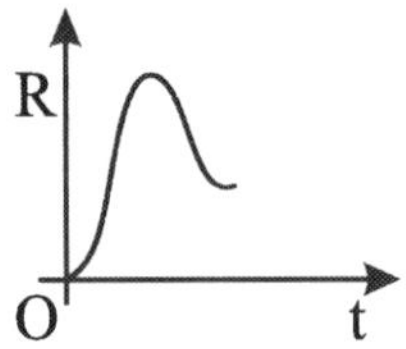

4) 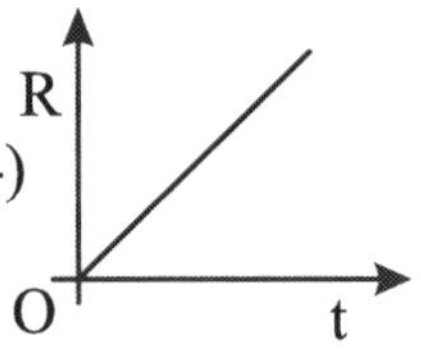

37. **The probability of survival of a radioactive nucleus for one mean life is**

1) $\frac{1}{e}$ 2) $1-\frac{1}{e}$ 3) $\frac{\ln 2}{e}$ 4) $1-\frac{\ln 2}{e}$

38. **A radioactive isotope is being produced at a constant rate A. The isotope has a half -life T initially there are no nuclei, after a time t>>T, the number of nuclei becomes constant. The value of this constant is**

1) AT 2) $\frac{A}{T}\ln(2)$ 3) $AT\ln(2)$ 4) $\frac{AT}{\ln(2)}$

LEVEL - III - KEY

1) 2	2) 2	3) 2	4) 1	5) 1	6) 2	7) 1	8) 4	9) 2	10) 3
11) 2	12) 3	13) 1	14) 2	15) 2	16) 1	17) 2	18) 3	19) 2	20) 3
21) 1	22) 3	23) 2	24) 3	25) 3	26) 1	27) 2	28) 3	29) 1	30) 1
31) 4	32) 3	33) 1	34) 1	35) 3	36) 1	37) 1	38) 4		

LEVEL - III - HINTS

1. $W = \frac{W_0}{2^n}$ $n = \frac{t}{t_{\frac{1}{2}}}$; $t_{1/2} = 5hrs$

2. $W = \frac{W_0}{2^n}$;no fo half lives $n = \frac{t}{t_{1/2}}$

3. $W = \frac{W_0}{2^n}; n = \frac{t}{t_{1/2}}$. initially value of Helium 22.4 lits no of half lives =2

4. $W = \frac{W_0}{2^n}$, $n = \frac{t}{t_{\frac{1}{2}}} = \frac{12}{3} = 4$

5. $W = \frac{W_0}{2^n}; n = 3$

7. $W = \frac{W_0}{2^n}$. Intial amount W_0, after one hour amount of remaining sample =80% W_0; $W_1 = \frac{80}{100} W_0$

8. $N = N_0 e^{-\lambda t}$

9. $\frac{dN}{dt} \propto \lambda N$; $\lambda = \lambda_1 + \lambda_2$

   $\lambda \propto \frac{1}{T}$

10. n = 4 half lifes

11. t = 1 half life

12. t = 2 half lifes

13. Given $N = \frac{N_o}{2}$; $\therefore \frac{N}{N_o} = \frac{1}{2}$

    that means t = 1 half life

14. $t = \frac{t_{1/2} \log \frac{N_0}{N}}{\log 2}$

15. At t_1 $\frac{2}{3} = \frac{1}{2^{t_1/20}}$ At t_2 $\frac{1}{3} = \frac{1}{2^{t_2/20}}$

    $t_2 - t_1 = 20$ mins.

16. $\frac{U_c}{A}$ = constant; $\frac{\frac{1}{2} C V_o^2}{\lambda N_o} = \frac{\frac{1}{2} C V^2}{\lambda N}$

    Where $V = V_o e^{-t/CR}$, $N = N_o e^{-\lambda t}$ and $\lambda = \frac{1}{t}$

17. t = 2 half lifes; $t = nT = 2 \times 5 \times \times 10^9$ *year*

    $= 10^{10}$ *year*

18. Activity $= \lambda N$ where $N = \dfrac{N_A m}{M}$

19. activity $= \lambda N_o e^{-\lambda t}$ where $N_o = \dfrac{N_A m}{M}$

20. Let the number of radioactive Potassium atoms present initially (t=0) is N_0 and the number of stable argon atoms at t=0 is zero. After time t the number of stable argon atoms is m and the radioactive potassium atoms is $N_0 - m$ given that

$$\frac{N_0 - m}{m} = \frac{1}{7}, m = \frac{7}{8} N_0 \; and \; N_0 - m = \frac{1}{8} N_0$$

since after one half-life N_0 reduces to $N_0 / 2$ after 2 half-lives $N_0 / 4$ and after 3 half-lives it reduces to $N_0 / 8$

Thus,
$$t = nT = 3 \times 2.5 \times 10^9 \; years$$
$$= 7.5 \times 10^9 \; year$$

21. Activity $R = \lambda N$

so that

$$R_1 = \lambda N_1 \; and \; R_2 = \lambda N_2$$
$$R_1 - R_2 = \lambda(N_1 - N_2) = \frac{0.6931}{T}(N_1 - N_2)$$
$$N_1 - N_2 = \frac{(R_1 - R_2)T}{0.6931}$$

$$N_1 - N_2 \propto (R_1 - R_2)T; \quad \therefore t_2 - t_1 \alpha (R_1 - R_2)T$$

22. $\dfrac{hc}{\lambda} + \dfrac{hc}{\lambda} = m_0 C^2 + m_0 C^2$

23. $\lambda = \lambda_1 + \lambda_2$

24. At closest approach, both protons willbe moving with same velocity.

$$\Delta K.E = \frac{1}{2} m v_0^2 - 2.\frac{1}{2} m \left(\frac{v_0}{2}\right)^2 = \frac{e^2}{4\pi \in_0 r}$$

$$\frac{1}{4} m v_0^2 = \frac{e^2}{4\pi \in_0 r} = \frac{K}{2}$$

$$Kr = \frac{e^2}{2\pi \in_o} = \text{constant}, \quad K \; \alpha \; \frac{1}{r}$$

25. $\dfrac{N}{N_0} = \left(\dfrac{1}{2}\right)^n = \left(\dfrac{1}{2}\right)^{\frac{t}{T}}$ $Here, t = \dfrac{T}{2} \; or \; \dfrac{t}{T} = \dfrac{1}{2}$

$$\frac{N}{N_0} = \left(\frac{1}{2}\right)^{\frac{1}{2}} = \frac{1}{\sqrt{2}}; \; \frac{N_0 - N}{N_0} = 1 - \frac{1}{\sqrt{2}} = \frac{\sqrt{2} - 1}{\sqrt{2}}$$

26. Total volume of blood,

$$V = \frac{\text{Total activity R}}{\text{Activity per cm}^3 (R_1)}$$

$$= \frac{R_0 e^{-\lambda t}}{R_1}, \qquad \text{(or)} \frac{VR_1}{R_0} = e^{-\lambda t}$$

$$-\ln\left(\frac{VR_1}{R_0}\right) = \lambda t = \frac{\ln(2)t}{T}; \; t = \frac{T}{\ln(2)} \ln\left(\frac{R_0}{VR_1}\right)$$

27. $K_1 + K_2 = 5.5\,MeV$

From conservation of linear momentum

$$[2K_1(216m)]^{1/2} = [2K_2(4m)]^{1/2}; \; K_2 = 54\,K_1$$

From the above $K_2 = 5.4\,MeV$

28. Since the intial activity is 50 times the activity for safe occupancy, therefore, $R_0 = 50R$ where $R = \lambda N$ since,

$$R \propto N$$

$$\frac{R}{R_0} = \frac{N}{N_0} = \left(\frac{1}{2}\right)^n = \left(\frac{1}{2}\right)^{t/T} \quad or$$

$$\left(\frac{1}{2}\right)^{t/10} = \frac{1}{50}$$

29. (a): Rest mass energy of U will be greater than the reat mass energy of nuclei into which it breaks. The consistituent nuclei and neutrons will have kinetic energy also, as a result of conservation of linear momentum.

30. The activity of radioactive element is given by $R = \lambda N = \frac{0.6931N}{T}$ *Thus*, $R \propto \frac{N}{T}$. The value of $\frac{N}{T}$ is largest for the element in the container A.

31. Energy of products must be more then the reactants to release energy.

32. ${}_3Li^7 + {}_1H^1 \rightarrow {}_4Be^8 + \gamma$

33: Let m is the total mass of the uranium mixture. The masses of the isotopes ${}_{92}U^{234}$, ${}_{92}U^{235}$ and ${}_{92}U^{238}$ in the mixture are $m_1 = \frac{0.006}{100}m$; $m_2 = \frac{0.71}{100}m$, and $m_3 = \frac{99.284}{100}m$.

If N_A is the Avogadro number, then number of atoms of threeisotopes are; $N_1 = \frac{m_1 N_A}{M_1}$, $N_2 = \frac{m_2 N_A}{M_2}$, and $N_3 = \frac{m_3 N_A}{M_3}$

Activity of radioactive sample $A = \lambda N$.

As $\lambda = \frac{0.693}{t_{1/2}}$, $\therefore \quad A = \frac{0.693}{t_{1/2}} N$

If t_1, t_2 and t_3 be the half lives, then

$$A_1 : A_2 : A_3 = \frac{N_1}{t_1} : \frac{N_2}{t_2} : \frac{N_3}{t_3}$$

or $$A_1 : A_2 : A_3 = \frac{m_1}{M_1 t_1} : \frac{m_2}{M_2 t_2} : \frac{m_3}{M_3 t_3}$$

$$= \frac{0.006}{234\left(2.5\times10^5\right)} : \frac{0.71}{235\left(7.5\times10^8\right)} : \frac{99.284}{238\left(4.5\times10^9\right)}$$

$$= 51.41\% : 2.13\% : 46.46\%$$

34. $A = A_0 e^{-\lambda t}$ $\Rightarrow \ln A = \ln A_0 - \lambda t$

$$\therefore \lambda = \frac{\ln A_0 - \ln A}{t} = \frac{6.2 - 0}{225} = 0.0275\,\text{min}^{-1}$$

$$\therefore T_{1/2} = \frac{0.693}{\lambda} = 25\,\text{min}$$

37. Probability of survival for any nucleus at time t is $P = \frac{N}{N_0} = \frac{N_0 e^{-\lambda t}}{N_0} = e^{-\lambda t}$

So, in one mean life, required probability is $e^{-\lambda \times \frac{1}{\lambda}} = \frac{1}{e}$;

38. $A = N\lambda$; $\therefore N = \frac{A}{\lambda} = \frac{AT}{\ln 2}$

LEVEL - IV

Comprehension-1:

The count rate meter is used to measure the activity of a given amount of a radio active element. At one instant, the meter shows 475 counts/minute. Exactly 5 minutes later, is shown 270 counts/minute then

1. **The decay constant is**

1) 0.82/ minute 2) 0.113/ minute 3) 0.166/ minute 4) 0.182/ minute

2. **Mean life of the sample is**

1) 6.35 minutes 2) 7.45 minutes 3) 8.85 minutes 4) 9.95 minutes

3. **Half life of the sample is**

1) 6.13 minutes 2) 8.42 minutes 3) 8.85 minutes 4) 9.92 minutes

Comprehension-2:

Atomic nucleus is central core of every atom in which the whole of positive charge and almost entire mass of atom is concentrated. It is a tiny sphere of radius R is given by $R = R_o A^{1/3}$, where $R_o = 1.4\times10^{-15}\,m$, a constant and A is the mass number of the nucleus

4. **A graph between '$\log\left(\frac{R}{R_o}\right)$', and 'log A'**

1) Is a parabola
2) Is a straight line passing through origin
3) Is a straight line have an intercept
4) Is an ellipse

5. **On increasing the value of 'A' the density of the nucleus**

1) Increases
2) Decreases
3) Remains constant
4) None

6. **The radius of the nucleus of mass number 125 is**

1) $175\times10^{-15}m$
2) $35\times10^{-15}m$
3) $70\times10^{-15}m$
4) $7\times10^{-15}m$

Comprehension-3:

The questions given below are based on the following paragraph

A nucleus of mass M + Δm is at rest and decays into two daughter nuclei of equal mass $\frac{M}{2}$ each. speed of light is c.

7. The binding energy per nucleon for the parent nucleus is E_1 **and that for the daughter nuclei is E_2 then**

1) $E_2 = 2E_1$
2) $E_1 > E_2$
3) $E_2 > E_1$
4) $E_1 = 2E_2$

8. **The speed of daughter nuclei is**

1) $c\frac{\Delta m}{M+\Delta m}$
2) $c\sqrt{\frac{2\Delta m}{M}}$
3) $c\sqrt{\frac{\Delta m}{M}}$
4) $c\sqrt{\frac{\Delta m}{M+\Delta m}}$

9. **Choose correct statement from the following.**

A)Large mass number nuclei undergo fission
B)Low mass number nuclei undergo fusion
C)For heavy nuclei the decrease in binding energy per nucleon shows the contribution of the increasing coulomb repulsion.

1) A, B are correct
2) A, B, C are correct
3) B , C are correct
4) A , C are correct

10. **Which of the following are not fundamental particles**

i)Electron **ii) Photon** **iii) α - Particle** **iv) Deutron**

1) Only i & ii are true
2) Only ii & iii are true
3) Only i &iii are true
4) Only iii & iv are true

11. **When a nucleus with atomic number Z and mass number A undergoes a radioactive decay process**

a) Both Z and A will decrease, if the process is α decay

b) Z will decrease but A will not change, if the process is β^+ decay

c) Z will increase but A will not change, if the process is β^- decay

d) Z and A will remain unchanged, if the process is γ decay

1) a & b are true
2) b & d are true
3) a, b & c are true
4) a, b, c, d are true

12. A nuclide A undergoes α decay and another nuclide B undergoes β^- decay

a) All the α–particle emitted by A will have almost the same speed

b) The β–particle emitted by A may have widely different speeds

c) All the β–particle emitted by B will have almost the same speed

d) The β–particle emitted by B may have widely different speeds

1) a, b are true 2) b, c are true

3) b, d are true 4) a, d are true

13. In the fission of U^{235}

i) Slow neutron is absorbed by U^{235}

ii) The products in the process are not same always, their atomic number varies from 34 to 58

iii) About 200 Mev energy is released per fission

iv) The product are always Ba and Kr

1)Only i, ii & iii are true 2)Only ii & iii are true

3)All are true 4)Only i , ii & iv are true

14. Which of the following statements are correct

i) Positron is predicted by Dirac and discovered by Anderson

ii) Liquid drop model of nucleus is developed by Bohr and Wheeler

iii) Carbon cycle was proposed by Bethe

iv) Fission reaction is first observed by OttoHahn and Strassman

1) All are true 2) Only i, ii & iv are true

3) Only i, iii & iv are true 4) Only iii & iv are true

15. Consider the following two statement A and B and identify the correct answer given below :

A) Nuclear density is same for all nuclei

B) Radius of the nucleus (R) and its mass number (A) are related as $\sqrt{A} \alpha R^{1/6}$

1) A and B are true 2) A and B are false

3) A is true but B is false 4) A is false but B is true

16. Consider the following statements A, B and identify the correct choice in the given answers

A: Density of a nucleus is independent of its mass number

B: Beryllium is used as moderator in nuclear reactors

1) A and B are correct 2) A and B are wrong

3) A is correct, B is wrong 4) A is wrong, B is correct

17. Consider the following statements (A) and (B) and identify the correct answer given below.

Statement (A) : Positive values of packing fraction implies a large value of binding energy.

Statement (B) : The difference between the mass of the nucleus and the mass number of the nucleus is called packing fraction

1) (A) and (B) are correct 2) (A) and (B) are false

3) (A) is true (B) is false 4) A is false, B is true

LEVEL - IV - KEY

1) 2	2) 3	3)1	4) 2	5) 3	6) 4	7) 3	8) 2	9) 2	10) 4
11) 4	12) 4	13) 1	14) 1	15) 3	16) 1	17) 2			

LEVEL - IV - HINTS

1. $A = A_0e^{-\lambda t}$; $270 = 475e^{-\lambda t}$

   On solving we get $\lambda = 0.113/\text{min}$

2. $\tau = \dfrac{1}{\lambda}$

3. $T = 0.693\,\tau$

6. $R = R_0A^{1/3}$

7. binding energy per nucleon of daughter nuclei is more than that of the parent nuclei.

8. According to conservation of momentum

   $0 = \frac{M}{2}V_1 - \frac{M}{2}V_2; V_1 = V_2$; $\Delta mc^2 = \frac{1}{2}.\frac{M}{2}V_1^2 + \frac{1}{2}.\frac{M}{2}V_2^2$

* * * * *

Chapter - 3
(PART - 1)
SEMICONDUCTOR DEVICES

SEMI CONDUCTOR DEVICES

SYNOPSIS

Classification of solids :

- Solids can be classified as metals, insulators and semiconductors basing on conductivity (or) band theory
- Classification according to conductivity is as follows
- Metals have high conductivity $\left(10^{8} to 10^{2}\ Sm^{-1}\right)$ (or) low resistivity
- Insulators have low conductivity $\left(10^{-19} to 10^{-11} Sm^{-1}\right)$ (or) high resistivity $\left(10^{+11} to 10^{+19} \Omega m\right)$
- Semiconductors have conductivity $\left(10^{5}\ to\ 10^{-6} Sm^{-1}\right)$ (or) resistivity $\left(10^{-5} to 10^{6} \Omega m\right)$, intermediate to metals and insulators

Energy Band theory in solids:

- An isolated atom has well defined energy levels and energy of an electron depends on its orbit (Principal quantum number)
- But in solids atoms are so close such that outer orbits are very close (or) overlaped to form energy band.
- Inside the crystal each electron has a unique position and no two electrons see exactly same pattern of surrounding charges and each electron has different energy level.
- Different energy level with continuous energy variation form energy bands (According to Pauli's principle)
- The energy band formed by a series of energy bands containing valance electrons is **valance band**.
- At 0 K, electrons start filling energy level in valance band starting from the lowest one.
- The highest energy level, occupied by an electron in the valance band at 0K is called **Fermi level**.
- The lowest unfilled energy band formed just above valance band is called **conduction band**.
- Depending on the forbidden energy gap between valance band and conduction band, the solids are classified as conductors, insulators and semiconductors.

1) **Conductors :** The energy band structure in conductors have two possibilities

- The valance band may be completely filled and the conduction band partially filled with an extremely small energy gap between them $E_g = 0$

  Ex: Sodium (Na)
- The valence band is completely filled and conduction band is empty but the two overlap each other. Eg: Zn

1) 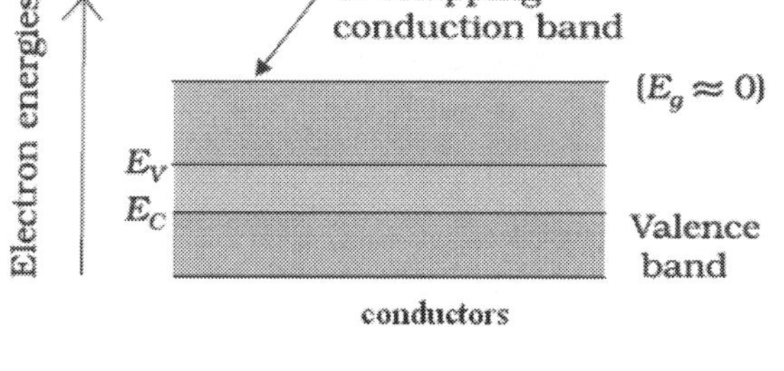

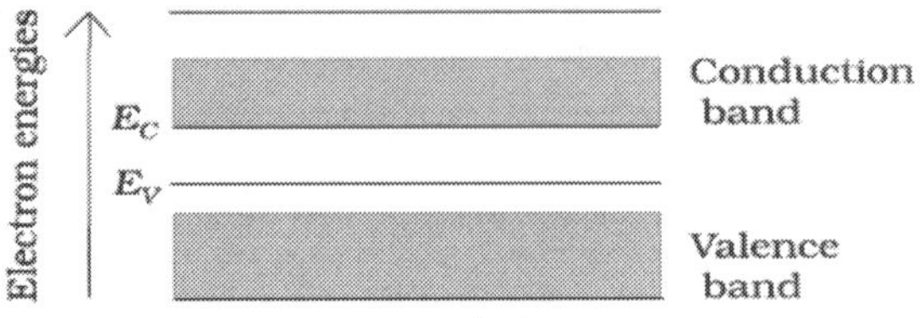

conductors

2) **Insulators :**

- In insulators forbidden energy gap is quite large. $E_g > 3ev$
- Ex. Energy gap for diamond is 5.5 eV.

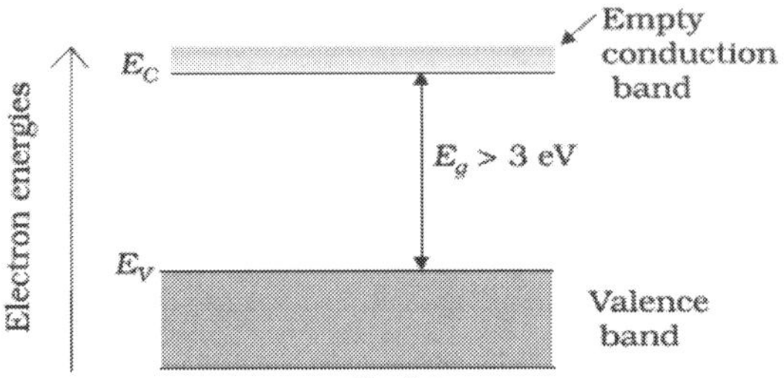

Insulators

3) **Semi conductors :**

- Semi conductors are the basic materials used in the present solid state devices like diode, transistor, Ic's.
- The energy band structure of the semiconductors is similar to insulators but in their case, the size of forbidden energy gap is much smaller than that of insulators. $E_g = 0.2eV \ to\ 3eV$

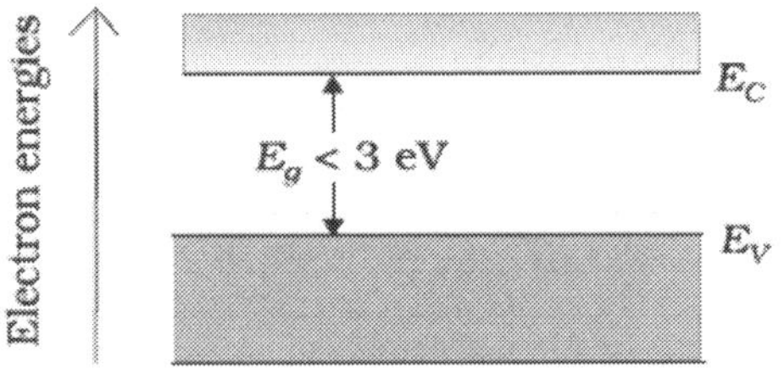

Semiconductors

- Eg: Forbidden energy gap for Ge is 0.67eV, for Si is 1.1 eV and for GaAs is 1.41eV
- The vacancy in the covalent band is called hole.
- The electrons and holes move in opposite directions.
- The difference in the resistivity of C, Si and Ge depends upon the energy gap between their conduction and valence bands. For C (diamond), Si and Ge, the energy gaps are 5.4 eV, 1.1 eV and 0.7 eV, respectively. Sn also is a group IV element but it is a metal because the energy gap in its case is 0 eV.

Intrinsic semi conductor :

- Semi conductors in their purest form having equal number of electrons in CB and holes in VB is called as intrinsic semi conductors.
- These electrons (free from covalent bonds) and holes are called intrinsic carriers and such a semiconductor is called intrinsic semiconductor.

Ex:- Germanium (Ge) and Silicon (Si) are the purest form of intrinsic semi conductors.

- In an intrinsic semiconductor if n_e denotes the electron number density in conduction band, n_h the hole number density in valence band and n_i the number density (or) concentration of intrinsic carriers, then

  $n_e = n_h = n_i$

Doping:

- The process of adding impurity to a pure semi conductor (in 1 : 10^6 ratio also called 1ppm) so as to improve its conductivity is called doping.
- Pentavalent impurity atoms are called donor impurity atoms.
- Trivalent impurity atoms are called acceptor impurity atoms.

Extrinsic semi conductor :

- The semi conductor formed by adding a small quantity of impurity (tri or penta valent) to pure semi conductor is called as extrinsic semi conductor.
- The extrinsic semi conductor formed by adding pentanalent impurity atoms is called **n-type semi conductor.**
- The extrinsic semi conductor formed by adding trivalent impurity atoms is called **p-type semi conductor.**
- Fermi level in n-type semi conductors (also known as donor energy level) lies in forbidden energy gap and is very close to conduction band (≈ 0.01eV below conduction band)
- Fermi level in p-type semi conductor (also known as acceptor energy level) lies in forbidden energy gap and is very close to valence band (≈ 0.01 to 0.05 eV above valence band)
- When ever a covalent bond breaks an electron hole pair is formed.
- When valency electron combines with a hole, a covalent bond formed and at the starting point of the electron a hole is formed.
- If a conduction electron combines with a hole a covalent bond is formed without the formation of a hole any where.
- In a doped (or) extrinsic semi conductor number density of electrons in conduction band (n_e), number density of holes in the valence band (n_h) and number density of electrons in conduction band (or) holes in valence band in a pure semi conductor (n_i) then they are related as $n_e n_h = n_i^2$.
- The intrinsic concentration n_i varies with T as

  $n_i^2 = A_o T^3 e^{-Eg/KT}$ (A_o is constant)
- The fraction of electrons of valence band present in conduction band is given by

  $f \alpha e^{-\frac{Eg}{KT}}$

  Where K is Boltzman's constant, E_g is forbidden energy gap and T is temperature

The Energy Gap : Experimentally it has been found that the forbidden energy region E_g depends on temperature.

For silicon $E_g(T) = 1.21 - 3.60 \times 10^{-4} T$

For germanium $E_g(T) = 0.785 - 2.23 \times 10^{-4} T$

At room temperature $(300K)$ for silicon $E_g = 1.1ev$ and for germanium $E_g = 0.72ev$

- Adding equal concentration of donor and acceptor or atoms to P type and N type semiconductor respectively results in an intrinsic semiconductor.
- When concentration of donor atoms exceeds the acceptor concentration in P type semiconductor, it changes from P type to N type semiconductor.
- In semi conductors the total current I is the sum of electron current I_e and holes current I_h

  $I = I_e + I_h$
- Electrical conductivity $\sigma = \sigma_e + \sigma_h$

  $\sigma = n_e\mu_e e + n_h\mu_h e$

  For intrinsic semi conductor $n_e = n_h = n_i$ so that $\sigma = n_i e(\mu_e + \mu_h)$

ELECTRICAL CONDUCTIVITY OF SEMI CONDUCTORS:(FORMULAE DERIVATION)

- Consider a block of semiconductor of length l, area of cross-section A and having number density of electrons and holes as n_e and n_h respectively .
- By that on applying a potential difference, say V, a current I flows through it as shown in figure.
- The electron current (I_e) and the hole current (I_h) constitute the total current I flowing through the semiconductor i.e ., $I = I_e + I_h$............(1)

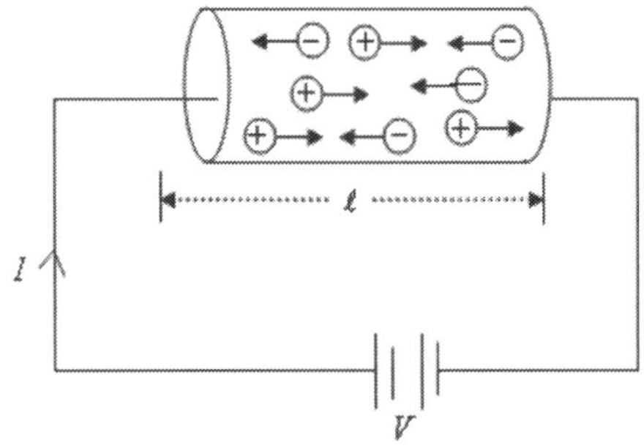

- If n_e is the density of conduction band electrons in the semiconductor and V_e is the drift velocity of electrons, then electron current is given by $I_e = en_eAv_e$........(2)

  Also, the hole current $I_h = en_hAv_h$ ......(3)

  Using the equation (2) and (3) , the equation (1) becomes

  $I = en_eAv_e + en_hAv_h$

  or $I = eA(n_ev_e + n_hv_h)$........(4)

  If ρ is the resistivity of the material of the semiconductor, then the resistance offered by the semiconductor to the flow of current is given by $R = \rho\frac{\ell}{A}$.........(5)

  Since V=RI, from the equations (4) and (5),

  we have $V = RI = \rho\frac{\ell}{A} \times eA(n_ev_e + n_hv_h)$

or $V = \rho \ell e(n_e v_e + n_h v_h)$.............(6)

If E is the electric field set up across the

semiconductor, then $E = \frac{V}{\ell}$........(7)

From the equations (6) and (7), we have

$$E = \rho e(n_e v_e + n_h v_h)$$

or $\frac{1}{\rho} = e\left(n_e \frac{v_e}{E} + n_h \frac{v_h}{E}\right)$.........(8)

On applying electric field, the drift velocity acquired by the electrons (or holes) per unit strength of electric field is called mobility of electrons (or holes)

$\mu_e = \frac{v_e}{E}$ and $\mu_h = \frac{v_h}{E}$

Therefore, the equation (8) becomes

$\frac{1}{\rho} = e(n_e \mu_e + n_h \mu_h)$.........(9)

Also, $\sigma = \frac{1}{\rho}$ is called the conductivity of the material of semiconductor.

$\therefore \sigma = e(n_e \mu_e + n_h \mu_h)$.......(10)

- Electrons mobility is greater than the hole mobility.
- Mobility is a property of the semiconductor itself. It does not depends on the doping concentration.
- The mobility of an electron or hole generally decreases with increase temperature.

• Resistance of semi conductors decrease with the increase in temperature so semi conductors are insulators at low temperature but becomes slightly conducting at room tempeture.

• P- type (or) n- type semi conductor material is elctrically neutral.

W.E-1: The number of silicon atoms per m^3 is 5×10^{28}. This is doped simultaneosuly with 5×10^{22} atoms per m^3 of Arsenic and 5×10^{20} per m^3 atoms of Indium. Calculate the number of electrons and holes. Given that $n_t = 1.5\times10^{16}\ m^{-3}$. Is the material - n type or p-type?

Sol: Arsenic is donor impurity

No. of donor atoms added, $N_D = 5\times10^{22} m^{-3}$,

Indium is acceptor impurity, no.of accepter atoms added

$N_A = 5\times10^{20} m^{-3}$

Therefore, no.of free electrons created

$n_e = N_D = 5\times10^{22}$

Now, $n_e > n_h$, therefore, net no.of free electrons created,

$n_e^1 = n_e - n_h = 5\times10^{22} - 5\times10^{20} = 4.95\times10^{22}\,m^{-3}$

Also net no. of holes created $n_h^1 = \dfrac{n_1^2}{n_e^1} = \dfrac{\left(1.5\times10^{16}\right)^2}{4.95\times10^{22}} = 4.55\times10^9\,m^{-3}$

AS $n_e^1 > n_h^1$, the resulting material is n-type semiconductor.

W.E-2: A semiconductor has an electron concentration of $0.45\times10^{12}\,m^{-3}$ and a hole concentration of $5.0\times10^{20}\,m^{-3}$. Calculate its conductivity . Given electron mobility $=0.135m^2V^{-1}s^{-1}$; hole mobility $=0.048m^2V^{-1}s^{-1}$,

Solution : The conductivity of a semiconductor is the sum of the conductivities due to electrons and holes and is given by $\sigma = \sigma_e + \sigma_h = n_e e\mu_e + n_h e\mu_h = e\left(n_e\mu_e + n_h\mu_h\right)$

As per given date , n_e is negligible as compared to n_h, so that we can write

$\sigma = en_h\mu_h = \left(1.6\times10^{-19}C\right)\left(5.0\times10^{20}m^{-3}\right)\left(0.048m^2V^{-1}s^{-1}\right) = 3.84\Omega^{-1}m^{-1} = 3.84Sm^{-1}$

W.E-3: An N-type silicon sample of width $4\times10^{-3}m$ thickness and length $6\times10^{-2}m$ carries a current of 4.8mA when the voltage is applied across the length of the sample. What is the current density ?

If the free electron density is $10^{22}m^{-3}$, then find how much time it takes for the electrons to travel the full length of the sample .

Soution : The current density J is given by

$$J = \frac{I}{A} = \frac{4.8\times10^{-3}}{\left(4\times10^{-3}\right)\left(25\times10^{-5}\right)} = \frac{4.8\times10^{-3}}{10^{-6}}$$

The drift velocity v_d given by $v_d = \dfrac{J}{ne} = \dfrac{4800}{10^{22}\times1.6\times10^{-19}} = 3m/s$

The time taken 't' is given by $t = \dfrac{L}{v_d} = \dfrac{6\times10^{-2}}{3} = 0.02\,\text{sec}$

W.E-4: The energy gap of pure Si is 1.1 eV. The mobilities of electrons and holes are respectively $0.135m^2V^{-1}s^{-1}$ and $0.048m^2V^{-1}s^{-1}$ and can be taken as independent of temperature. The intrinsic carrier concentration is given by $n_i = n_0 e^{-Eg/2kT}$. Where n_0 is a constant, E_g The gap width and k The Boltzmann's constant whose value is $1.38\times10^{-23}JK^{-1}$. The ratio of the electrical conductivities of Si at 600K and 300K is.

Solution: The total electrical conductivity of a semiconductor is given by

$\sigma = e\left(n_e\mu_e + n_h\mu_h\right)$

For an intrinsic semiconductor,

$n_e = n_h = n_i$

We can thus write for the conductivity

$\sigma = e\left(\mu_e + \mu_h\right)n_i$

or $\sigma = e(\mu_e + \mu_h) n_0 e^{E_g/2kT}$

As the mobilities μ_e, μ_h are independent of temperature, they can be regarded as constant. The ratio of the conductivities at 600 K and 300 K is then,

$$\frac{\sigma_{600}}{\sigma_{300}} = \frac{e(\mu_e + \mu_h) n_0 e^{-E_g/2k\times 600}}{e(\mu_e + \mu_h) n_0 e^{-E_g/2k\times 300}} = e^{-E_g/1200k}$$

As per given data $E_g = 1.1eV$

$$k = 1.38\times 10^{-23} JK^{-1} or \left(\frac{1.38\times 10^{-23}}{1.6\times 10^{-19}}\right) eVK^{-1}$$

$$\therefore k = 8.625\times 10^{-5} eVK^{-1}$$

Solving we get the ratio of electrical conductivities is 4×10^4

W.E-5: In a p-n junction diode, the current I can expressed as $I = I_0 \exp\left(\frac{eV}{2k_B T} - 1\right)$ where I_0 is called the reverse saturation current, V is the voltage across the diode and is positive for forward bias and negative for reverse bias, and I is the current through the diode, K_B is the Boltzmann constant $(8.6\times 10^{-5} eV/K)$ and T is the absolute temperature. If for a given diode $I_o = 5\times 10^{-12}$ A and T= 300K, then

(a) What will be the forward current at a forward voltage of 0.6V?

(b) What will be the increase in the current if the voltage across the diode is increased to 0.7V?

(c) What is the dynamic resistance?

(d) What will be current if reverse bias voltage changes from 1V to 2V?

Solution:

$I_o = 5\times 10^{-12} A, k = 8.6\times 10^{-5} eVk^{-1}$

$= 8.6\times 10^{-5} \times 1.6\times 10^{-19} Jk^{-1}$

a) $I = I_0\left(e^{ev/2kT} - 1\right)$,

For V=0.6V,

$$I = 5\times 10^{-12}\left(e^{\frac{1.6\times 10^{-19}\times 0.6}{2\times 8.6\times 10^{-5}\times 1.6\times 10^{-19}\times 300}} - 1\right) = 5\times 10^{-12}\left(e^{23.52} - 1\right)$$

$$= 5\times 10^{-12}\left(1.256\times 10^{10} - 1\right) = 0.0628.A$$

b) For v=0.7v, we have

$$I = 5\times 10^{-12}\left(e^{\frac{1.6\times 10^{-19}\times 0.7}{2\times 8.6\times 10^{-5}\times 1.6\times 10^{-19}\times 300}} - 1\right)$$

$I = 5\times10^{-12}\left(e \quad -1\right)$

$= 5\times10^{-12}\left(6.054\times10^{11} - 1\right) = 3.0271A$

$\therefore \Delta I = 3.271 - 0.0628 = 2.9643A$

c) $\Delta I = 2.9643, \Delta v = 0.7 - 0.6 = 0.1V$ dynamic resistance

$$= \frac{\Delta v}{\Delta I} = \frac{0.1}{2.9643} = 0.0337\Omega$$

d) For change in voltage from 1 to 2v, the current will remain equal to $I_0 = 5\times10^{-12} A$. It shows that the diode possesses practically infinite resistance in reverse biasing

W.E-6: The energy of a photon of sodium light $(\lambda = 589nm)$ equal to the band gap of a semiconducting material (a) Find the minimum energy E required to create a hole-electron pair (b) Find the value of E/kT at a temperature of 300K

Solution :

(a) The energy of the photon in $eV = \frac{12400}{\lambda} = \frac{12400}{5890} = 2.1eV$

(Wavelength of photon =589nm=5890 A^0)

Thus the band gap is 2.1ev. This is also the minimum energy E required to push an electron from the valence band into the conduction band. Hence the minimum energy required to create a hole-electron pair is 2.1eV

(b) At T=300K, kT= $\left(8.62\times10^{-5} eV/K\right)\left(300K\right)$

$= 25.86\times10^{-3} eV$ Thus, $\frac{E}{kT} = \frac{2.1eV}{25.86\times10^{-3} eV} = 81$

The availale thermal energy is nearly 81 times less than that of the required energy to create electron hole pair. So it is difficult for the thermal energy to create the hole-electron pair but a photon of light can do it easily

P-n junction :

- In a single Ge (or) Si crystal by doping one part with trivalent impurity and the other part with pentavalent impurity , a p - n junction is formed
- At p-n junction migration of majority charges i.e. holes form p side to n side and electrons from n side to p - side due to concentration difference is called diffussion
- At p-n junction , the donor impurity atoms become +ve ions and the acceptor impurity atoms become -ve ions, so that the ions act as two electrodes and the junction is called p-n junction diode.
- The ions at junction develop a potential difference called potential barrier and this barrier prevents further diffusion of charges across junction
- Around junction electron combine with holes to form a small region, free of charge carriers is called depletion region.
- The large electric field of intensity (E) is directed from n-type semi conductor to p-type at the junction.
- If potential barrier is V_b, width of depletion layer is d then $E = \frac{V_b}{d}$.
- Only inside depletion layer the +ve and -ve immobile ions exist whereas outside this layer the material is neutral

- Potential barrier depends on nature of semi conductor crystal, temperature and amount of doping
- The symbol of p-n junction diode is given below

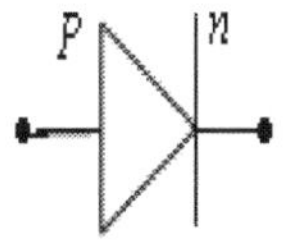

Semi conductor diode :

1) **Forward bias** :

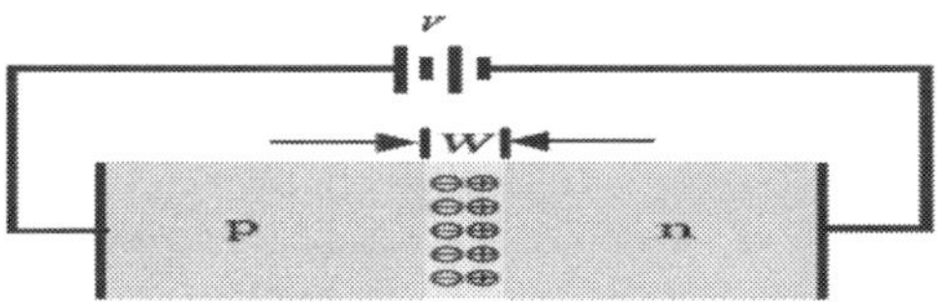

- When an external voltage V is applied across a semiconductor diode such that p-type is connected to +ve terminal and n-type to -ve terminal of battery (in general p - type to high voltage and n-type to low voltage), the diode is said to be forward biased.
- External voltage V is greater and opposite to barrier potential V_b. So width of depletion layer and resistance decrease.
- Effective barrier voltage under forward bias is V_b-V.

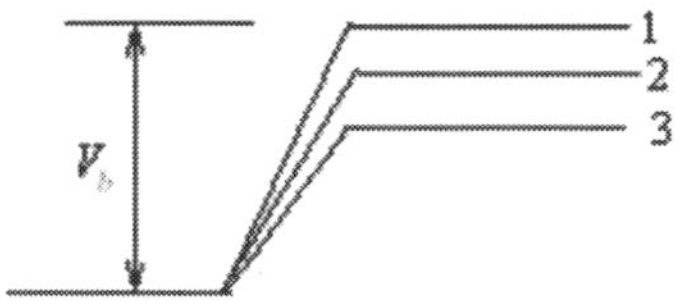

Fig. p-n junction diode under forward bias. Barrier potential (1) without battery, (2) low battery voltage, and (3) high voltage battery.

- Resistance of ideal diode in forward bias is zero
- If external voltage (V) is greater than barrier voltage then majority charge carriers diffuse across the junction and constitute diffusion current ($I = I_e + I_h$)
- Direction of diffusion current is from p- type to n-type
- The current flows through the diode in the below circuit is

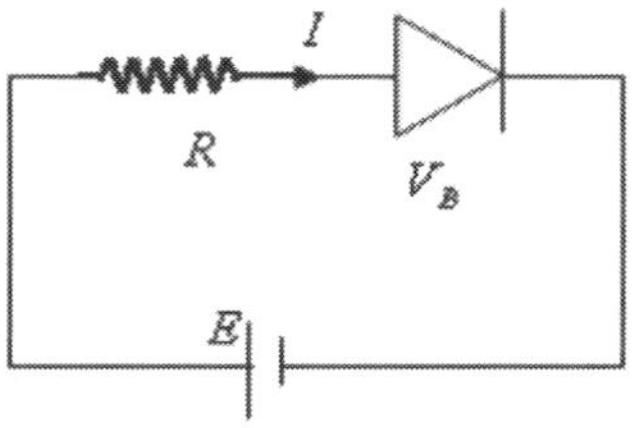

$$I = \frac{E - V_B}{R + r_f}$$

Where R= external resistance,

r_f = resistance of diode in forward bias,

V_B =barrier potential.

i) Power developed across the diode $= V_B I$

ii) Power developed across the resistor $= (E - V_B)I$,

- The external voltage beyond which diode current start increasing rapidly is called knee voltage (V_{knee})
- The below diagram show forward bias of junction diode.

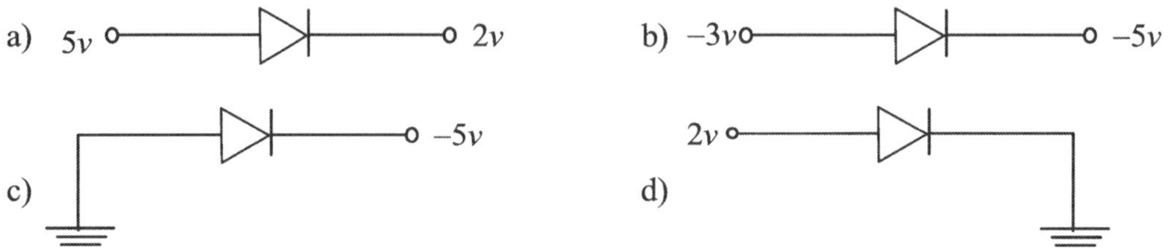

2) **Reverse bias :**

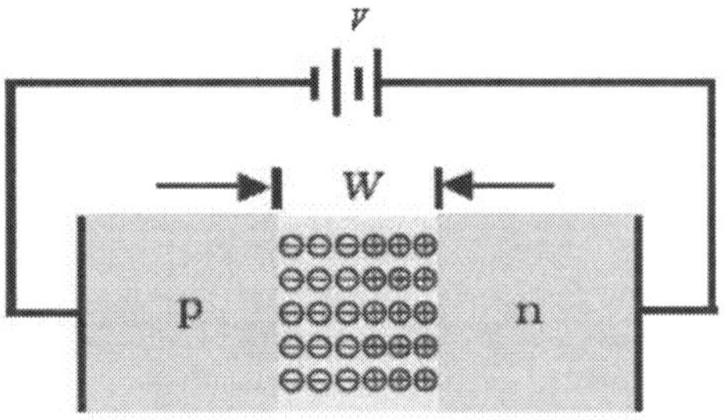

- When an external voltage V is applied across a semiconductor diode such that p-type is connected to -ve terminal and n-type to +ve terminal of battery (in general p - type to low voltage and n-type to high voltage), the diode is said to be reverse biased.
- External voltage V is greater and in same direction to barrier potential V_b, so width of depletion layer and resistance increase.
- Effective barrier voltage under reverse bias is $V_b + V$.

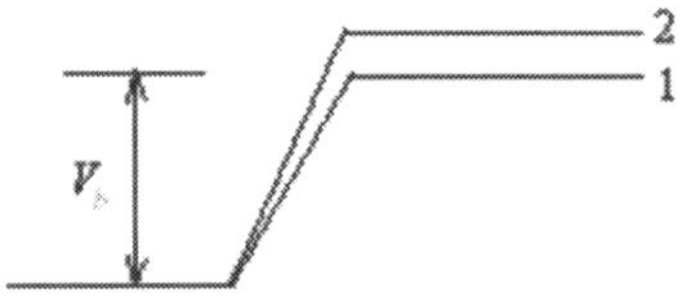

Fig : Diode under reverse bias, Barrier potential under reverse bias.

- Resistance of ideal diode in reverse bias is infinity upto a large external voltage (V).
- At lower external voltage few covalent bonds are broken to liberate electrons and holes and these constitute reverse saturation current
- At very high reverse bias voltage all the covalent bonds are broken to liberate large number of electrons called as breakdown voltage.
- The below diagram show reverse bias of p n junction diode.

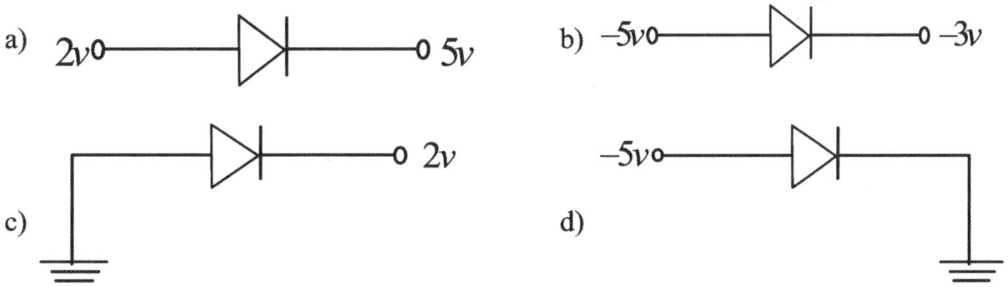

- Thus there is an unexpected release (avalanche) of large number of electrons and holes there by sharp increase in current takes place at a voltage called avalanche breakdown voltage.

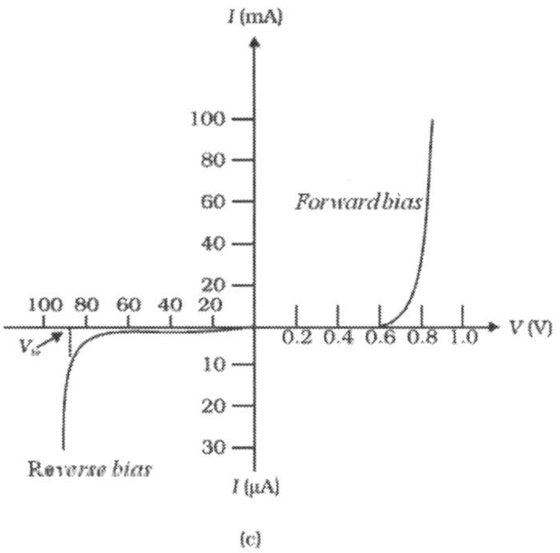

Fig : Avalanche breakdown

NOTE:

The potential barrier existing across an unbiased p-n junction is V_B volt

i) The minimum kinetic energy required by a hole to diffuse from the p-side to the n-side is 'eV_B'

ii) If the junction is forward biased at V volt, then the minimum kinetic energy required by a hole to diffuse from the p-side to the n-side is $e(V_B - V)$

iii) If the junction is reverse biased at V volt, then the minimum kinetic energy required by a hole to diffuse from the p-side to the n-side is $e(V_B + V)$

W.E-7:The V-I characteristic of a silicon diode is shown in the Fig. Calculate the resistance of the diode at

a) $I_D = 15mA$ *and* $(b) V_D = -10V$

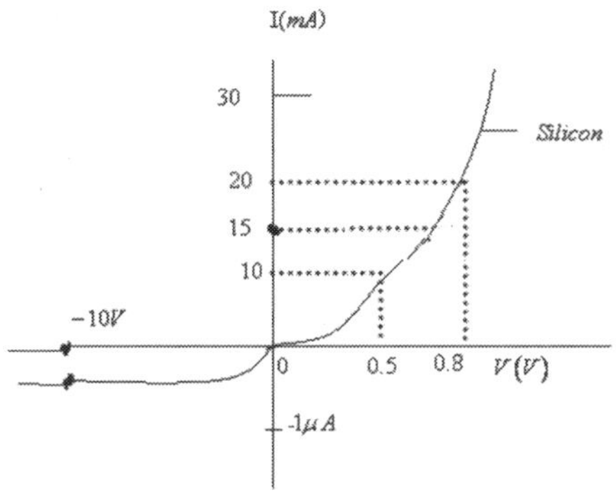

Solution: Considering the diode characteristics as a straight line between I=10mA to I=20mA passing through the origin, we can calculate the resistance using Ohm's law

a) From the curve at I=20mA, V=0.8V

I=10mA, V=0.7V

$r_{fb} = \Delta V / \Delta I = 0.1V / 10mA = 10\Omega$

b) From the curve at $V = -10V,\ I = -1\mu A$

Therefore, $r_{rb} = 10V / 1\mu A = 1.0\times10^{7}\Omega$

W.E-8: Two junction diodes, one of germanium (Ge) and other of silicon (Si) are connected as shown in fig to a battery of 12V and a load resistance $10k\Omega$. The germanium diode conducts at 0.3V and silicon diode at 0.7V. When current flows in the circuit, the potential of terminal Y will be

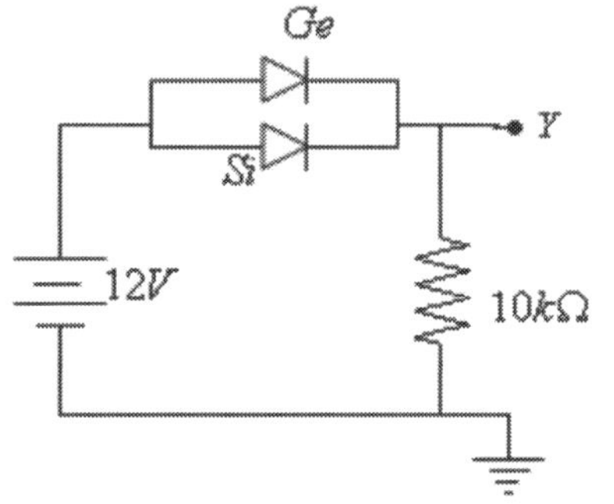

Solution : The Ge diode conducts for a p.d of 0.3V, therefore the current passes through it and the Si diode do not conduct. Hence the potential of terminal $Y = 12 - 0.3 = 11.7V$

W.E-9: Find maximum voltage across AB in the circuit shown in Fig. Assume that diode is ideal

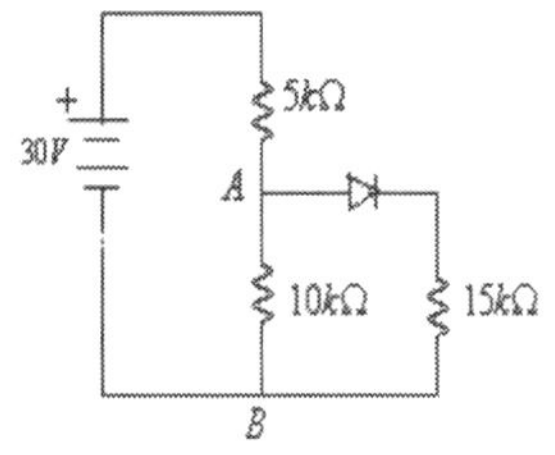

Solution: As the diode is treated ideal, its forward resistance $R_f = zero$. It acts as short circuit. So $10\,k\Omega$ is in parallel with $15\,k\Omega$ and the effective resistance across AB is

$$R_{AB} = \frac{10\times15}{10+15} = \frac{10\times15}{25} = 6k\Omega$$

$6k\Omega$ is in series with $5k\Omega$

$\therefore$ total resistance $= R_T = 6k\Omega + 5k\Omega = 11k\Omega$,

V=30V. Current drawn from the battery is $I = \frac{V}{R_T} = \frac{30V}{11k\Omega} = 2.72mA$

$V_{AB} = IR_{AB} = 2.72mA \times 6k\Omega = 16.32V$

S **W.E-10:Find the voltage V_A in the circuit shown in figure . The potential barrier for Ge is 0.3 V and for Si is 0.7 V**

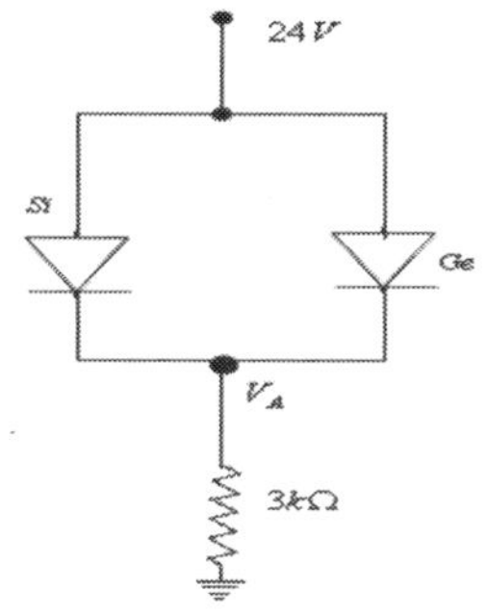

Solution: In the situation given, germanium diode will turn on first because potential barrier for germanium is smaller. The silicon diode will not get the opportunity to flow the current and so remains in open circuit. The equivalent circuit is as in figure

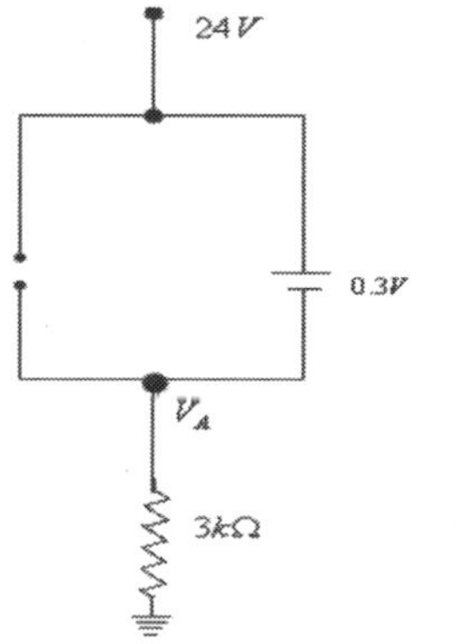

$V_A = 24 - 0.3 = 23.7V$

W.E-11:Considering the circuit and data given in the diagram, calculate the currents flowing in the diodes D_1 and D_2 Forward resistance of D_1 and D_2 is 20Ω

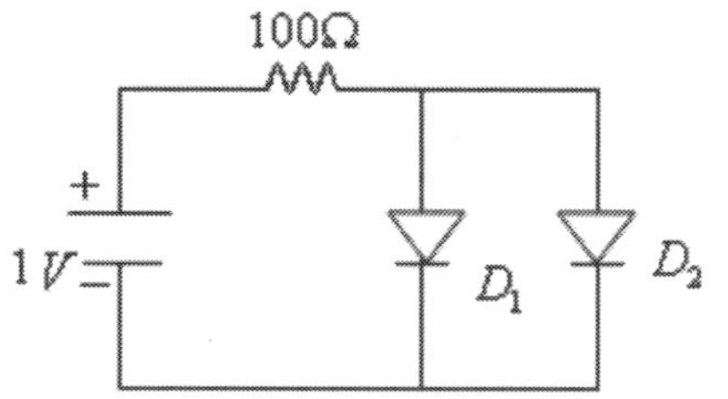

Solution: Since the positive terminal of battery is connected to P-type of both diodes D_1 and D_2, they are forward biased. These diodes are replaced by with their forward resistance as shwon is Fig:

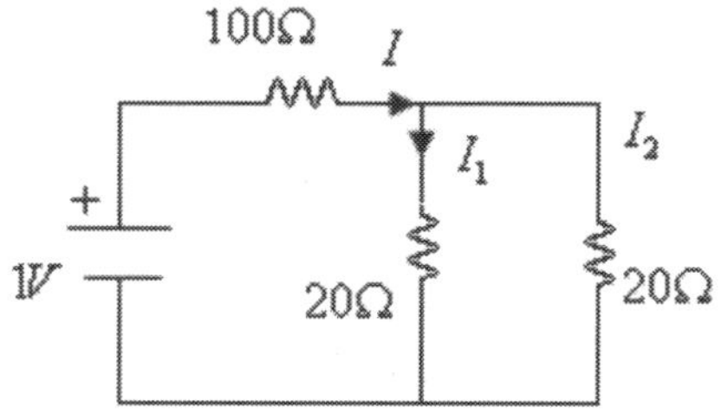

he resistance of 20Ω and $20\,\Omega$in

parallel, $\frac{1}{R}=\frac{1}{20}+\frac{1}{20}(or)R=\frac{20}{2}=10\Omega$

Therefore, total current I in the circuit

$I=\frac{1}{100+10}=\frac{1}{110}$ amp and $I_1=I_2=\frac{1}{2}\times\frac{1}{110}=\frac{1}{220}$ amp

W.E-12:In the circuit shown, the potential drop across each capacitor is (assuming the two diodes are ideal)

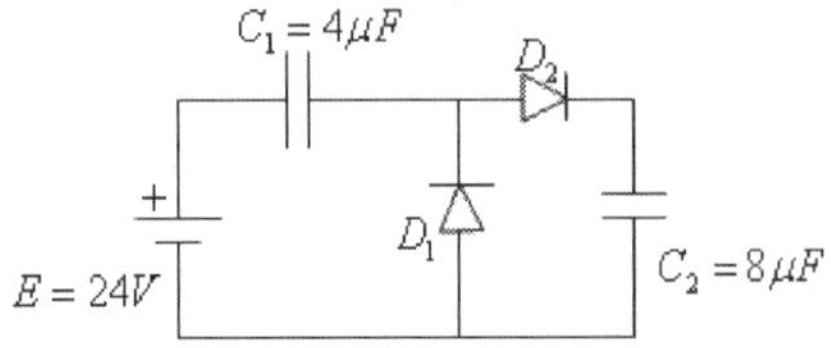

Solution: The diode D_1 is reverse biased (open circuit), but the diode D_2 is forward (short circuit).

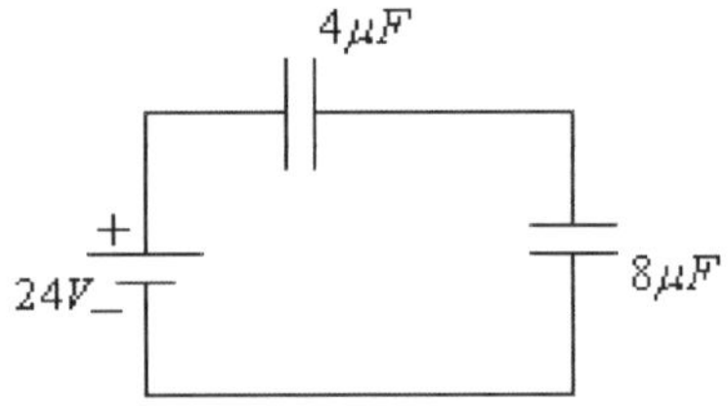

$\therefore$ the potential of the battery divides across the two capacitors in the inverse ratio of their capacities.

i.e, $\frac{V_1}{V_2}=\frac{C_2}{C_1}=\frac{8}{4}=\frac{2}{1}$

$V_1=\frac{2}{3}E=\frac{2}{3}\times 24=16V$ $\quad$ $V_2=\frac{1}{3}E=\frac{1}{3}\times 24=8V$

W.E-13:A potential barrier of 0.50V exists across a p-n junction

a) **If the depletion region is $5.0\times10^{-7}m$ wide,what is the intensity of the electric field in this region?**

b) **An electron with a speed of $5.0\times10^{5}m/s$ approaches the p-n junction from the n-side. With what speed will it enter the p-side ?**

Solution :

a) The electric field is E=V/d $=\frac{0.50V}{5.0\times10^{-7}m}=1.0\times10^{6}V/m$

b)

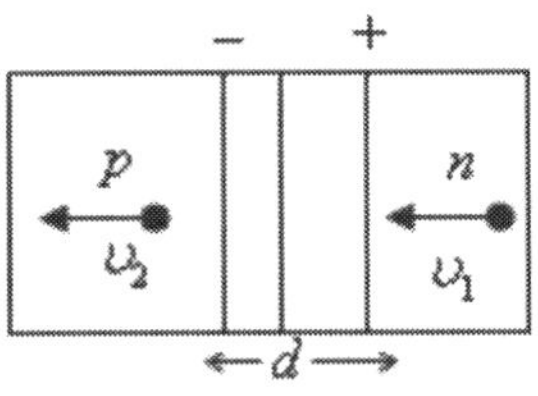

Let the electron has a speed v_1 when

it enters the depletion layer and v_2 when it comes out of it. As the potential energy

increases by $e \times 0.50V$. From the principle of conservation of energy $\frac{1}{2}m\upsilon_1^2 = e \times 0.50 + \frac{1}{2}m\upsilon_2^2$

$$\frac{1}{2}(9.1\times10^{-31})(5\times10^5)^2 = 1.6\times10^{-19}\times0.5+\frac{1}{2}\times9.1\times10^{-31}\times\upsilon_2^2$$

Solving this. $\upsilon_2 = 2.7\times10^5 m/s$

- **Application of junction diode as a RECTIFIER**
- The process of conversion of ac to dc is called rectification, tha arrangement is called rectfier. They are
- **Half wave rectifier**

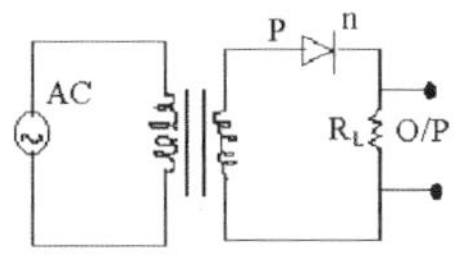

Half Wave Rectifier

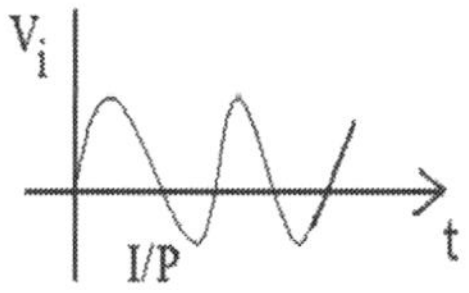

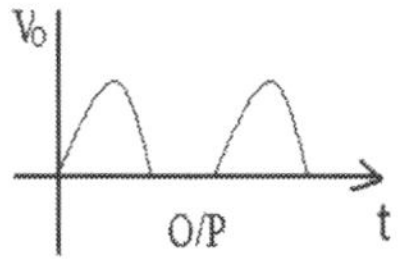

- In halfwave rectification we need atleast one semicondcutor diode.
- In half wave rectifier the efficiency is $\eta = 0.406\frac{R_L}{r_f + R_L}$ (where $r_f \rightarrow$ forward resistance of diode and $R_L \rightarrow$ load resistance.)

  It's maximum value is 40.6%.
- Maximum current $I_m = \frac{V_m}{R_L + r_f}$ (V_m is maximum voltage)
- Mean dc current $I_{dc} = \frac{I_m}{\pi}$
- $I_{rms} = \frac{I_m}{2}$
- Value of ripple factor $= 1.21$
- The ripple frequency is equal to the frequency of applied emf.

- The value of dc component in out put voltage is less than the ac.
- Input ac power $= I_{rms}^2 \left(R_L + r_f\right)$
- Out put dc power $= I_{dc}^2 R_L$

Efficiency of a half-wave rectifier
(Formula derivation)

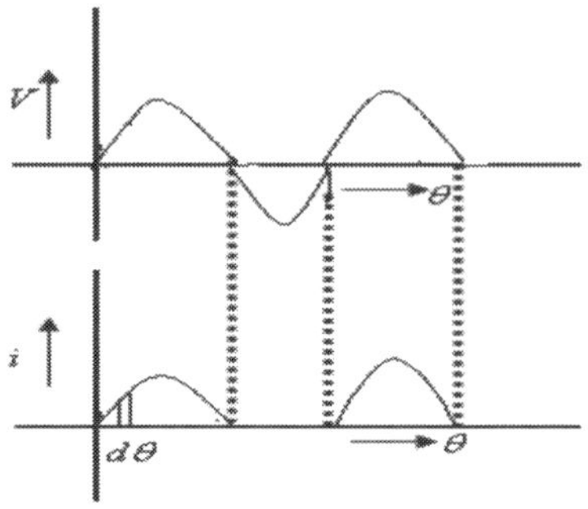

From fig

$$I_{average} = \frac{Area\ under\ the curve\ for\ a\ cycle}{base}$$

$$= \frac{\int_0^{\pi} i d\theta}{2\pi}$$

$$= \frac{1}{2\pi}\int_0^{\pi} \frac{V_m \sin\theta}{r_f + R_L} d\theta$$

$$= \frac{V_m}{2\pi\left(r_1 + R_L\right)}\left[-\cos\theta\right]_0^{\pi}$$

$$= \frac{V_m}{r_1 + R_L} \times \frac{1}{\pi} = \frac{I_m}{\pi}$$

$$\left[where\, I_m = \frac{V_m}{r_f + R_L}\right]$$

Where r_f and R_L denote diode resistance and load resistance respectively

Hence d.c. power

$$= I_{d.c}^2 \times R_L = \left(\frac{I_m}{\pi}\right)^2 \times R_L$$

a.c. power input= $I_{r.m.s}^2\left(r_f + R_L\right)$

For a half-wave rectifier

$$I_{r.m.s} = \frac{I_m}{2}$$

Hence $P_{a.c} = \frac{I_m^2}{2} \times \left(r_f + R_L\right)$

$$\therefore \text{ Rectifier efficiency } = \frac{P_{dc}}{P_{ac}} = \frac{\left(I_m / \pi\right)^2 \times R_L}{\left(I_m / 2\right)^2 \left(r_f + R_L\right)} = \frac{0.406 R_L}{r_f + R_L}$$

$$\eta = \frac{0.406}{1 + \frac{r_f}{R_L}}$$

The efficiency is maximum when is neglible.

- **Full wave rectifier:**

  $\eta \quad =$

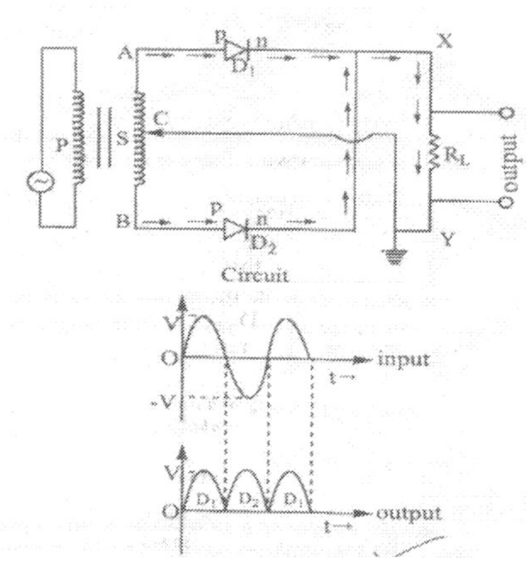

- In fullfwave rectification we need at least two semicondcutor diodes.

- In full wave rectifier the efficiency is $\eta = 0.812 \frac{R_L}{r_f + R_L}$. Its maximum value is 81.2%

  (where $r_f \rightarrow$ forward resistance of diode and $R_L \rightarrow$ load resistance.)

- Maximum current

  $$I_m = \frac{V_m}{R_L + r_f} \left(V_m \text{ is maximum voltage}\right)$$

- Mean dc current $I_{dc} = \frac{2I_m}{\pi}$

- $I_{rms} = \frac{I_m}{\sqrt{2}}$
- Value of ripple factor $= 0.482$
- The ripple frequency is twice to the frequency of applied emf.
- The value of dc component in out put voltage is more than the ac.
- Input ac power $= I_{rms}^2 \left(R_L + r_f\right)$
- Out put dc power $= I_{dc}^2 R_L$
- **Efficiency of full-wave rectifier:**

(Formula derivation)

Efficiency of full-wave rectifier is given by

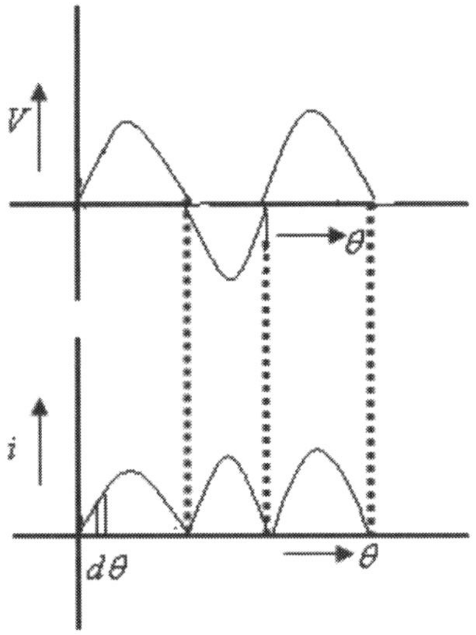

$$\eta = \frac{P_{d.c}}{P_{a.c}}$$

Instantaneous current is $i = \frac{\upsilon}{r_f + R_L}$

Average d.c. current is $I_{d.c} = \frac{2I_m}{\pi}$

$\therefore$ d.c . power output $= I^2{}_{d.c} \times R_L = \left(\frac{2I_m}{\pi}\right)^2 \times R_L$

a.c. input power is given by $P_{d.c} = I_{r.m.s}^2 \left(r_f + R_L\right)$

For a full-wave rectified cycle, $I_{r.m.s} = I_m / \sqrt{2}$

$$P_{a.c} = \left(\frac{I_m}{\sqrt{2}}\right)^2 \left(r_f + R_L\right)$$

$\therefore$ Full-wave rectification efficiency,

$$\eta = \frac{P_{d.c.}}{P_{a.c.}} = \frac{(2I_m/\pi)^2 R_L}{\left(\frac{I_m}{\sqrt{2}}\right)^2 (r_f + R_L)}$$

$$= \frac{0.812 R_L}{r_f + R_L} = \frac{0.812}{1 + \frac{r_f}{R_L}}$$

The efficiency will be maximum if r_f is zero

$\therefore$ maximum efficiency $= 81.2\%$

The efficiency of a full-wave rectifier is double the efficiency of a half-wave rectifier

- **Full-wave bridge rectifier**: In a centre tap full-wave rectifier, it is difficult to locate the centre tap on the secondary winding, which can be overcome in bridge rectifier. The circuit is shown in Figure Four diodes D_1, D_2, D_3 and D_4 are used in the circuit.

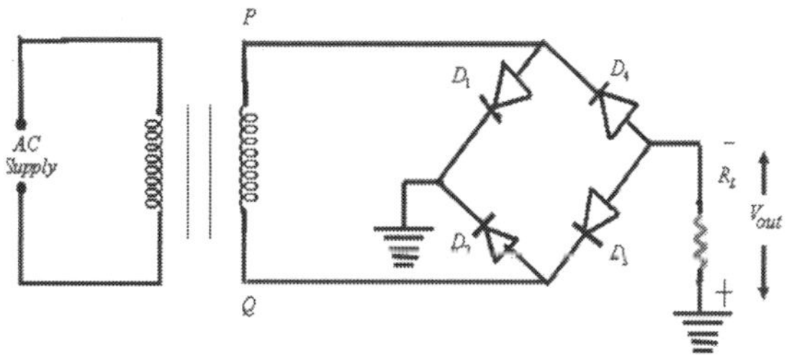

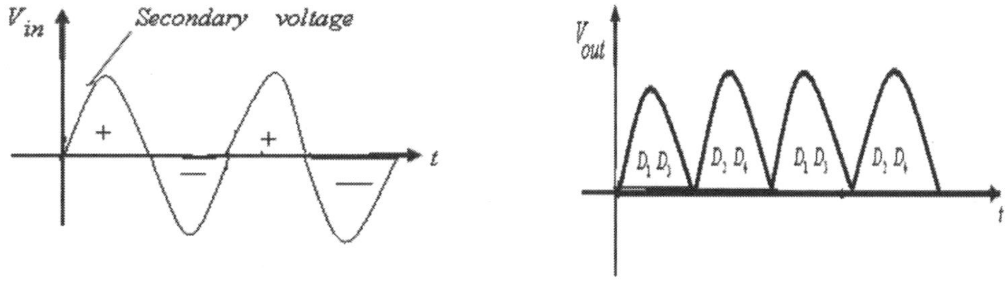

- **Ripple factor**: The ratio of r.m.s value of a.c component to d.c. component in the rectifier output is called ripple factor.

$$\text{Ripple factor} = \frac{I_{a.c.}}{I_{d.c.}} = \sqrt{\left(\frac{I_{r.m.s}}{I_{d.c.}}\right)^2 - 1}$$

Ripple factor decides the effectiveness of a rectifier. The smaller value of ripple factor shows lesser a.c. component; hence more effectiveness of rectifier.

i) For half-wave rectification, $I_{r.m.s.} = \frac{I_m}{2}$; $I_{d.c.} = \frac{I_m}{\pi}$

$$\text{Ripple factor} = \sqrt{\left(\frac{I_m/2}{I_m/\pi}\right)^2 - 1} = 1.21$$

ii) For full-wave rectification

In full-wave rectification,

$$I_{r.m.s.} = \frac{I_m}{\sqrt{2}}; \qquad I_{d.c.} = \frac{2I_m}{\pi}$$

$$\text{Ripple factor} = \sqrt{\left(\frac{I_m/\sqrt{2}}{2I_m/\pi}\right)^2 - 1} = 0.48$$

- **Form factor**: It is ratio of $I_{r.m.s.}$ and $I_{d.c.}$

i) For half-wave rectification:

$$F = \frac{I_{r.m.s}}{I_{d.c.}} = \frac{\pi}{2} = 1.57$$

$$\left[as\ I_{r.m.s} = \frac{I_m}{2}\ and\ I_{d.c} = \frac{I_m}{\pi}\right]$$

ii) For Full-wave rectification:

$$F = \frac{I_{r.m.s.}}{I_{d.c.}} = \frac{\pi}{2\sqrt{2}}$$

$$\left[as\ I_{r.m.s} = \frac{I_m}{\sqrt{2}}\ and\ I_{d.c} = \frac{2I_m}{\pi}\right]$$

The role of capacitor in filtering:

- When the voltage across the capacitor is rising, it gets charged. If there is no external load, it remains charged to the peak voltage of the rectified output.
- When there is a load, it gets discharged through the load and the voltage across it begins to fall. In the next half-cycle of rectified output it again gets charged to the peak value in figure.
- The rate of fall of the voltage across the capacitor depends upon the inverse product of capacitor C and the effective resistance R_L used in the circuit and is called the time cosntant.
- To make the time constant large value of C should be large. So capacitor input filters use large capacitors. The output voltage obtained by using capacitor input filter is nearer to the peak volatge of the rectified voltage. This type of filter is most widely used in power supplies.

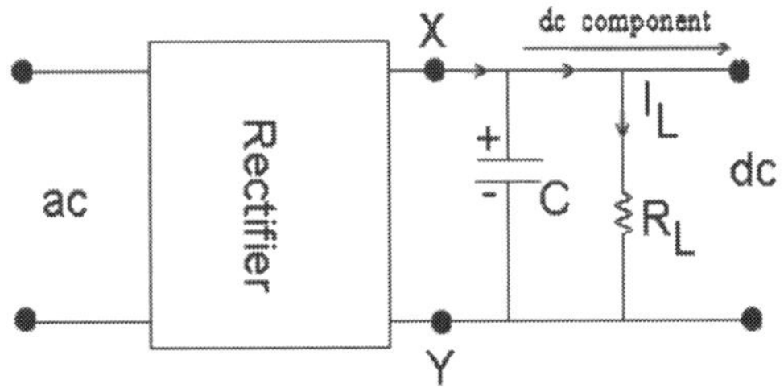

(a)

a)Full wave rectifier with capacitor filter

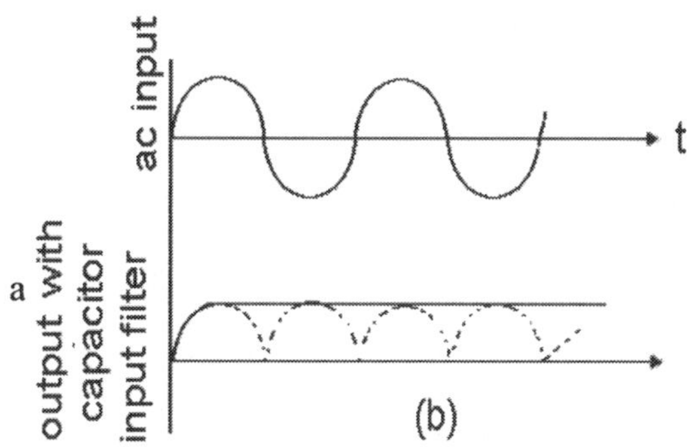

b) Input and out put voltage of rectifier in (a)

WE:14:A p-n diode is used in a half wave rectifier with a load resistance of 1000Ω. If the forward resistance $\left(r_f\right)$ of diode is 10Ω, calculate the efficiency of this halfwave rectifier.

Solution: Load resistance $R_L = 1000\Omega$

Forward resistance of the diode $= r_f = 10\Omega$

Efficiency of halfwave rectifier $\left[\dfrac{0.406R_L}{r_f + R_L}\right] = \dfrac{0.406 \times 1000}{1010} = 0.4019$

The percentage efficiency of the half wave rectifier $\eta = 40.19\%$

WE:15:Afull wave rectifier uses two diodes with a load resistance of 100Ω. Each diode is having negligible forward resis tance. Find the efficiency of this full wave rectifier.

Solution: Forward resistance of the diode $r_f = 0$,

Load resistance, $R_L = 100\Omega; \eta = ?$

efficiency of full wave rectifier $= \dfrac{0.812 \times 100}{100} = 0.812$

The percentage efficiency of the full wave rectifier $= 81.2\%$

WE:16:If a p-n junction diode, a square input signal of 10V is applied as shwon

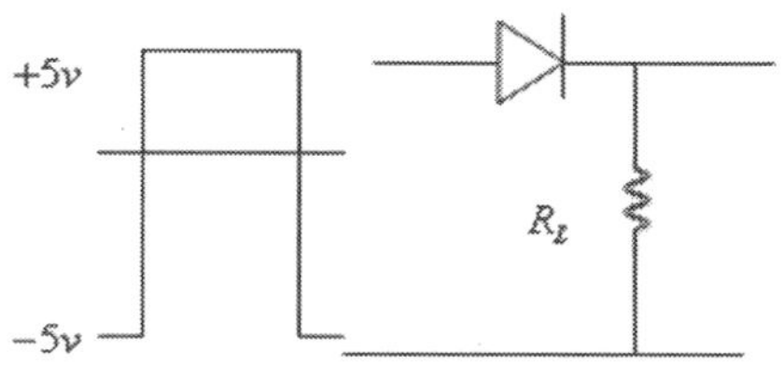

Then the out put signal across R_L will be

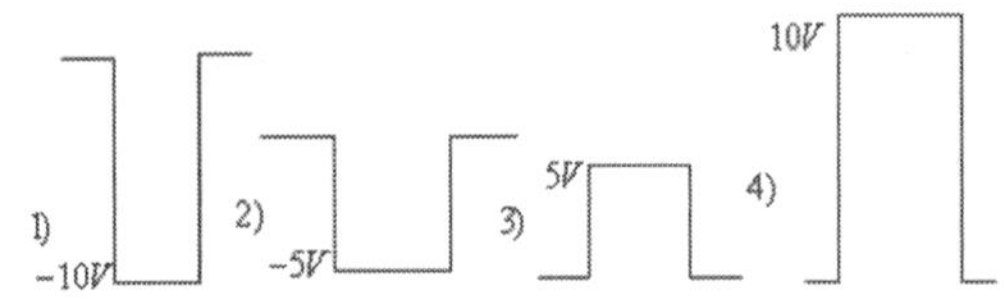

Solution : The junction diode will conduct when it is forward biased. Therefore, the output voltage will be obtainded during positive half cycle only . So option is (3)

- **p-n junction diodes are specially used as :**

  1) Zener diode 2) Opto electronic junction devices

1) Zener diode :

- A heavily doped p-n junction diode used to operate in reverse bias is called zener diode.
- In Zener diode even at low reverse voltage direct breakage of covalent bonds due to very strong ... hole pairs. This voltage where current increases sharply

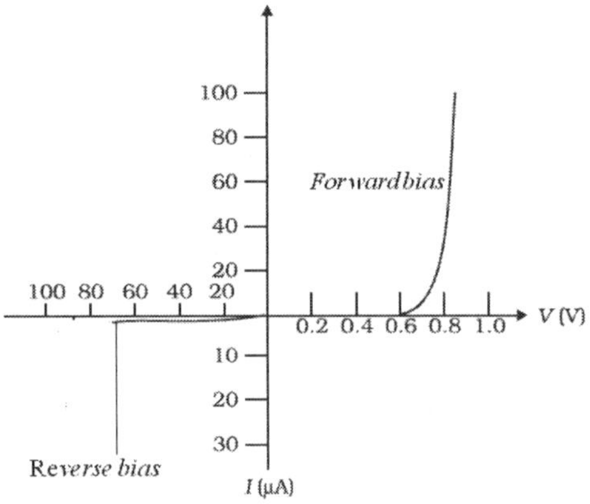

(c)

- Symbol of zener diode is 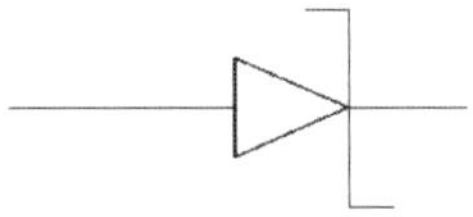

- Zener diode is used as voltage regulator its circuit diagram is

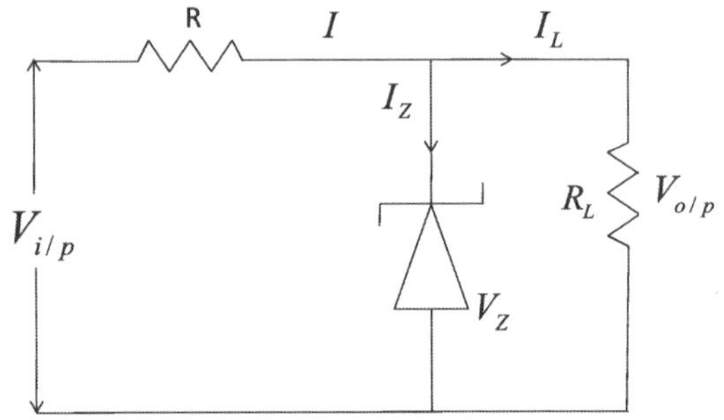

1) $I = I_Z + I_L$ 2) $V_{O/P} = V_Z = I_L R_L$ 3) $V_{I/P} = IR + V_Z$

$$V_{I/P} = IR + V_Z \Rightarrow V_{O/P} = V_{I/P} - (I_Z + I_L)R$$

As $V_{I/P}$ changes current through zener diode change so that $V_{O/P}$ (or) V_Z remain constant.

WE:17:For the circuit shown in figure, Find

1)the output voltage; 2)the voltage drop across series resistance;

3)the current through Zener diode.

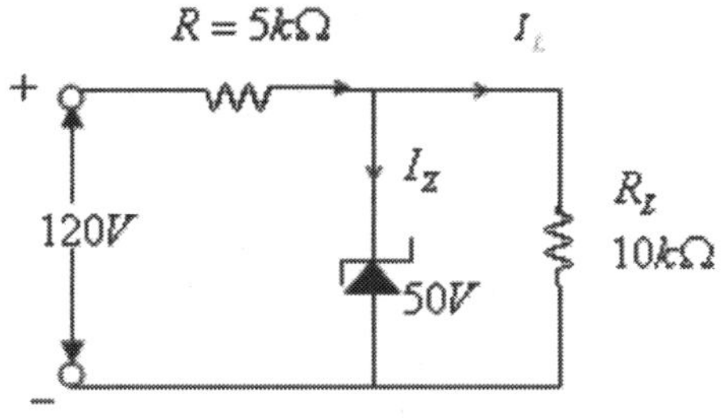

Solution: From the figure $R = 5k\Omega = 5\times10^3\Omega$;

input voltage $V_{in} = 120V$;

zener voltage, $V_z = 50V$

1) Output voltage $V_z = 50V$

2) Voltage drop across series tesistance R= $V_{in} - V_z = 120 - 50 = 70V$

3) Load current $I_L = \frac{V_z}{R_L} = \frac{50}{10\times10^3} = 5\times10^{-3}A$

Current through R= $i = \frac{V_{in} - V_z}{R} = \frac{70}{5\times10^3} = 14\times10^{-3}A$

According to Kirchoff's first law $I = I_L + I_z$

$\therefore$ Zener current

$I_Z = I - I_L = 14\times10^{-3} - 5\times10^{-3}$

$= 9\times10^{-3} = 9mA$

2) Opto electronic junction devices

PHOTODIODE

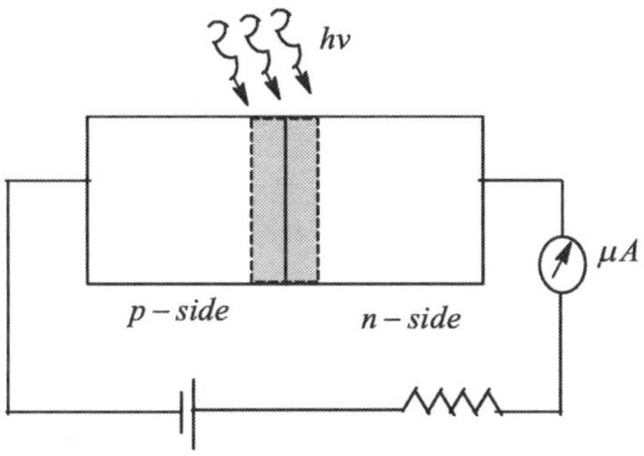

Fig : An illuminated photodiode under reverse bias

- Photocurrent is proportional to incident light intensity.
- Photodiode can be used as a photodetector to detect optical signals.

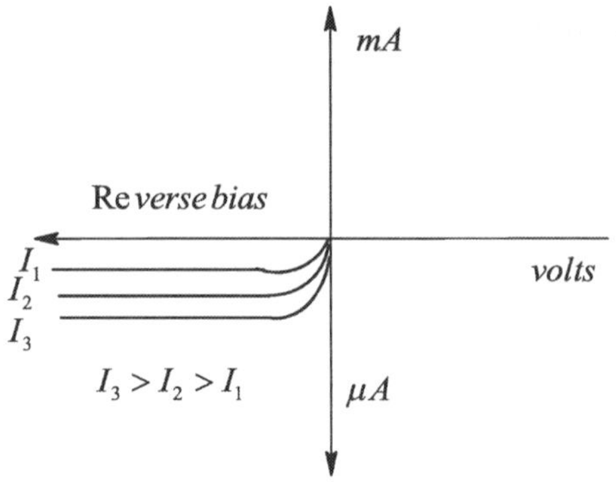

Fig : I-V characteristics of a photodiode for different illumination intensity $I_3 > I_2 > I_1$.

WE:18 The current in the forward bias is known to be more (mA) than the current in the reverse bias (μA). What is the reason then to operate the photodiodes in reverse bias ?

Solution: Consider the case of an n-type semiconductor. The majority carrier density (n) is considerably larger than the minority hole density p(n>>p) On illumination, let the excess electrons and holes generated be Δn and Δp, respectively

$n^1 = n + \Delta n$

$p^1 = p + \Delta p$

Here n^1 and p^1 are the electron and hole concentrations at any particular illumination and n and p are carriers concentration when there is no illumination. Remember $\Delta n = \Delta p$ and $n >> p$. Hence, the fractional change in the majority carriers (i.e., $\Delta n / n$)would be much less than that in the minority carrier dominated reverse bias current is more easily measurable than the fractional change in the forward bias current. Hence,photodiodes are preferably used in the reverse bias conditionfor measuring light intensity

- **LIGHT EMITTING DIODE (LED)**

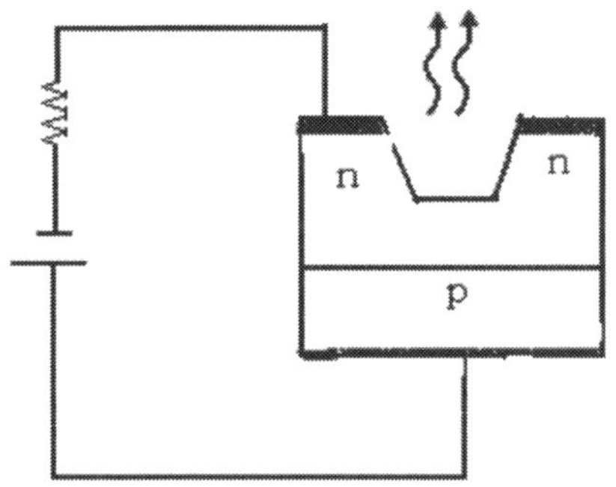

- Light-emitting diode (LED) is a forward-biased p-n junction diode which emits visible light when energised .
- The energy of radiation emitted by LED is equal to or less than the band gap of the semiconductor.
- The band width of emitted light is $100 A^0$ to $500 \overset{0}{A}$ or in other words it is nearly (but not exactly) monochromatic.

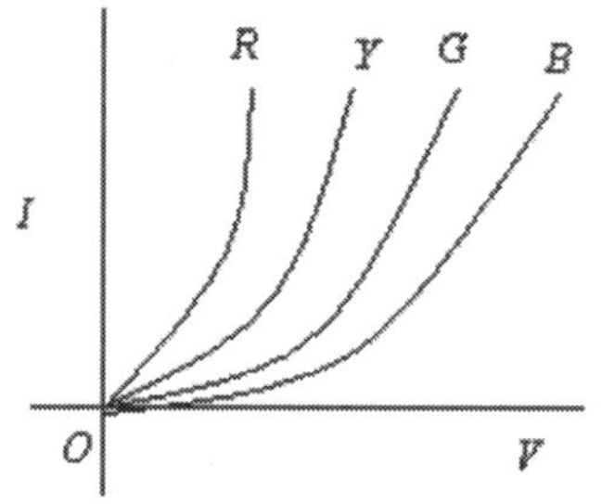

The I-V characterstics of L.E.D:

- **SOLAR CELL :**

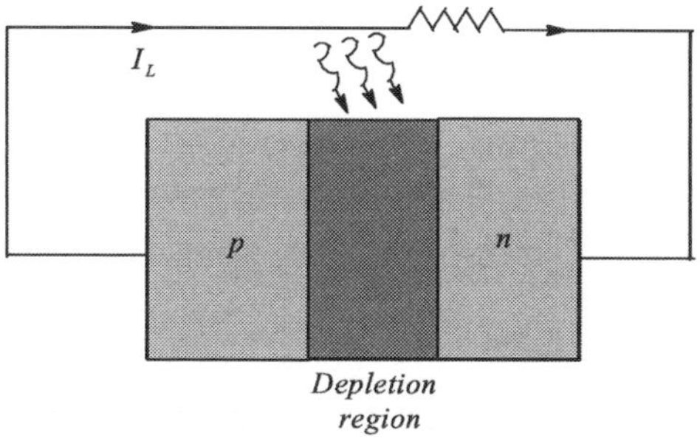

Fig : A typical illuminated p-n junction solar cell

- Unlike a photodiode, a solar cell is not given any biasing. It supplies emf like an ordinary cell.
- Sunlight is not always required for a solar cell.
- Semiconductors with band gap close to 1.5ev. are ideal materials for solar cell fabrication
- Si and GaAs are preferred material for solar cells.

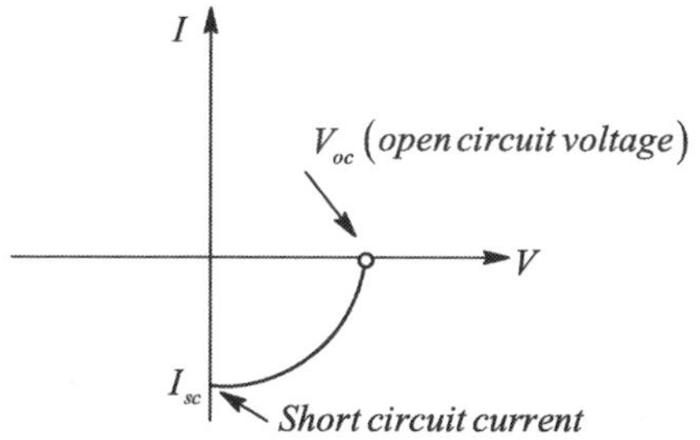

Fig : I-V characteristics of a solar cell

- **SELECTION OF SOLAR MATERIAL:**
- i) band gap (1.0 to 1.8eV) ii) high optical absorption ($10^4 cm^{-1}$)
  iii) good electrical conductivity iv) availability of the raw material and (v) cost.

TRANSISTORS:

- Transfer +resistor= Transistor
- Transistors are current operated solid-state devices.
- Silicon is the element from which most of the transistors and other semiconductor components are made today.
- Transistor has three regions known as the emitter (E), base (B) and collector (C)
  1) Emitter is heavily doped and medium in size.
  2) Base of a transistor is lightly doped and very thin.
  3) Collector of a transistor is moderately doped and large in size.

- Transistor has two junctions
  1) emitter-base junction 2) Collector-base junction
- In a transistor the emitter base junction is forward biased and the collector base junction is reverse biased.
- **In an n-p-n Transistor:**

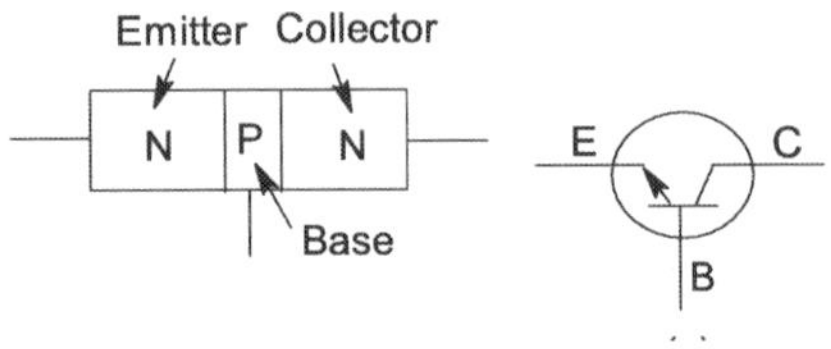

- The current is due to electrons inside and outside the n-p-n transistor and they are the majority charge carriers
- The conventional current flows from base to emitter.
- The emitter junction is forward biased and the collector junction is reverse biased.
- The emitter current (I_E) is the sum of base current (I_B) and collector current (I_C), i.e.,$I_E = I_B + I_C$
- I_C is 97 to 98% of I_E and I_B is 2 to 3% of I_E.
- **In a p-n-p Transistor:**

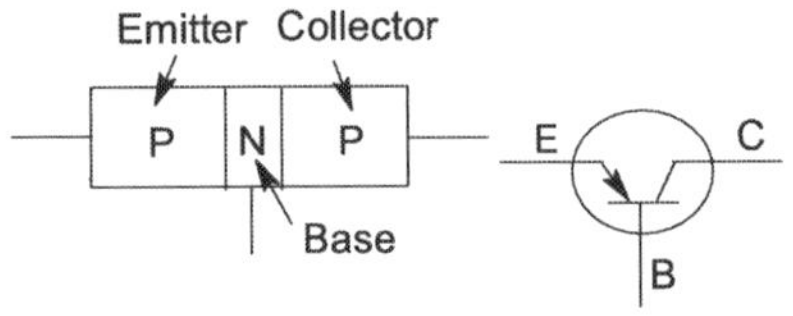

- The current is due to holes inside as they are the majority charge carriers and due to electons outside the p-n-p transistor.
- The conventional current is from emitter to base.
- The emitter junction is forward biased and the collector junction is reverse biased.
- Here also $I_E = I_C + I_B$
- I_C is 97 to 98% of I_E and I_B is 2 to 3% of I_E.
  Thus the collector current is less than the emitter current ($I_C < I_E$)
- **TRANSISTOR CONFIGURATIONS**
  In electronic circuits transistors are connected in three ways. They are
  1) Common base configuration ,
  2) Common emitter configuration,
  3) Common collector configuration
- **COMMON BASE CONFIGURATION**
- In this configuration base is common to both input and output.
- Base terminal is earthed and input is given across base - emitter and output is taken across base - collector as shown in figure
- This mode is called grounded base configuration.

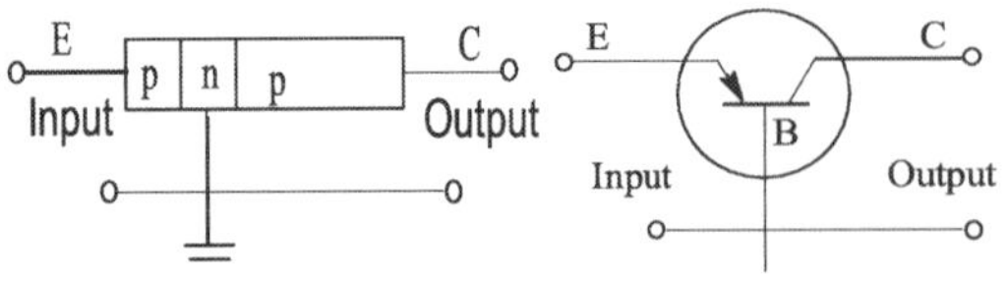

- Current amplification factor of common base configuration for ac $\alpha = \left(\frac{\Delta I_C}{\Delta I_E}\right)_{\text{constant } V_{CB}}$
- Current amplification factor of common base configuration for dc $\alpha = \frac{I_c}{I_E}$
- Values of α range from 0.95 to 0.99.
- Input resistance of transistor in CB configuration is $R_{in} = \left(\frac{\Delta V_{BE}}{\Delta I_B}\right)_{V_{CB}}$
- Output resistance of transistor in CB configuration is $R_{out} = \left(\frac{\Delta V_{CB}}{\Delta I_C}\right)_{I_E}$
- Voltage gain = current gain x resistance gain. $A_v = \alpha \times \frac{R_{out}}{R_{in}}$
- Power gain = Voltage gain x current gain. $A_p = A_v \times \alpha = \alpha^2 \times \frac{R_{out}}{R_{in}}$
- **COMMON EMITTER CONFIGURATION**
- In this configuration emitter is common to both input and output.
- The emitter is earthed and input is given across base - emitter and output is taken across collecter - emitter as shown in fig
- This mode is called grounded emitter configuration.

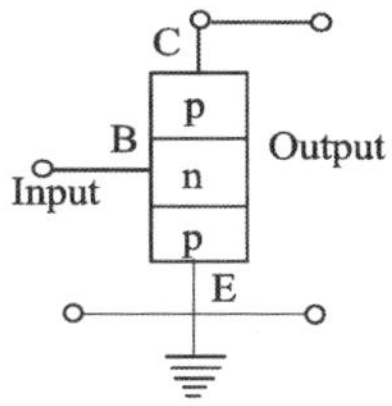

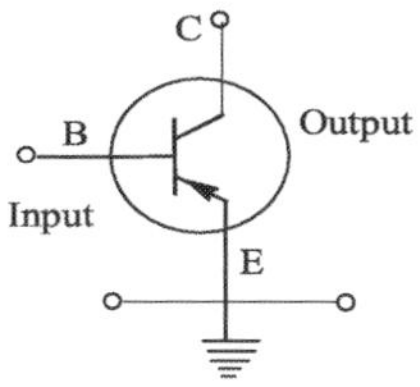

- Current amplification factor of common emitter configuration for ac $\beta = \left(\frac{\Delta I_C}{\Delta I_B}\right)_{\text{constant } V_{CE}}$
- Current amplification factor of common emitter configuration for dc $\beta = \frac{I_c}{I_B}$
- Values of β range from 20 to 500.
- Input resistance of transistor in CE configuration is $R_{in} = \left(\frac{\Delta V_{BE}}{\Delta I_B}\right)_{V_{CE}}$
- Output resistance of transistor in CE configuration is $R_{out} = \left(\frac{\Delta V_{CE}}{\Delta I_C}\right)_{I_B}$

- Voltage gain = current gain x resistance gain.

$$A_v = \beta \times \frac{R_{out}}{R_{in}}$$

- Power gain = Voltage gain x current gain. $A_p = A_v \times \beta = \beta^2 \times \frac{R_{out}}{R_{in}}$
- **COMMON COLLECTOR CONFIGURATION**
- In this configuration collector is common to both input and output.
- The collector is earthed and input is given across base - collector and output is taken across emitter -collector as shown in figure
- This mode is called grounded collector configuration.

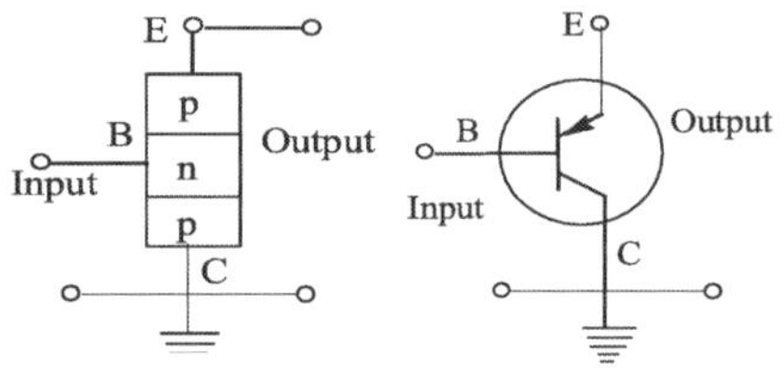

- Current amplification factor of common collector configuration for ac $\gamma = \left(\frac{\Delta I_E}{\Delta I_B}\right)_{\text{constant } V_{CE}}$
- Current amplification factor of common collector configuration for dc $\gamma = \frac{I_E}{I_B}$
- Input resistance of transistor in CC configuration is $R_{in} = \left(\frac{\Delta V_{BE}}{\Delta I_B}\right)_{V_{CE}}$
- Output resistance of transistor in CCconfiguration is $R_{out} = \left(\frac{\Delta V_{CE}}{\Delta I_E}\right)_{I_B}$
- **Relation between α & β :-**

ac current gain in C.B , $\alpha = \frac{\Delta I_C}{\Delta I_E}$

ac current gain in C.E , $\beta = \frac{\Delta I_C}{\Delta I_B}$

$$\beta = \frac{\Delta I_C}{\Delta I_B} = \frac{\Delta I_C}{\Delta I_E - \Delta I_C} = \frac{\frac{\Delta I_C}{\Delta I_E}}{1 - \frac{\Delta I_C}{\Delta I_E}} = \frac{\alpha}{1-\alpha} \Rightarrow \beta = \frac{\alpha}{1-\alpha} \text{ (or) } \alpha = \frac{\beta}{1+\beta}$$

- **Relation between α, β and γ :-** $\gamma = \frac{\Delta I_E}{\Delta I_B} = \frac{1}{1-\alpha} = \frac{\beta}{\alpha} = \beta + 1$

- **Comparative study of CB, CE, CC configurations**

	parameter	CB configuration	CE configuration	CC configuration
1	Input resistance	Minimum (50-20 Ω)	Medium (1-2K Ω)	Maximum (150-180K Ω)
2	Output resistance	Maximum (1-2M Ω)	More (≈50K Ω)	Minimum (≈1K Ω)
3	Current gain	Minimum $\alpha = 0.95-0.99$	More $\beta = 20-500$	Maximum $\gamma = 20-500$
4	Voltage gain	Medium	maximum	minimum
5	Power gain	Medium (20-30)	Maximum (30-40)	Minimum (≈10)
6	Useful in application of	Current	power	Impedance matching
7	Phase difference	0	π rad (or) 180°	0

- **Characterestics of a transistor:-**

n–p–n Transistor (C.E)

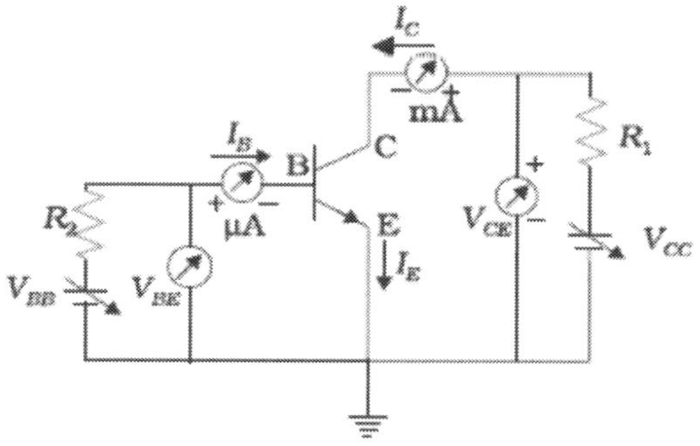

1) Input characterestics

Input characterstics are drawn between V_{BE} verses I_B at constant V_{CE}. The output voltage V_{ce} is fixed (say at zero volts). The input voltage V_{be} is changed in steps (say 0.1V) upto 1 volt and the corresponding base current I_b is noted down. This process is repeated for different values (say 10V, 20V etc.) of V_{ce}.

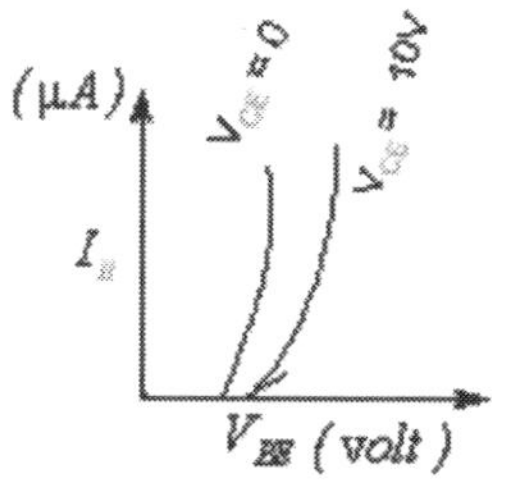

2) Output characterestics

Out put characterstics are drawn between V_{CE} verses I_C at constant I_B

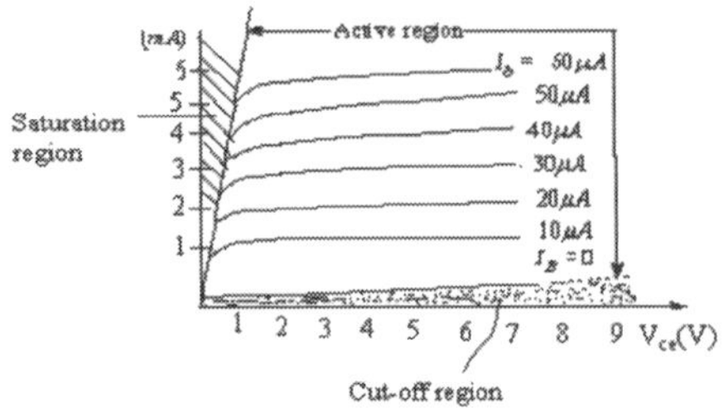

- **Saturation region :-** In this region the collector current becomes almost independent of the base current. This happens when both junctions are forward biased.
- **Cut off region:-** In this region the collector curent is almost zero. This happens when both the junctions are reverse biased.
- **Active (or) Linear region:-** In this region collector current (I_c) is many times greater than base current (I_b). A small change in input current (ΔI_b) produces a large change in the output current (ΔI_c). This happens when emitter junction is forward biased and collector junctions is reverse biased.
- The transistor works as an amplifier when operated in the active region.
- When the transistor is used in the cut off (or) saturation state it acts as a switch

- **Transistor as a switch**

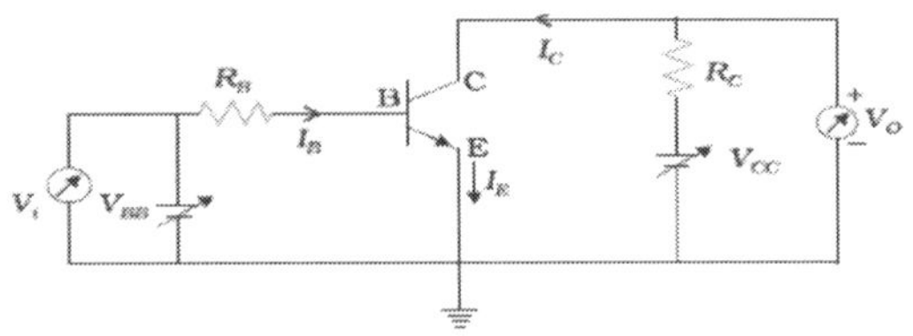

- Appling KVL to the input and output sides of the circuit, We get

  $V_i = I_B R_B + V_{BE}$ and $V_O = V_{CC} - I_C R_C$
- In the case of Si transistor, as long as input Vi is less than 0.6 V, the transistor will be in cut off state and current I_C will be zero. Hence $V_o = V_{CC}$ When Vi becomes greater than 0.6 V the transistor is in active state ,so V_o decreases linearly till its value becomes less than about 1.0 V.
- **graph:**

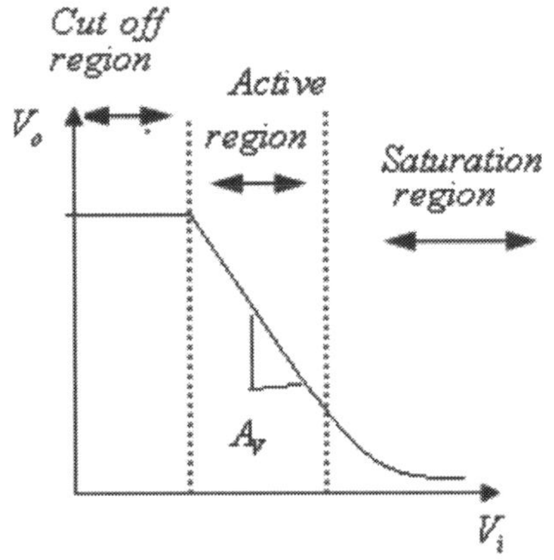

- If we plot the V_0 vs V_i curve [also called the transfer characterstics of the base biased transistor in figure,we see that between cut off state and active state and also between active state and saturation state there are regions of non-linearty showing that the transition from cut off state to active state and from active state to saturation state are not sharply defined.
- **TRANSISTOR AS AN AMPLIFIER (CE CONFIGURATION)**

The process of raising the strength of a weak input signal to a strong output signal is called 'amplification'.

Amplifier : It is a device which increases the weak input signal into strong output signal

Amplifier has wide applications in industries, T.V, radio and communication systems.

Amplifiers are of two types

1) Power amplifiers 2) Voltage amplifiers

1) **Power amplifier:** Amplifier which is used to raise the power level is known as "Power amplifier."

2) **Voltage amplifier:** The amplifier which is used to raise voltage level is known as voltage amplifier.

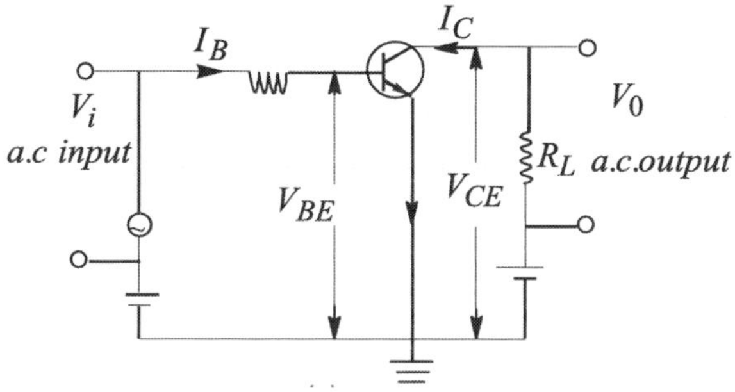

Fig:N-P-N TRANSISTOR AS AMPLIFIER

i) **Voltage gain :**

It is defined as the ratio of change in output voltage to the change in input voltage.

$$A_v = \frac{\Delta V_{CE}}{\Delta V_{BE}} = -\frac{R_L(\Delta I_c)}{R_i(\Delta I_b)} = -\beta\frac{R_L}{R_i}$$

Negative sign indicates input and output voltages are in opposite phase.

ii) **Power gain :**

It is defined as the ratio of output power to the input power.

$$\text{power gain} = \frac{output\ \ power}{input\ \ power} = \frac{I_{out}V_{out}}{I_{in}V_{in}}$$

Power gain= current gain × voltage gain $\Rightarrow A_p = \beta \times A_v = \beta^2 \times \frac{R_{out}}{R_{in}}$

NOTE:1 In common base amplifier, the phase difference between the input and output signals is zero

NOTE:2 In common emitter amplifier, the phase difference between input and output signals is π

WE:19:Current amplification factor of a common base configuration is 0.88. Find the value of base current when the emitter current is 1 m A.

Solution: In a common -base arragement, the current amplification factor

$$\alpha = \left(\frac{\Delta I_C}{\Delta I_E}\right)_{V_{CB}} = \frac{I_C}{I_E}$$

Given $\alpha = 0.88, I_E = 1mA$

$\therefore$ Collector current

$I_C = \alpha I_E = 0.88 \times 1 = 0.88mA$

Now since $I_E = I_B + I_C$

$\therefore$ Base current $\quad I_B = I_E - I_C = 1 - 0.88 = 0.12mA$

WE:20:In a transistor, the emitter circuit resistance is 100Ω and the collector resistance is 100Ω. The power gain, if th emitter and collector currents are as sumed to be equal, will be

Solution:

If $I_C \approx I_B \Rightarrow \beta \approx 1$

$$\therefore\ A_P = \beta^2 \left(\frac{R_L}{R_i}\right) = \left(\frac{R_L}{R_i}\right)$$

$$= \frac{100 \times 10^3}{100} = 1000$$

WE:21:From the output characteristics shown in Fig .Calculate the values of β_{ac} *and* β_{dc} of the transistor when V_{CE} is 10V and $I_c = 4.0mA$

Solution:

Consider any two characteristics for two values of I_B which are above and below the given value of I_C, Here $I_C = 4.0mA.$

(Choose characteristics for $I_B = 30$ and $20\mu A$) At $V_{CE} = 10V$ we read the two values of I_C from the graph Then

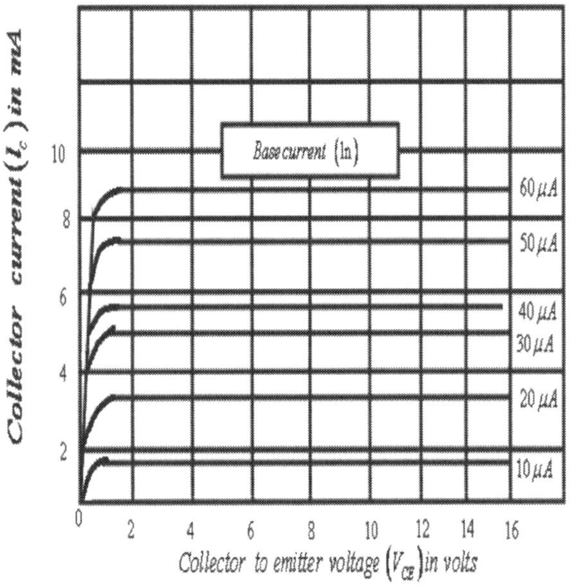

$\Delta I_B = (30 - 20)\,\mu A$

$\Delta I_C = (4.5 - 3.0)\,mA = 1.5mA$

Therefore

$$\beta_{ac} = \left(\frac{\Delta i_c}{\Delta I_B}\right)_{V_{CE}} = 1.5mA/10\mu A = 150$$

For determining β_{dc} calculate the two values of β_{dc} for the two characteristics chosen and find their mean . Therefore for $I_C = 4.5mA$ *and* $I_B = 30\mu A$ $\beta_{dc} = \frac{I_C}{I_B} = 4.5mA\ /30\mu A = 150$ and for

$$I_C = 3.0mA \text{ and } I_B = 20\mu A$$

$\beta_{dc} = 3.0mA/20\mu A = 150$

Hence $\beta_{dc} = (150+150)/2 = 150$

WE:22:In Figure the V_{BB} supply can be varie form 0V to 5.0V. The Si transistor has $\beta_{dc} = 250$ and $R_c = 1K\Omega, V_{CC} = 5.0V$. Assume that when the transistor is saturated, $V_{CE} = 0V$ and $V_{BE} = 0.8V$. Calculate (a) the minumum base current, for which the transistor will reach saturation. Hence, (b) determine V_1 when the transistor is switched on(c) find the ranges of V_1 for which the transistor is switched off and switched on.

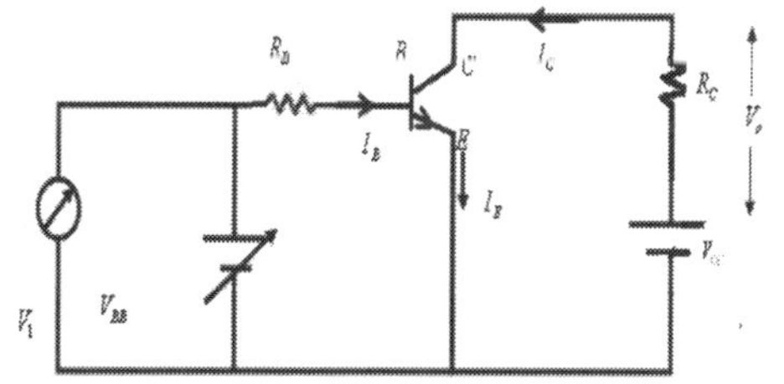

Solution:

Given at saturation $V_{CE} = 0V$, $V_{BE} = 0.8V$ $V_{CE} = V_{CC} - I_C R_C \Rightarrow I_C = V_{CC}/R_C$
$= 5.0V/1.0K\Omega = 5.0mA$

Therefore $I_B = I_C/\beta = 5.0mA/250 = 20\mu A$

The input voltage at which the transistor will go into saturation is given by

$V_{IH} = V_{BB} = I_B R_B + V_{BC} = 20\mu A \times 100K\Omega + 0.8V = 2.8V$

The value of input voltage below which the transistor remains cutoff is given by $V_{IL} = 0.6V, V_{IH} = 2.8V$. Between 0.0V and 0.6V, the transistor will be in the 'switched off state. Between 2.8V and 5.0V,it will be in 'switched on' state Note that the transistor is in active state when I_B varies from 0.0mA to 0.20 mA.

In this range, $I_C = \beta I_B$ is valid. In the

In this range, $I_C = \beta I_B$ is valid. In the

saturation range, $I_C \leq \beta I_B$

WE:23:Two amplifiers are connected one after the other in series (cascaded). The first amplifier has a voltage gain of 10 and the second has a voltage gain of 20. If the input signal is 0.01 volt, calculate the out put ac signal.

Solution: When the amplifiers are Connected in series, the net voltage gain is equal to the product of the gains of the individual amplifiers.

$\therefore Av = Av^1 \times Av^{11}$,hear $Av^1 = 10$,and $Av^{11} = 20$

also $Av = \dfrac{V_{output}}{V_{input}}$

$\therefore$ we can write $\dfrac{V_{output}}{V_{input}} = Av^1 \times Av^{11} : V_{in} = 0.01V$

$V_{out} = V_{in} \times Av^1 \times Av^{11} = 0.01 \times 10 \times 20 = 2V$

WE:24:In a single state transistor amplifier, when the signal changes by 0.02V, the base current by $10\mu A$ and collector current by 1mA. If collector load $R_C = 2k\Omega$ and $R_L = 10k\Omega$, Calculate:(i)

Current Gain (ii)Input impedance,(iii) Effective AC load,(iv) Voltage gain and (v) Power gain.

Solution :

i) Current Gain $\beta = \dfrac{\Delta i_c}{\Delta i_b} = \dfrac{1mA}{10\mu A} = 100$

ii) Input impedance $R_i = \dfrac{\Delta V_{BE}}{\Delta i_b} = \dfrac{0.02}{10\mu A} = 2000\Omega = 2k\Omega$

iii) Effective (a,c) load $R_{AC} = R_C \parallel R_L \therefore R_{AC} = \dfrac{2\times 10}{2+10} = 1.66k\Omega$

iv) Voltage gain $A_V = \beta \times \dfrac{R_{AC}}{R_{in}} = \dfrac{100\times 1.66}{2} = 83$

v) Power gain, A_p = Current gain$\times$ Voltage gain $= 100 \times 83 = 8300$

WE:25:An n-p-n transistor in a common emitter mode is used as a simple voltage amplifier with a collector current of 4mA. The terminal of a 8V battery is connected to the collector through a load resistance R_L and to the base through a resistance R_B. The collector-emitter voltage $V_{CE} = 4V$, base-emitter voltage $V_{BE} = 0.6V$ and base current amplification factor $\beta_{d.c} = 100$.Calculate the values of R_L and R_B

Solution :

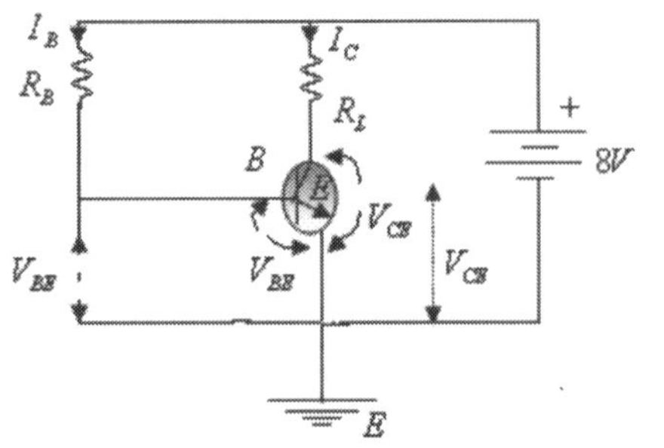

Potential difference across R_L

$= 8V - V_{CE} = 8V - 4V = 4V$

Now $I_C R_L = 4V$

$$R_L = \frac{4}{4\times10^{-3}} = 10^3\Omega = 1k\Omega$$

Further for base emitter equation

$V_{CC} = I_B R_B + V_{BE}$

or $I_B R_B$ = Potential difference across $R_B = V_{CC} - V_{BE} = 8. - 0.6 = 7.4V$

$$\text{Again},\ I_B = \frac{I_C}{\beta} = \frac{4\times10^{-3}}{100} = 4\times10^{-5}A$$

$$\therefore R_B = \frac{7.4}{4\times10^{-5}} = 1.85\times10^5\Omega = 185k\Omega$$

- **CONCEPT OF FEEDBACK**
- When a part of the output voltage (or current) of an amplifier is injected back into the input circuit, feedback is said to exist.
- If the voltage feedback is in phase with the applied voltage, the feedback is said to be positive or regenerative;
- If the voltage feedback is in opposite phase to the incoming signal, the feedback is said to be negative or degenerative.
- Fig. illustrates the principles of feedback.

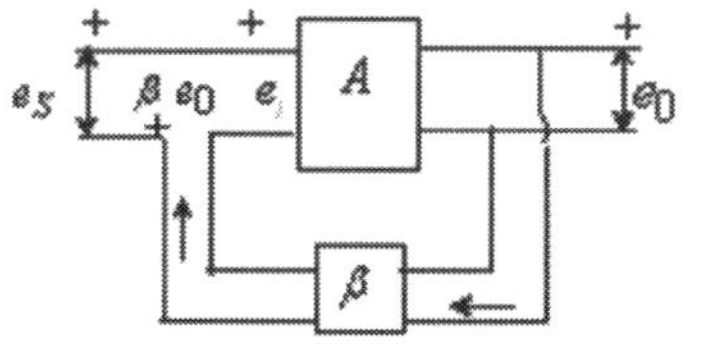

- The gain of the amplifier without feedback is A. If a signal e_S is applied at the input terminals of the amplifier, then let the output voltage be e_0.
- If a fraction β ($\beta = \frac{e_f}{e_0}$) of this output voltage is fedback into the input in phase with the applied signal, then the actual input voltage of the amplifier,

$e_i = e_s + e_f$

$e_i = e_s + \beta e_0$ (for positive feed back)

- This total input voltage multiplied by the gain A of the amplifier must be equal to the output voltage i.e. $e_0 = Ae_i \Rightarrow e_0 = A(e_s + \beta e_0) = Ae_s + A\beta e_0$

$\Rightarrow e_0 - A\beta e_0 = Ae_s \Rightarrow e_0(1-\beta A) = Ae_s$ from which the gain A_f with feedback is, $A_f = \frac{e_0}{e_s} = \frac{A}{1-\beta A}$

- The positive feedback thus, increases the gain of the amplifier. If too much positive feedback is applied so that $1-\beta A=0$, the gain of the amplifier becomes infinite.
- For stable oscillation $\beta A=1$ (Barkhausen's criteria)
- In this case the amplifier becomes unstable and the output can be obtained with no external input signal, i.e., the amplifier becomes an oscillator.
- In the case of negative feedback $e_i = e_s - e_f$
- In the case of negative feedback the voltage feedback βe_0 is in opposite phase to the applied voltage

  e_S, so that gain with negative feedback becomes, $A_f = \dfrac{e_0}{e_s} = \dfrac{A}{1+\beta A}$
- The negative feed back also reduces noise & distortion in an amplifier.

NOTE:

i) When $|1+A\beta| > 1, |A_f| < |A|$, feed back is negative

ii) When $|1+A\beta| < 1, |A_f| > |A|$, feed back is positive

- **Transistor as an oscillator:**
- The simplest eletrical oscillating system consists of an inductance L and capacitor C connected in parallel.

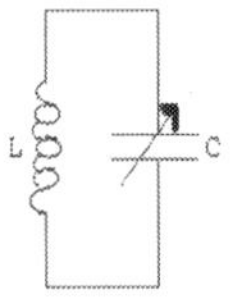

- Once an eletrical energy is given to the circuit, this energy oscillates between capacitance (in the form of eletrical energy) and inductance (in the form of magnetic energy) with a frequency

  $$\nu = \frac{1}{2\pi\sqrt{LC}}$$
- The amplitude of oscillations is damped due the presence of inherent resistance in the circuit
- In order to obtain oscillations of constant amplitide, an arrangement of regenerative or positive feedback from an output circuit to the input circuit is made so that the circuit losses may be compensated.

WE:26:In a negative feedback amplifier, the gain without feedback in 100, feed back ratio is 1/25 and input voltage is 50m V. Calculate (i)gain with feedback (ii) feedback factor(iii) output voltage (iv) feedback voltage (v) new input voltage so that output voltage with feedback equals the output voltage without feedback

Solution :

i) Gain with feedback

$$A_f = \frac{A}{1+\beta A} = \frac{100}{1+(1/25)\times 100} = 20$$

ii) $\beta = \dfrac{1}{25}$

iii) Out put voltage

$$V_0' = A_f V_i = 20 \times 50mV = 1volt$$

iv) Feedback voltage $\beta V_0' = \frac{1}{25} \times 1 = 0.04\ volt$

v) New increased input voltage

$$V_i^1 = V_i(1+\beta A) = 50\left(1+\frac{1}{25}\times 100\right) = 250mV$$

Digital Eletronics:

1) Binary number system:

- It is a two-valued system developed by George Boole. Only two digits 0 and 1, called bits, are used in binary system. But binary addition (in this 1+1=10) is different from addition in Boolean algebra (in this 1+1=1 and remaining will be same i.e., 0+0=0, 1+0=1, 0+1=1).
- A number in a decimal system can be converted into binary by the successive division of 2 until the quotient is zero. The remainders obtained in the successive divisions, taken in the reverse order (from bottom to top shown below) give the binary representation of that number.

WE:27: Convert the decimal number 23 into binary number.

Solution:

2|23
2|11–1
2| 5–1
2| 2–1
2| 1–0
2| 0–1

Arranging the remainders in the reverse order(from bottom to top shown below) the binary equivalent of 23 is found to be 10111. i.e., $23_{(10)} = 10111_{(2)}$

- In binary representation of any number, the first bit is the 'most significant bit' (MSB) and the last bit is the 'least significant bit' (LSB).
- Binary number can be converted into decimal number by multiplying each bit with 2^n and adding them as mention below. where n is positon of the bit from right side (LSB) to leftside (MSB).

WE:28: Convert the binary number 10111 into decimal number.

Solution:

4 3 2 1 0

$$10111_{(2)} = 1\times 2^4 + 0\times 2^3 + 1\times 2^2 + 1\times 2^1 + 1\times 2^0$$

$$= 16 + 0 + 4 + 2 + 1 = 23_{(10)}$$

$$10111_{(2)} = 23_{(10)}$$

2) Boolean Algebra:

- Only two states or values of a varible are allowed in Boolean algebra. In logic these two states correspond to 'on' and 'off' and 'saturation' of electronic devices.
- The two allowed states in Boolean alegbra are represented by the digits 0 and 1.

  0 can also represent Off, low, false,No, 0 V

  1 can also represent ON, high, true, Yes, 5 V
- The variables of Boolean algebra are subjected to three operations.
- The OR addition indicated by a plus (+) sign.
- The AND multiplication indicated by a cross (X) or a dot (.) sign.
- The NOT operation indicated by a bar over a variable.
- **OR Addition:** The + sign in Boolean algebra represents OR addition. The equation Y = A+B is read as 'Y equals A OR B'. The
- **AND multiplication:** In Boolean algebra the equation Y=AXB or Y=A.B or Y=AB is read as 'Y equals A AND B'
- **NOT operation:** The NOT operation on a variable A is represented by $\overline{A}$.

  The equation $Y = \overline{A}$ is read as 'Y equals NOT A'.
- Rule for OR, AND and NOT functions in Boolean algebra

OR	AND	NOT
0+0=0	0.0=0	$\overline{0} = 1$
0+1=1	0.1=0	$\overline{1} = 0$
1+0=1	1.0=0	
1+1=1	1.1=1	

- Some useful laws of Boolean algebra
- **Commulative laws:**

  A+B=B+A

  A.B=B.A
- **Associative laws:**

  A+(B+C) = (A+B) +C

  A.(B.C) = (A.B).C
- **Distributive laws:**

  A.(B+C) = A.B +A.C
- **Basic OR and AND Relations:**

OR	**AND**
A+0=A	A.0=0
A+1=1	A.1=A
A+A=A	A.A=A
$A+\overline{A}=1$	$A.\overline{A}=0$

- **Double complement function:**

  $\overline{\overline{A}} = A$

- **De Morgan's theorems:**

  DeMorgan's theorems (or rules) are very useful in simplifying complicated logical expressions and can be stated as under :

  Theorem 1 : The complement of the sum of two (or more) variables is equal to the product of complements of the variables i.e. $\overline{A+B} = \overline{A} \bullet \overline{B} \Rightarrow \overline{A+B} = \overline{A}.\overline{B}$

  Theorem 2 : The complement of the product of two (or more) variables is equal to the sum of complements of the variables i.e

  $\overline{A \bullet B} = \overline{A} + \overline{B}$. $\Rightarrow \overline{A.B} = \overline{A} + \overline{B}$

- **LOGIC GATES :-**
- A digital circuit having a certain logical relationship between the input and the output voltages is called a logic gate.
- There are three basic logic gates

  (1) OR gate (2) AND gate (3) NOT gate

- **The OR gate**

  1) It is a logic gate which has two or more inputs and one output

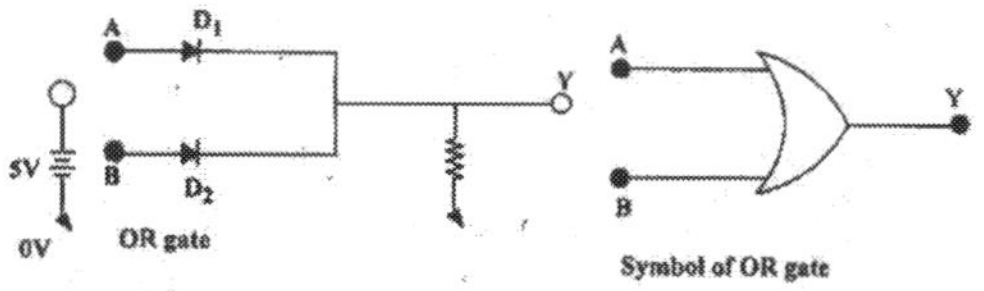

OR gate

Symbol of OR gate

- If its one or more inputs are high, then iis output will be high. Therefore it has a logic of OR.
- **Boolean expression for OR gate**

  $A + B = Y$

  which reads as A OR B is equal to Y.

- Truth table for OR gate -

A	B	Y = A+B
0	0	0
0	1	1
1	0	1
1	1	1

- **The AND gate**
- It is logic gate which has two or more inputs and one output

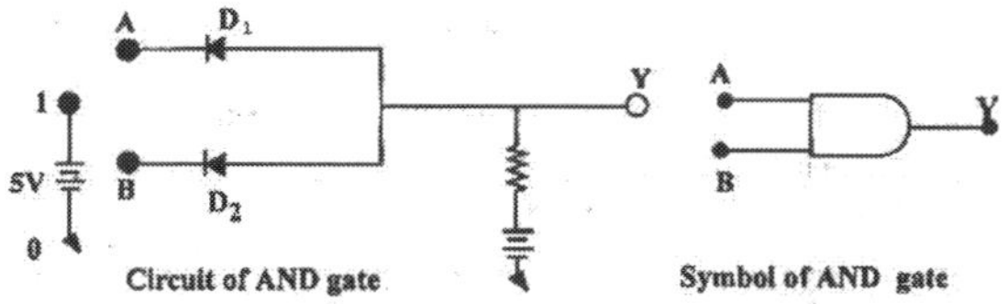

Circuit of AND gate

Symbol of AND gate

- If its all inputs are high, then it is output will be high (1). Therefore it has a logic of AND.
- **Boolean expression for AND gate -**

  A. B = Y reads as A AND B is equal to Y.
- Truth table for AND gate

A	B	Y = A . B
0	0	0
0	1	0
1	0	0
1	1	1

- **The NOT gate -**
- It is the logic gate which has one input and one output

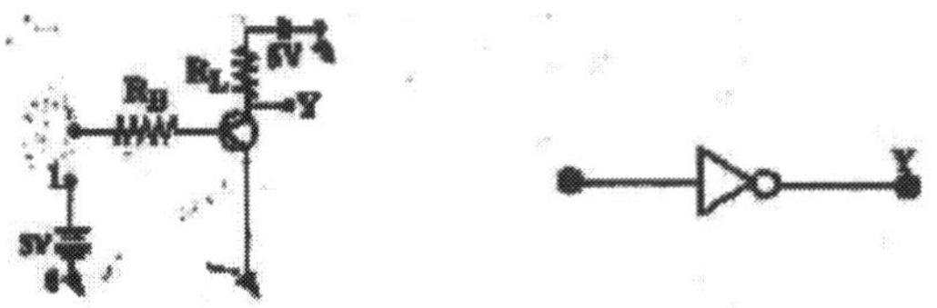

Circuit of NOT gate Symbol of NOT gate

- If its input is high (1), then its output will be low(0). Therefore it has a logic of NOT.
- **Boolean expression for NOT gate** $\overline{A} = Y$ reads as 'A NOT is equal to Y'.
- Truth table for NOT gate

A	$Y = \overline{A}$
0	1
1	0

- **Combination of gates -**

  In complicated digital circuits used in calculators, computers etc. the different types of combination of three basic logic gates are used.
- **NOR gate**
- This logic gate is the combination of OR gate and NOT gate.

  OR + NOT = NOR
- In this logic gate the output of OR gate is given to the input of NOT gate as shown in the below figure.

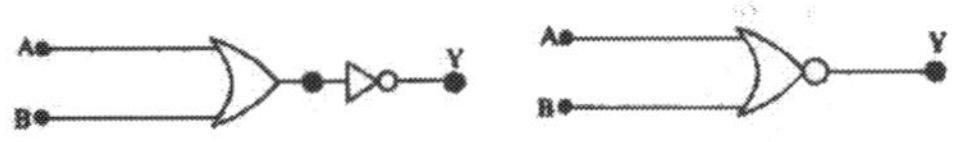

OR NOT NOR gate

- **Boolean expression for NOR gate**

  $Y = \overline{A + B}$

  Which reads as A OR B negated.

- Truth table for NOR gate

A	B	A+ B	$Y = \overline{A+B}$
0	0	0	1
0	1	1	0
1	0	1	0
1	1	1	0

- **NAND gate -**
- This logic gate is the combination of AND gate and NOT gate

  AND + NOT $\rightarrow$ NAND
- In this logic gate the output of AND gate is given to the input of NOT gate as shown below

AND NOT NAND gate

- Boolean expression for NAND gate $Y = \overline{A.B}$
- Truth table for NAND gate

A	B	A.B	$\overline{A.B}$
0	0	0	1
0	1	0	1
1	0	0	1
1	1	1	0

- **Uses of NOR gate and NAND gate -**
- The NAND gate and NOR gates are the building blocks of digital circuits. All the basic gates (OR, AND and NOT) can be obtained by the repeated use of NAND or NOR gates.
- **NOT gate from NAND gate -**

1) Diagram

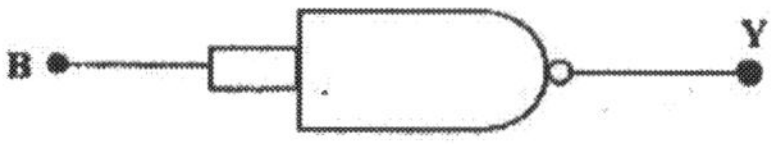

2) **Truth table**

A	B	$Y = \overline{A.A} = \overline{A}$
0	0	1
1	1	0

- **AND gate from NAND gate**

1) Diagram -

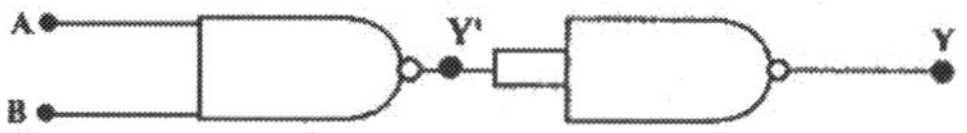

2) Truth table

A	B	Y
0	0	0
0	1	0
1	0	0
1	1	1

- **OR gate from NAND gate -**

1) Diagram -

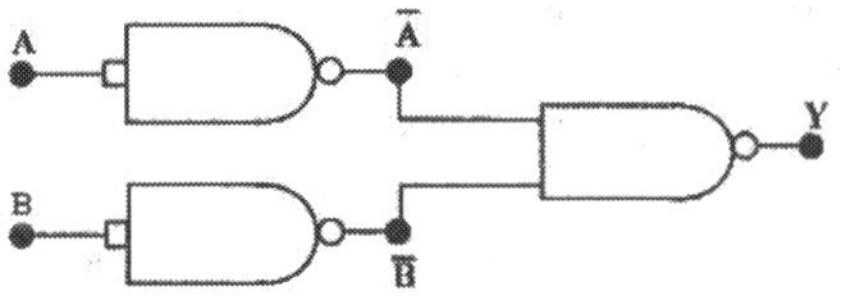

2) Truth table

A	B	$\overline{A}$	$\overline{B}$	$Y = (A + B)$
0	0	1	1	0
1	0	0	1	1
0	1	1	0	1
1	1	0	0	1

- **NOT gate from NOR gate -**

1) Diagram

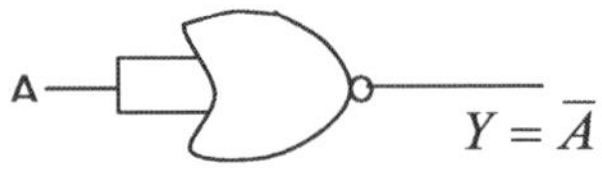

2) Truth table

A	B	$Y = \overline{A.A} = \overline{A}$
0	0	1
1	1	0

- **AND gate from NOR gate**

1) Diagram -

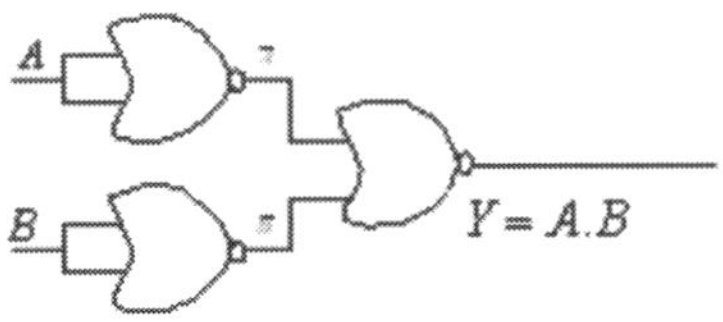

2) Truth table

A	B	Y
0	0	0
0	1	0
1	0	0
1	1	1

- **OR gate from NOR gate -**

  1) Diagram -

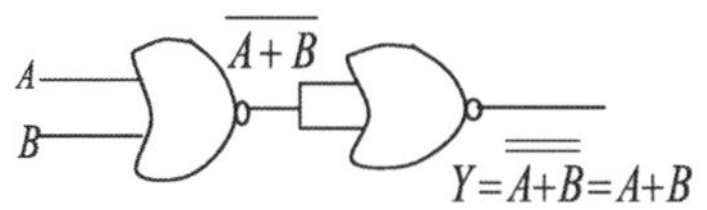

  2) Truth table

A	B	Y
0	0	0
0	1	1
1	0	1
1	1	1

- **XOR GATE**
- XOR gate is obtained by using OR, AND and NOT gates. It is also called exlusive OR gate.
- The output of a two input XOR gate is 1 only when the two inputs are different.
- The Boolean equation is $Y = A.\overline{B} + B.\overline{A}$

  1) two input XOR gate

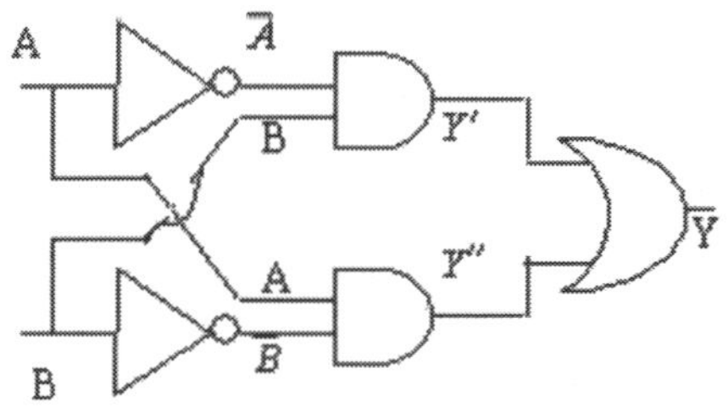

  2) circuit symbol

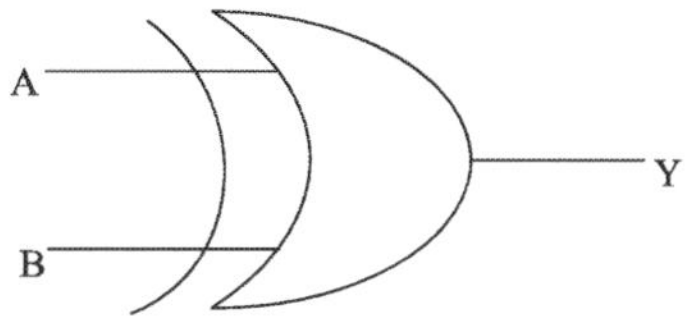

  3) truth table

A	B	Y
0	0	0
0	1	1
1	0	1
1	1	0

XNOR GATE

- XNOR gate is obtained by using OR, AND and NOT gates.
- It is also called exlusive NOR gate.
- The output of a two input XNOR gate is 1 only when both the inputs are same.
- The Boolean equation is $Y = A.B + \overline{A}.\overline{B}$

XNOR gate is inverse of XOR gate.

1) Diagram -

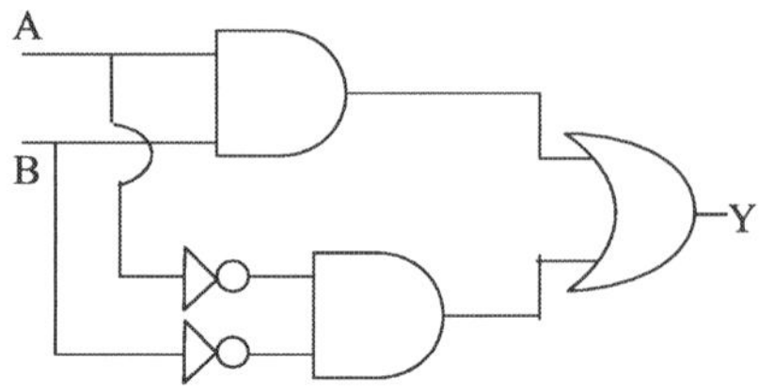

2) circuit symbol

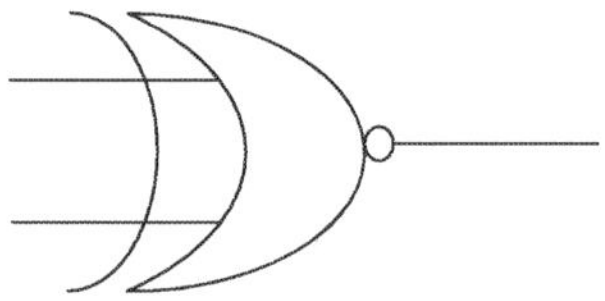

3) truth table

A	B	Y
0	0	1
0	1	0
1	0	0
1	1	1

WE:29:The Boolean expression of the output Y of the inputs A and B for the circuit shown in the fig:

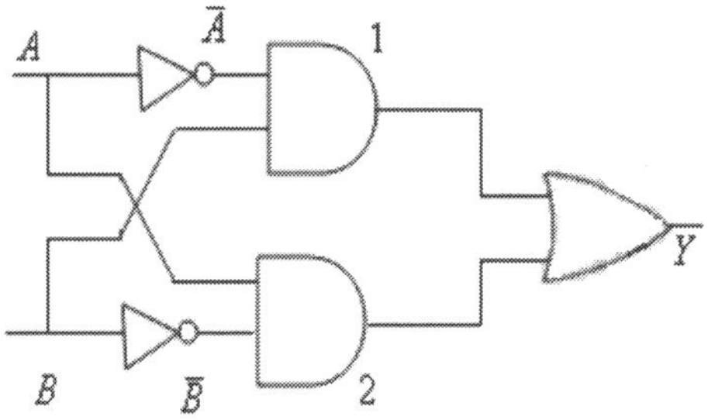

Solution : The output of AND gate 1 is $\overline{A}B$

The output of AND gate 2 is $A\overline{B}$

$\therefore$ the output of OR gate is Y= $= \overline{A}B + A\overline{B}$

WE:30:The diagram of a logic circuit is given below. The output of the circuit is represented by

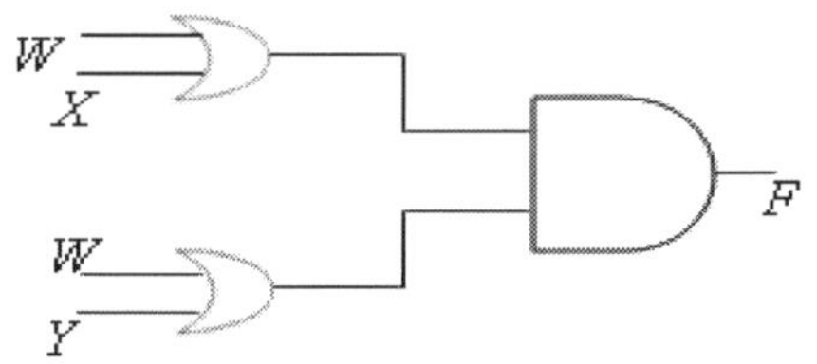

Solution :

$$F=(W+X).(W+Y)=W.W+W.Y+X.W+X.Y$$

$$=W+W.Y+X.W+X.Y=W(1+Y)+X.W+X.Y$$

$$=W+XW+X.Y=W(1+X)+X.Y=W+X.Y$$

WE:31:The logic circuit and its truth table are given, what is the gate X in the diagram

A	B	Y
1	1	1
1	0	1
0	1	1
0	0	0

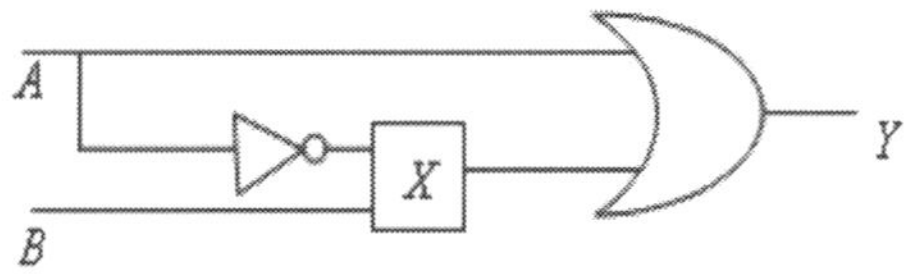

Solution: From the truth table we note that $Y=A+B$

i.e., it is for OR gate (or)

$$A+X=A+B=A+B.(A+\bar{A})(\because A+\bar{A}=1)$$

then, $A+X=A+B.A+B.\bar{A}=A.(1+B)+\bar{A}.B=A+\bar{A}.B$

So $X=A+\bar{A}.B$**, which is AND gate with inputs as** $\bar{A}$ **and B**

WE:32:You are given two circuits as shown in Figure which consist of NAND gates. Identify the logic operation carried out by the two circuits.

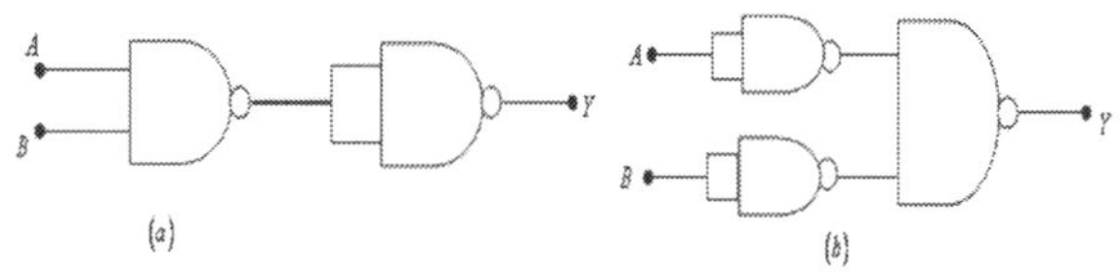

Solution: From fig(a). The output of NAND gate is connected to NOT gate (obtained from NAND gate) Let Y^1 be the output of NAND gate and the final output of the combination of two gates is Y. The output of a NAND gate is O only when both the inputs are zero, while in NOT gate, the input gets inverted.

Truth table for the arrangement

A	B	Y^1	Y
0	0	1	0
1	0	1	0
0	1	1	0
1	1	0	1

=

A	B	Y
0	0	0
1	0	0
0	1	0
1	1	1

It is the truth tables of AND gate. Therefore the given circuit acts as AND gate.

b) The output of two NOT gates are Connected to NAND gate Let Y_1 and Y_2 be the outputs of the two NOT gates and the final output of the combination of three gates be Y. In a NOT gate. the input gets inverted, while the output of a NAND gate is 'O' only when both the inputs are zero. Truth table for given arrangement.

A	B	Y
0	0	0
1	0	1
0	1	1
1	1	1

WE:33:Justify the output waveform (Y) of the OR gate for the following inputs A and B given is Fig

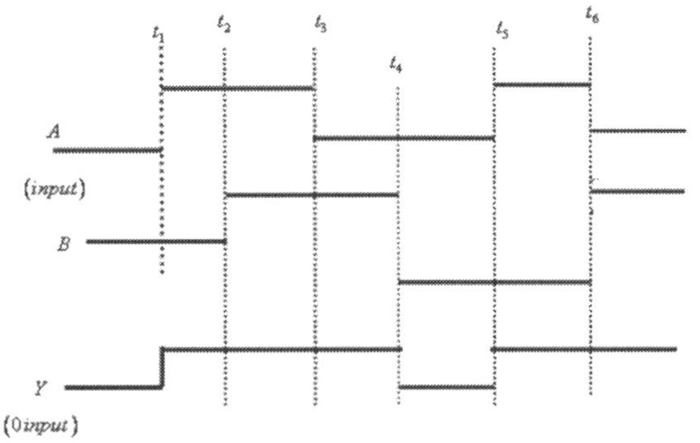

Solution: Note the following:

. At $t < t_1$; $A = 0, B = 0$; $Hence Y = 0$

. For t_1 $to t_2$; $A = 1, B = 0$; $Hence Y = 1$

. For t_2 $to t_3$; $A = 1, B = 1$; $Hence Y = 1$

. For t_3 $to t_4$; $A = 0, B = 1$; $Hence Y = 1$

. For t_4 $to t_5$; $A = 0, B = 0$; $Hence Y = 0$

. For t_5 $to t_6$; $A = 1, B = 0$; $Hence Y = 1$

. For $t > t_6$; $A = 0, B = 1$; $Hence Y = 1$

Therefore the wave form Y will be as shown in the Fig

WE:34:Draw logic diagrams for the Boolean expressions given below.

i) $A \bullet \overline{B} + \overline{A} \bullet B = Y$ **ii)** $\left(A + \overline{B}\right) \bullet \left(\overline{A} + B\right) = Y$

Solution: (i) The required logic diagram for the given Boolean expression is given in figure. Here the input B before applying to first AND gate and input A before applying to second AND gate have been inverted. The output of these gates are, therefore, $A\overline{B}$ and $\overline{A}B$ respectively. These outputs are fed to

OR gate which gives $Y = A\overline{B} + \overline{A}B$ as shown in figure.

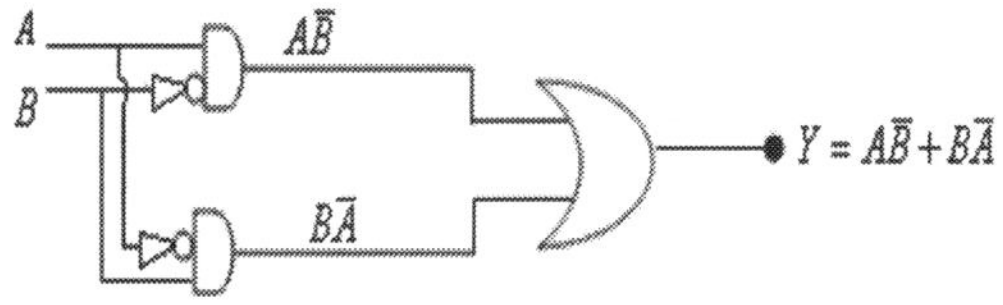

ii) The required logic diagram for the given Boolean expression is shown in figure. The input B to first OR gate and input A to second OR gate have been inverted. The output of these gates are, therefore, $A+\overline{B}$ and $\overline{A}+B$ as shown. These inputes when applied to AND gate give the required output.

$Y = \left(A+\overline{B}\right).\left(\overline{A}+B\right)$

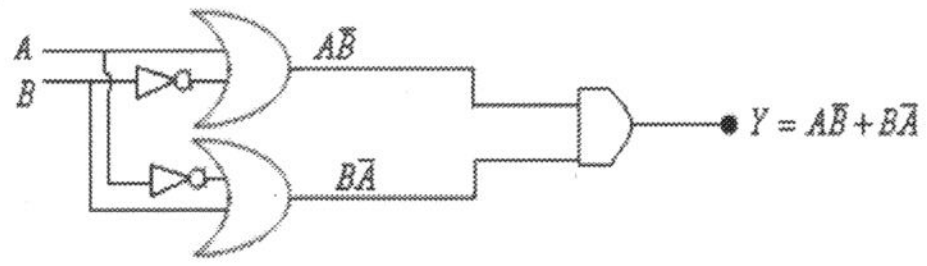

INTEGRATED CIRCUITS:

An entire circuit (consisting of many passive components like R and C active devices like diode and transistor) on a small single block (or chip) of a semiconductor is known as Integrated Circuit (IC). The most widely used technology is the Monolithic Integrated Circuit.

- The chip dimensions are as small as $1mm \times 1mm$ or it could even be smaller.
- Depending on nature of input signals, IC's can be grouped in two categories:
- (a) linear or analogue IC's and
- (b) digital IC's. The linear IC's process analogue signals varies linearly with the input.
- One of the most useful linear IC's is the operational amplifier.
- The digital IC's process signals that have only two value.
- They contain circuits such as logic gates. Depending upon the level of integration (i.e,. the number of circuit components or logic gates), the IC's are termed as

i) Small Scale Integration, SSI(logic gates ≤ 10)

ii) Medium Scale Integration, MSI (logic gates < 100)

iii) Large Scale Integration, LSI (logic gates <1000)

iv) very large scale integration VLSI (logic gates > 1000).

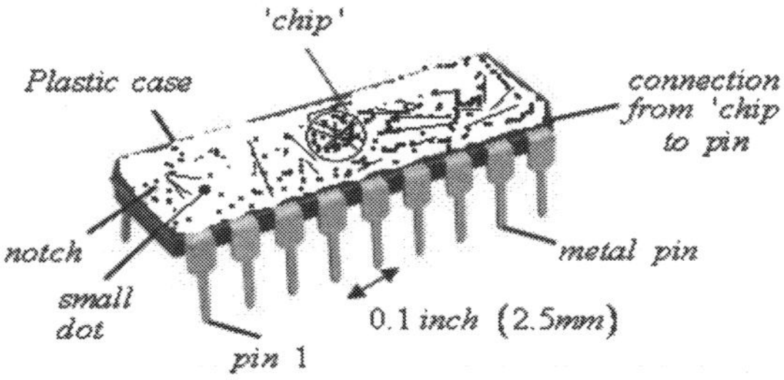

The casing and connection of a 'chip'

Chapter - 4
(PART - 2)
SEMICONDUCTOR DEVICES

CONCEPTUAL QUESTIONS

INTRINSIC AND EXTRINSIC SEMICONDUCTORS

1. **Which property makes the crystalline solids to have sharp melting point?**
   1) Equal strength of all the interatomic bonds 2) anisotropic
   3) long range order 4) short range order

2. **Which of the following is not a property of crystalline substance**
   1) sharp melting point 2) bounded by flat surface 3) isotropic 4) long range order

3. **A strip of copper and another of germanium are cooled from room temperature to 80K. The resistance of:**
   1) each of these decreases
   2) copper strip decreases and that of germanium decreases
   3) copper strip decreases and that of germanium increases
   4) each of these increases

4. **The difference in the variation of resistance with temperature in a metal and a semiconductor arises essentially due to the difference in the :**
   1) crystal structure
   2) variation of the number of charge carriers with temperature
   3) type of bonding
   4) variation of scattering mechanism with temperature

5. **The manifestation of band structure in solids is due to :**
   1) Heisenberg's uncertainty principle 2) Pauli's exclusive principle
   3) Bohr's correspondence principle 4) Boltzman's law

6. **A solid which is not transparent to visible light and whose conductivity increases with temperature is formed by:**
   1) ionic bonding 2) covalent bonding 3) vander Wall's bonding 4) metallic bonding

7. **To a germanium sample, traces of gallium are added as an impurity. The resultant sample would behave like :**
   1) A conductor 2) A p-type semiconductor
   3) A n-type semiconductor 4) An insulator

8. **In semiconductors at a room temperature**
   1) The valence band is partially empty and the conduction band is partially filled.
   2) The valence band is completely filled and the conduction band is partially filled
   3) The valence band is completely filled 4) The conduction band is completely empty

9. **Identify the property which is not characteristic for a semiconductor?**
   1) at a very low temperatures, it behaves like an insulator
   2) at higher temperatures two types of charge carriers will cause conductivity
   3) the charge carriers are electrons and holes in the valence band at higher temperatures
   4) the semiconductor is electrically neutral.

10. **Carbon, silicon, and germanium have four valence electrons each. At room temperature, which one of the following statements is most appropriate?**
    1) The number of free conduction electrons is negligibly small in all the three.
    2) The number of free electrons for conduction is significant in all the three.
    3) The number of free electrons for conduction is significant only in Si and Ge but small in C.
    4) The number of free conduction electrons is significant in C but small in Si and Ge

11. **The relation between number of free electrons (n) in a semiconductor and temperature (T) is given by**

1) $n \propto T$ 2) $n \propto T^2$ 3) $n \propto \sqrt{T}$ 4) $n \propto T^{3/2}$

12. **An electrically neutral semi conductor has**

1) equal amounts of negative and positive charges 2) no minority charge carriers

3) no majority charge carriers 4) no free charges

13. **There is no hole current in conductors because they have**

1) high conductivity 2) high electron density 3) no valence band

4) overlapping of valence and conduction bands.

14. **In the insulators**

1) the valence band is partially filled with electrons

2) the conduction band is partially filled with electrons

3) the conduction band is partially filled with electrons and valence band is empty

4) the conduction band is empty and the valence band is filled with electrons.

15. **In semiconductors the forbidden energy gap between V.B. and C.B. is of the order of**

1) 1eV 2) 5eV 3) 1 keV 4) 1MeV

16. **The level formed due to impurity atom, in the forbidden energy gap, very near to the valence band in a p-type semicoductor is called**

1) acceptor level 2) donar level 3) conduction level 4) forbidden level

17. **The bond in semiconductors is**

1) covalent 2) ionic 3) metallic 4) hydrogen

18. **On increasing temperature, the conductivity of pure semiconductors**

1) decreases 2) increases 3) remains unchanged 4) becomes zero

19. **The mobility of free electrons is greater than that of free holes because**

1) they carry negative charge 2) they are light

3) their mutual collisions are less 4) they require low energy to continue their motion

20. **A semiconductor at 0 K behaves as**

1) conductor 2) insulator

3) super conductor 4) extrinsic semiconductor

21. **Carbon, Silicon, and Germanium have four valence electrons each. These are characterised by valence and conduction bands separated by energy band gap respectively equal to $(E_g)_C$, $(E_g)_{Si}$ and $(E_g)_{Ge}$. Which of the following statement is true?**

1) $(E_g)_{Si} < (E_g)_{Ge} < (E_g)_C$ 2) $(E_g)_C < (E_g)_{Ge} > (E_g)_{Si}$

3) $(E_g)_C > (E_g)_{Si} > (E_g)_{Ge}$ 4) $(E_g)_C = (E_g)_{Si} = (E_g)_{Ge}$

22. **The valency of impurity element for making p-type semiconductor is**

1) 5 2) 4 3) 3 4) 7

23. **In n-type semiconductors the electron concentration is equal to**

1) density of donor atoms
2) density of acceptor atoms
3) density of both type of atoms
4) neither density of acceptor atoms nor density of donor atoms

24. **Which of the following statements is not true?**

1) the resistance of intrinsic semiconductors decreases with increase of temperature.
2) doping pure Si with trivalent impurities give p-type semiconductors
3) the majority charge carries in n-type semiconductors are holes
4) a p-n junction can act as a semiconductor diode

25. **p-type semi conductor is**

1) negatively charged
2) positively charged
3) neutral
4) may be positive or negative

26. **n-type semi conductor is**

1) negatively charged
2) positively charged
3) neutral
4) may be positive or negative

27. **An electric field is applied across a semi conductor. Let n be the number of charge carriers. As temperature increases, n will**

1) increase
2) decrease
3) does not change
4) may increase or decrease

28. **In a n - type semiconductor , the fermi energy level lies**

1) in the forbidden energy gap nearer to the conduction band.
2) in the forbidden energy gap nearer to the valence band.
3) in the middle of forbidden energy gap
4) outside the forbidden energy gap

29. **An n-type and p-type silicon can be obtained by doping pure silicon with**

1) Arsenic and phosphrous
2) Indium and aluminium
3) Phosphorous and indium
4) aluminium and boron

30. **The width of forbidden gap in silicon crystal is 1.1 ev. When the crystal is converted into a n-type semiconductor the distance of fermi level from conduction band is**

1) Greater than 0.55eV
2) Equal to 0.55 eV
3) lesser than 0.55eV
4) Equal to 1.1 eV

31. **In extrinsic semiconductors**

1) the conduction band and valence band overlap
2) the gap between conduction band and valence band is near about 16eV
3) the gap between conduction band and valence band is near about 1eV
4) The gap between conduction band and valence band will be 100eV and more

32. **The element that can be used as acceptor impurity to dope silicon is**

1) Antimony
2) Arsenic
3) Boron
4) phosphorous

33. **Among the following, the wrong statement in the case of semiconductor is**

1) Resistivity is in between that of a conductor and insulator
2) Temperature coefficient of resistance is negative
3) Doping increases conductivity
4) At absolute zero temperature it behaves like a condutor

34. The value indicated by fermi energy level in an intrinsic semiconductor is

1) the average energy of electrons and holes 2) the energy of electrons in conduction band
3) the energy of holes in valence band 4) the energy of forbidden region

35. The conduction band and valency band of a good conductors are

1) well separated 2) just touch 3) very close 4) overlap

36. Two pieces one of germinium and the other of aluminium are cooled from T_1K to T_2K. The resistance of

1) aluminium increases and that of germanium decrease
2) each of them decreases
3) aluminium decreases and that of germanium increases
4) each of them increases

37. In intrinsic semiconductor at room temperature the no. of electrons and holes are

1) equal 2) zero 3) unequal 4) infinite

38. Band gap in insulator is of the order

1) 6 eV 2) 0.60 eV 3) –6 eV 4) 0eV

39. In p-type semiconductor conduction is due to

1) greater number of holes and less number of electrons
2) only electrons 3) only holes
4) greater number of electrons and less number of holes

40. In an intrinsic semiconductor, the fermi nergy level is

1) nearer to valence band than conduction band
2) equidistant from conduction band and valence band
3) nearer to conduction band than valence band 4) bisecting the conduction band

41. With increase in temperature in an intrinsic semiconducter the ratio of conduction electrons and holes is

1) 1:1 2) 1: 2 3) 2 :1 4) 1 : 3

42. To obtain n-type extrinsic semiconductor, the impurity element to be added to germanium should be of valency

1) 2 2) 5 3) 4 4) 3

43. The majority carriers in a p-type semi-conductor are........

1) Electrons 2) Holes 3) Both 4) Impurities

44. The objective of adding impurities in the extrinsic semiconductor is

1) to increase the conductivity of the semiconductor
2) to increase the density of total current carries
3) to increase the density of either holes or electrons but not both
4) to eliminate the electron -hole pairs produced in intrinsic semiconductor.

45. In intrinsic semiconductor conductivity is

1) low 2) average 3) high 4) very low

46. In intrinsic semiconductor conductivity is due to

1) doping 2) breaking of covalent bonds 3) free electrons 4) holes

47. When the conductivity of a semi conductor is only due to breaking of covalent bonds, the semi conductor is called

1) n-type 2) p-type 3) intrinsic 4) extrinsic

48. **Assertion (A) : C,Si and Ge have 4 Valancy each but C is an insulator where as Si and Ge are semi conductors**

Reason (R) : Energy gap is least for Ge, less for Si compared to C, So that free electrons for conduction in Ge and Si are significant but negligible small for C.

1) A, R are true and R explains A correctly 2) A, R are true and R do not explain A correctly

3) A is true, but R is false 4) R is true, but A is false

JUNCTION DIODE

49. **The potential barrier at PN junction is due to**

1) fixed acceptor and donor ions on either side of the junction

2) minority carriers on either side of the junction 3) majority carriers on either side of the junction

4) both majority and minority carriers on either side of junction

50. **A PN junction diode cannot be used**

1) as rectifier 2) for converting light energy to electric energy

3) for getting light radiation 4) for increasing the amplitude of an ac signal

51. **A full wave rectifier along with the output is shown in fig. the contributions from the diode (2)are**

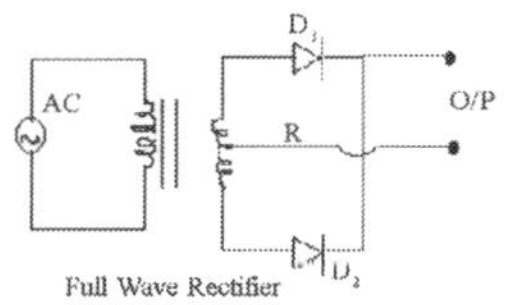

Full Wave Rectifier

1) C 2) A,C 3) B, D 4) A,B,C,D

52. **A full-wave rectifier is used to convert 'n' Hz a.c into d.c,then the number of pulses per second present in the rectified voltage is.**

1) n 2) n/2 3)2n 4)4n

53. **If the input frequency of half-wave rectifier is n Hz ac, then its output is**

1) a constant dc 2) n/2 Hz pulsating dc

3) n Hz pulsating dc 4) 2n Hz pulsating dc

54. **p-n junction diode acts as**

1) ohmic resistance 2) non-ohmic resistance 3) both 1 and 2 4) amplifier

55. **The process of converting alternating current into direct current is known as**

1) modulation 2) amplification 3) detection 4) rectification

56. **On increasing reverse voltage in a p-n junction diode the value of reverse current will**

1) gradually increases

2) first remains constant and then suddenly increase

3) remains constant 4) gradually decrease

57. **In forward bias the depletion layer behaves like**

1) an insulator 2) a conductor 3) a semiconductor 4) capacitor

58. p-n junction in reverse bias behaves like

1) an inductor 2) a condenser 3) amplifier 4) an off switch

59. The main cause of avalenche breakdown is

1) collision by ionisation 2) high doping

3) recombination of electrons and holes 4) low doping

60. The main cause of Zener breakdown is

1) the base semiconductor being germanium

2) production of electron - hole pairs due to thermal exitation

3) low doping 4) high doping

61. When p-n junction is foward biased, the current across the junction is mainly due to

1) diffusion of charges 2) drifting of charges

3) both diffusion and drifting of charges 4) holes only

62. The current through any p-n junction is due to

a) drift of charge carriers **b) diffusion of charge carriers**

c) different concentrations of same type of charge carriers in different regions

d) same concentrations of same type of charge carriers in different regions

1) a,b and c 2) a and b only 3) only d 4) a,b,c,d

63. The thickness of depletion layer is approximately

1) 1 μm 2) 1mm 3) 1cm 4) 1m

64. The depletion region is

1) region of opposite charges 2) neutral region

3) region of infinite energy 4) region of free current carriers

65. The diffusion current in a p-n junction is greater than the drift current when the junction is

1) forward biased 2) reverse biased 3) un biased

4) both forward and reverse biased

66. Germanium diode

1) may be used as rectifier because it offers a relatively low resistance for forward bias and very high resistance for reverse bias.

2) may be used as a rectifier because it offers a relatively high resistance for forward bias and very low resistance for reverse bias.

3) cannot be used as a rectifier 4) may be used as an amplifier

67. The output of the given circuit in Fig.

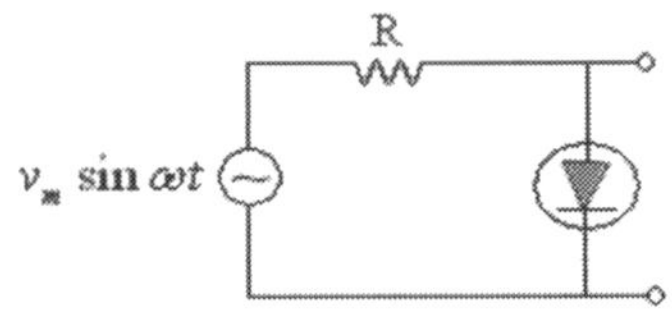

1) would be zero at all times

2) would be like a half-wave rectifier with positive cycles in output

3) would be like a half-wave rectifier with negative cycles in output

4) would be like that of a full-wave rectifier

68. Diode is forward biased and the applied voltage is greater than the potential barrier then

I) resistance of the junction in the forward bias decreases

II) potential barrier remains same **III) width barrier remains decreases**

IV)p-type is at higher potential than the n-type

1) all are true 2)all are false 3) I,III,IV are true 4)I,II, III are true

69. When a junction diode is reverse biased, the current called drift current is due to

1) majority charge carriers of both n&p sides 2) minority charge carriers of both n&p sides

3) holes of both n &p sides 4) conduction band electrons of n-side only

70. Among the following one statement is not correct when a junction diode is in forward bias

1) the width of depletion region decreases

2) free electron on n- side will move towards the junction

3) holes on p -side move towards the junction

4) electron on n- side and holes on p-side will move away from junction

71. Consider a.p-n junction as a capacitor, formed with p and n material acting as thin metal electrodes and depletion layer width acting as seperation between them. Basing on this assume that a n-p-n transistor is working as an amplifier in CE configuration. If C_1 and C_2 are base-emiter and collector emitter junction capacitances, then :

1) $C_1 > C_2$ 2) $C_1 < C_2$ 3) $C_1 = C_2$ 4) $C_1 = C_2 = 0$

72. Consider the following statements A and B and identify the correct answer

1) : Germanium is preferred over silicon in the construction of zener diode.

2) : Germanium has high thermal stability than silicon in the construction of Zener diode

1) Both (1) and (2) are true 2) Both (1) and (2) are false

3) (1) is true but (2) is false 4) (1) is false but (2) is true

73. A Zener diode when used as a voltage regulator is connected

(a) in forward bias **(b) in reverse bias**

(c) in parallel to the load **(d) in series to the load**

1) (a) and (b) are correct 2) (b) and (c) are correct

3) (a) only is correct 4) (d) only is correct

74. Consider the following statements A and B and identify the correct answer

(A) A Zener diode is always connected in reverse bias to use it as voltage regulator

(B) The potential barrier of a p - n junction lies between 0.1 to 0.3V, approximately

1) A and B are correct 2) A and B are wrong

3) A is correct but B is wrong 4) A is wrong but B is correct

75. Consider the following statements A and B identify the correct of the give answer.

A) The width of the depletion layer in a p-n junction diode increases in forward bias.

B) In an intrinsic semiconductor the fermi energy level is exactly in the middle of the forbidden gap

1) A is true and B is false 2) Both A and B are false

3) A is false and B is true 4) Both A and B are true

76. The potential in the depletion layer due to

1) Electrons 2) Holes 3)Ions 4) Forbidden band

77. **Pickout the incorrect statement regarding reverse saturation current in the p-n junction diode**
1) this current doubles for every 100^0C rise of temperature
2) this current is due to minority carriers
3) the current carriers are produced by thermal agitation
4) reverse saturation current is also known as leakage current

78. **When the p-n junction diode is reverse biased, the thickness of the depletion layer**
1) increases 2) decreases 3) becomes zero 4) remains constant

79. **p-n junction diode can be used as**
1) amplifer 2) detector 3) oscillator 4) capacitor

80. **A p-n junction diode is reverse biased.Then**
1) more current flows 2) the barrier potential decreases
3) the barrier potential increases 4) resistance offered is low

81. **In the middle of the depletion layer of reverse biased p-n junction, the :**
1) electric field is zero 2) potential is maximum
3) electric field is maximum 4) potential is zero

82. **When p-n junction diode is forward biased then**
1) the depletion region is reduced and barrier height is increased
2) the depletion region is widened and barrier height is reduced
3) both the depletion region and barrier height are reduced
4) both the depletion region and barrier height are increased

83. **In a p-n junction photo cell, the value of the photo electromotive force produced by monochromatic light is proportional to**
1) The barrier voltage at the p-n junction. 2) The intensity of the light falling on the cell
3) The frequency of the light falling on the cell. 4) The voltage applied at the p-n junction

84. **In Fig. V_0 is the potential barrier across a p-n junction, when no battery is connected across the junction:**

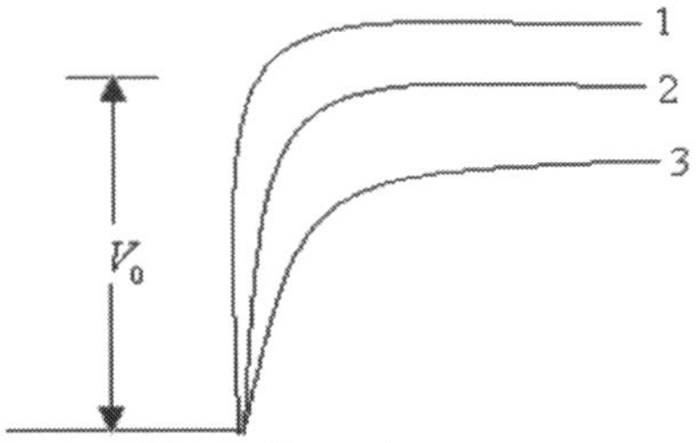

1) 1 and 3 both correspond to forward bias of junction
2) 3 corresponds to forward bias of junction and 1 corresponds to reverse bias of junction
3) 1 corresponds to forward bias and 3 corresponds to reverse bias of junction
4) 3 and 1 both correspond to reverse bias and 3 corresponds to reverse bias of junction

85. **There is a sudden increase in current in zener diode is**
1) Due to rupture of bonds 2) Resistance of deplection layer becomes less
3) Due to high doping 4) Due to less doping

86. **Application of a forward bias to a p-n junction:**
1) increases the number of donors on the n-side
2) increases the electric field in the depletion zone
3) increases the potential difference across the depletion zone 4) widens the depletion zone

87. In a p-n junction diode, not connected to any circuit

1) the potential is the same everywhere
2) the p-type side is at a higher potential than the n-type side.
3) there is an electric field at the junction directed from the n-type side to the p-type side.
4) there is an electric field at the junction directed from the p-type side to the n-type side.

88. Select the correct statement from the following:

1) A diode can be used as a rectifier
2) A triode cannot be used as a rectifier
3) The current in a diode is always proportional to the applied voltage
4) The linear portion of the I -V characteristic of a triode is used for amplification without distortion

89. When a p-n junction diode is forward-biased, energy is released at the junction due to the recombination of electrons and holes. This energy is in

1) Visible region 2) Infrared region 3) UV region 4) X-ray region

90. In Fig. assuming the diodes to be ideal:

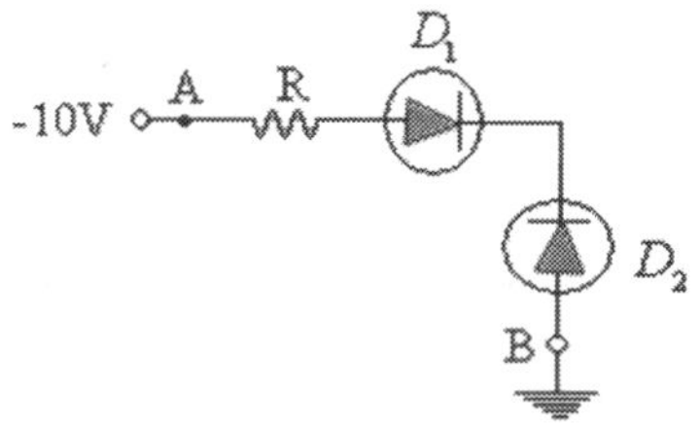

1) D_1 is forward biased and D_2 is reverse biased and hence current flows from A to B

2) D_2 is forward biased and D_1 is reverse biased and hence no current flows from B to A and vice versa

3) D_1 and D_2 are both forward biased and hence current flows from A to B

4) D_1 and D_2 are both reverse biased and hence no current flows from A to B and vice-versa

91. **A Si and a Ge diode has identical physical dimensions.The band gap in Si is larger than that in Ge.An identical reverse bias is applied across the diodes.**

1)The reverse current in Ge is larger than that in Si
2)The reverse current in Si is larger than that in Ge
3)The reverse current is identical in the two diodes
4)The relative magnitude of the reverse currents cannot be determined from the given data only

92. **The correct curve between potential(V) and distance(d) near p-n junction is**

1) 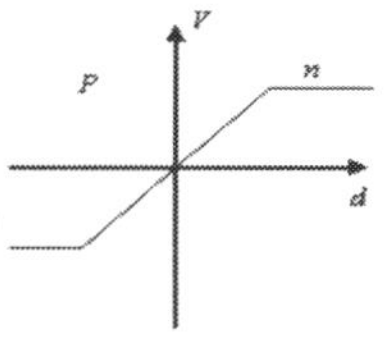

2) 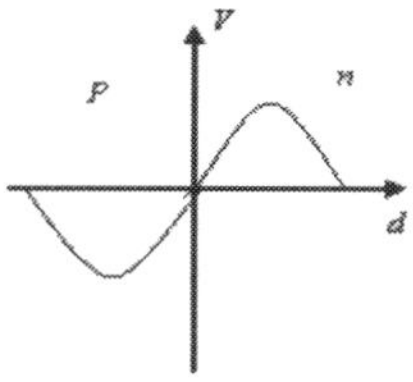

3) 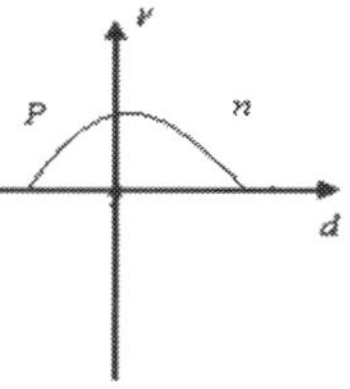

4) 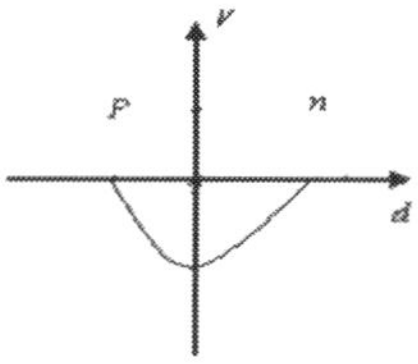

93. **The graph given below represents the I-V characterstics of a zener diode.Which part of the characterstics curve is most relavant for its operation as a voltage regulator.**

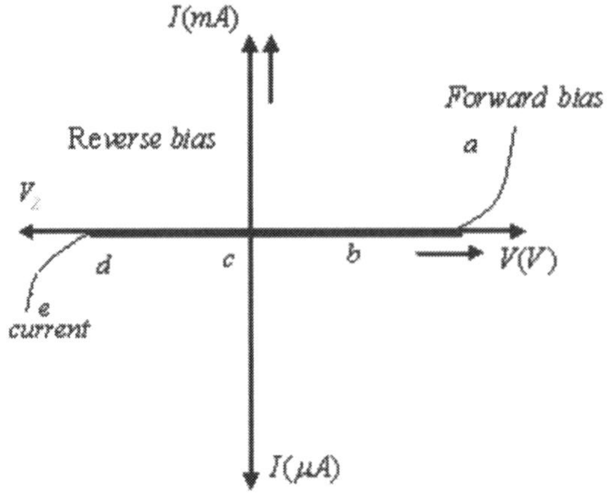

1)ab　　2)bc　　3)cd　　4)de

94. **Assertion (A) : Si and GaAs are preferred materials for solar cells**
Reason (R) : Energy gap of Si is 1.1eV and that of GaAs is 1.53eV which give maximum irradiance where as other materials like CdS or CdSe $\left(E_g = 2.4\,eV\right)$ and PbS $\left(E_g = 0.4\,eV\right)$ give minimum irradiance.
1) A, R are true and R explains A correctly
2) A, R are true and R do not explain A correctly
3) A is true, but R is false
4) R is true, but A is false

95. **Assertion (A) : Putting p-type semiconductor slabs directly in physical contact can not form p-n junction**
Reason (R) : The roughness at contact will be much more than inter atomic crystal spacing $\left(=2\overset{0}{A}\,to\,3\overset{0}{A}\right)$ and continuous flow of charge carriers is not possible
1) A, R are true and R explains A correctly
2) A, R are true and R do not explain A correctly
3) A is true, but R is false
4) R is true, but A is false

96. **Assertion (A) : Eventhough forward bias is known to be more (mA) than the current in the reverse bias (μA), photodiodes are still used in reverse bias.**
Reason (R) : The fractional change in majority charge carriers would be much less than that in minority charge carriers due to photo-effects so that minority carrier reverse bias current is more easily to measure with photo diodes than majority carrier forward current.
1) A, R are true and R explains A correctly
2) A, R are true and R do not explain A correctly
3) A is true, but R is false
4) R is true, but A is false

97. **Match List-I with List-2**

LIST-1	**LIST-2**
A) Intrinsic	**E) Donor level semiconductor**
B) N-type	**F) Acceptor level semiconductor**
C) p-type	**G) Immobile ions semiconductor**
D) Depletion layer centre of forbidden	**H) Fermi energy levels is at the energy gap**
	I) Fermi energy level is near the conduction band

	A	B	C	D		A	B	C	D
1)	F	E,G	H	I	2)	H	E,I	F	G
3)	E	F	G	H,I	4)	I	H	G	E,F

98. Match the following

LIST - I	**LIST - II**
a. Antimony	**e. P - type impurity**
b. Indium	**f. N - type impurity**
c. Carbon	**g. not semiconductor**
d. Silicon	**h. semiconductor**

1)a-f, b-e, c-g, d-h 2)a- e, b- g, c- h, d- f 3) a-g, b-h, c-f, d-e 4) a-h, b-f, c-e, d- g

TRANSISTORS

99. The correct relation between current gains α and β is

1) $\beta=\frac{\alpha}{1-\alpha}$ 2) $\beta=\frac{\alpha}{1+\alpha}$ 3) $\beta=\alpha(1-\alpha)$ 4) $\beta=\frac{1-\alpha}{\alpha}$

100. Transistors are made of

1) insulators 2) conductors

3) alloys 4) doped semi-condcutors

101. In n-p-n transistor the arrow head on emitter represents that the conventional current flows from

1) base to emitter 2) emitter to base 3) emitter to collector 4) base to collector

102. In a junction transistor the emitter, base and collector are made of

1) extrinsic semi conductors 2) intrinsic semi conductors

3) both 1 and 2 4) metal

103. In a transistor

1) both emitter and the collector are equally doped

2) base is more heavily doped than collector

3) collector is more heavily doped than the emitter

4) the base is made very thin and is lightly doped

104. In a transistor

1) length of emitter is greater than that of collector 2) length of collector is greater than that of emitter

3) both emitter and collector have same length

4) any one of emitter and collector can have greater length

105. In transistor the emitter current is

1) slightly more than the collector current 2) slightly less than the collector current

3) equal to the collector current 4) equal to the base current

106. In the use of transistor as an amplifier

1) the emitter - base junction is reverse biased and the collector base junction is also reverse biased

2) the emitter - base junction is forward biased and the collector - base junction is reverse biased

3) both the junctions are forward biased

4) any of the two junctions may be forward biased.

107. One way in which the operation of an npn transistor differ from that of a pnp transistor is that
1) the emitter junction is reverse biased in npn
2) the emitter junction injects minority carriers into the base region of the pnp.
3) the emitter injects holes into the base of the pnp and electrons into the base region of npn
4) the emitter injects holes into the base of npn

108. npn transistors are preferred to pnp transistors because they have
1) low cost
2) low dissipation energy
3) capable of handling large power
4) electrons have high mobility than holes and hence high mobility of energy

109. A CE transistor amplifies a weak current signal because collector current is
1) β times I_b
2) β times I_C
3) α times I_b
4) α times I_C

110. When a positive voltage signal is applied to the base of a common emitter npn amplifier
1) The emitter current decreases
2) The collector voltage becomes more positive
3) The collector voltage becomes less positive
4) The collector current decreases

111. In case of common emitter p-n-p transistor input characteristic is a graph drawn
1) With I_C on y-axis and V_{CE} on x-axis keeping I_B constant
2) With I_B on y-axis and V_{BE} on x-axis keeping V_{CE} constant
3) With I_C on y-axis and I_B on x-axis keeping V_{CE} constant
4) With V_{BE} on y-axis and V_{CE} x-axis keeping I_B constant

112. The output characterstics of an n-p-n transistor represent, [I_C = Collector current, V_{CE} = potential difference between collector and emitter, I_B = Base current, V_{BB} = voltage given base; V_{BE} = the potential difference between base and emitter]
1)change in I_C *as* I_B and V_{BB} are changed
2)Changes in I_C with changes in V_{CE} (I_B = cosntant)
3)changes in I_B with changes in V_{CE}
4)Change in I_C as V_{BE} is changed

113. In a transistor the base is made very thin and is lightly doped with an impurity, because
1) to enable the collector to collect about 95% of the holes or electrons coming from the emitter side
2) to enable the emitter to emit small number of holes or electrons
3) to save the transistors from high current effects
4) to enable the base to collect about 95% of holes or electrons coming from the emitter side

114. A p-n-p transistor is said to be in active region of operation, When
1) Both emitter junction and collector junction are forward biased
2) Both emitter junction and collector junction are reverse biased
3) Emitter junction is forward biased and collector junction is reverse biased
4) Emitter junction is reverse biased and collector junction is forward biased

115. An n-p-n transistor power amplifier in C-E configuration gives
1) Voltage amplification only
2) Current amplification only
3) Both current and voltage amplification
4) Only power gain of unity

116. When n-p-n transistor is used as an amplifier:
1) electrons move from base to collector
2) holes moves from emitter to base
3) holes move from collector to base
4) holes move from base to emitter

117. The part of a trasistor which is heavily doped to produce large number of majority carriers is :

1) emitter 2) base 3) collector
4) can be any of the above three

118. When n-p-n transistor is used as an amplifier :

1) electrons move from collector to base 2) holes move from collector to base
3) electrons move from base to emitter 4) holes move from base to emitter

119. A n-p-n transistor conducts when

1) both collector and emitter are positive with respect to the base
2) collector is positive and emitter is negative with respect to the base
3) collector is positive and emitter is at same potential as the base
4) both collector and emitter are negative with respect to the base

120. In a common -base amplifier, the phase difference between the input signal voltage and output voltage is :

1) $\frac{\pi}{4}$ 2) π 3) zero 4) $\frac{\pi}{2}$

121. In a common emitter amplifier, the phase difference between the input signal voltage and output voltage is

1) $\frac{\pi}{2}$ 2) 0 3) π 4) $\frac{\pi}{4}$

122. When n-P-n transistor is used as an amplifer

1) Electrons move from emitter to collector 2) Holes move from emitter to base
3) Electrons move form collector to base 4) Holes move from base to collector

123. In a PNP transistor the base is the N-region. Its width relative to the P-region is

1) smaller 2) larger 3) same 4) not related

124. A three terminal device with one terminal common to both the output and input is called

1) rectifier 2) transistor 3) diode 4) triode

125. Input and output signal of an amplifier in CE configuration are always

1) Equal 2) Inphase 3) Having a phase difference 4) Out of phase

126. Ge transistor can be operated at a temperature

1) upto 90^0C 2) upto 40^0C 3) upto 100^0C 4) upto 20^0C

127. Transistor amplifier circuit with a feed back circuit is called

1) oscillator 2) detector 3) modulator 4) all

128. A pulsating voltage is a mixture of an a.c componet and a d.c component. the circuit used to seperate a.c and d.c component is called

1) an oscillator 2) an amplifier 3) a rectifier 4) a filter

129. The α and β of a transistor are always

1) $\alpha > 1$, $\beta < 1$ 2) $\alpha < 1$, $\beta > 1$ 3) $\alpha = \beta$ 4) $\alpha\beta = 1$

130. In case of NPN transistor, emitter current is always greater than collector current, because:

1) Collector side is revese biased and emitter side is forward biased

2) Collector being reverse biased attracts more electrons

3) Some electrons are lost in base

4) Collector side is forward biased and emitter side is reverse biased.

131. When a transistor amplifier having current gain of 75 is given an input signal, $V_I = 2\sin(157t + \pi/2)$, the output signal is found to be $V_O = 200\sin(157t + 3\pi/2)$. The transistor is connected as:

1) A common collector amplifier
2) A common base amplifier
3) A common emitter amplifier
4) An oscillator

132. An oscillator is an amplifier with

1) A large gain
2) Negative feedback
3) Positive feedback
4) No feedback

133. In which of the transistor configurations, the voltage gain is highest?

1) Common-base
2) Common-emitter
3) Common-collector
4) Same in all threee

134. A working transistor with its three legs marked P, Q, and R is tested using a multimeter to R and the other (positive) terminal to P or Q. Some resistance is seen on the multimeter. Which of the following is true for the transistor?

1) It is a pnp transistor with R as emitter
2) It is an npn transistor with R as collector
3) It is an npn transistor with R as base
4) It is pnp transistor with R as collector

135. Match the following Column I with Column II

Column I | **Column II**

i n -p-n transistor — A 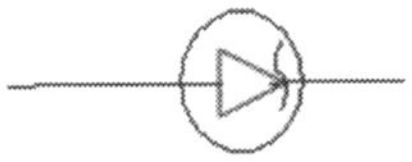

ii p-n-p transistor — B 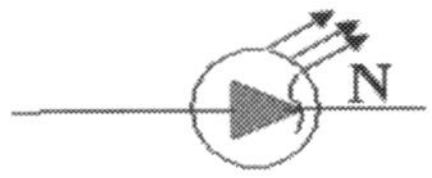

iii Light emmiting diode — C 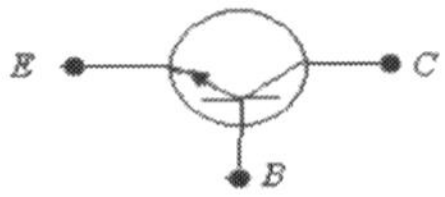

iv Zener diode — D 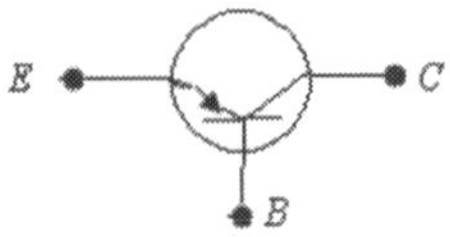

1) I-A,II-B,III-C, IV-D
2) I-D, II-A, III-B, IV-C
3) I-C, II-D, III-B, IV-A
4) I-B, II-A, III-C, IV-D

136. Match List-I with List-2

List-1	**List-2**
A) Emitter	E) Transistor
B) Base	F) Moderately doped
C) Collector	G) lightly doped
D) Transfer of	H) Heavily doped resistance
	I) Largest physical size

	A	B	C	D		A	B	C	D
1)	F	E	H,I	G	2)	G	F,I	E	H
3)	H	G	F,I	E	4)	E	H,I	G	F

137. Match the following for C E transistor amplifier

LIST - I	**LIST - II**
a. Current gain	**e.** $\frac{\Delta I_c}{\Delta I_b}$
b. Voltage gain	**f.** $\frac{\Delta I_c}{\Delta I_b}\frac{R_L}{R_i}$
c. Power gain	**g.** $\left(\frac{\Delta I_c}{\Delta I_b}\right)^2 \frac{R_L}{R_i}$
d. Resistance gain	**h.** $\frac{R_L}{R_i}$

1) a-e, b-f, c-g, d-h
2) a- f, b- g, c- h, d-e
3) a-g, b-h, c-e, d-f
4) a- h, b- e, c- f, d- g

138. Assertion (A) :- Common emiter mode of a transistor is widely used

Reason (R) :-Current gain, voltage gain, and power gain are maximum in C.E mode of a transistors.

1) If both A and R are correct and R is the correct explanation of A
2) If both A and R are correct and R is not the correct explanation of A
3) If 'A' is correct and 'R'is incorrect
4) If 'A' is incorrect and 'R'is correct

139. *Assertion (A)* :- Transistor in C.E. mode can be used as amplifier

***Reason (R)* :- A small change in base current produces a relatively large change in collector current**

1) If both A and R are correct and R is the correct explanation of A
2) If both A and R are correct and R is not the correct explanation of A
3) If 'A' is correct and 'R'is incorrect
4) If 'A' is incorrect and 'R'is correct

LOGIC GATES

140. In the Binary number system the number 100 represents :

1) one
2) three
3) four
4) hundred

141. Among the following is not the function of NOT gate is

1) stop a signal
2) invert an input signal
3) complement a signal
4) change the logic in a digital circuit

142. Digital circuit can be made by repetitive use of

1) OR gates 2) AND gates 3) NOT gates 4) NAND gates

143. Among the following one gives output 1 in the AND gate

1) A = 0, B = 0 2) A = 1, B = 1 3) A = 1, B = 0 4) A = 0, B = 1.

144. Person who use Boolean algebra for describing the operation of logic gates first was

1) Boole 2) Shannon 3) Schottky 4) Zener

145. NAND and NOR gates are called universal gates because they

1) are universally available

2) can be combined to produce OR, AND and NOT gates

3) are widely used in the Integrated circuits 4) can be easily manufactured.

146. In positive logic, the logic state 1 corresponds to

1) positive voltage 2) zero voltage

3) lower voltage level 4) higher voltage level.

147. In Boolean algebra A + B = Y implies that :

1) sum of A and B is Y

2) Y exists when A exists or B exists or both A and B exist

3) Y exists only when A and B both exist

4) Y exists when A or B exist but not when both A and B exist.

148. In the Boolean algebra, the following one is wrong

1) 1+0 = 1 2) 0+1=1 3) 1+1=1 4) 0+0 =1

149. In Boolean algebra A.B = Y implies that:

1) product of A and B is Y 2) Y exists when A exists or B exists

3) Y exists when both A and B exist but not when only A or B exists

4) Y exists when A or B exists but not both A and B exist.

150. In the Boolean algebra, the following one is wrong

1) 1.0 = 0 2) 0.1 = 0 3) 1.1 = 0 4) 1.1 = 1

151. The following truth table is for

A	B	Y
1	1	0
1	0	1
0	1	1
0	0	1

1) NAND gate 2) AND gate 3) XOR gate 4) NOT gate

152. The output of a 2-input OR gate is zero only when its

1) both inputs are 0 2) either input is 1 3) both inputs are 1 4) either input is zero.

153. Boolean algebra is essentially based on

1) symbols 2) logic 3) truth 4) numbers

154. The value of $\overline{A}+A$ in the Boolean algebra is

1) A 2) $\overline{A}$ 3) 0 4) 1

155. The value of $A.\overline{A}$ in Boolean algebra is

1) 0 2) 1 3) A 4) $\overline{A}$

156. The following is NOT equal to 0 in the Boolean algebra is

1) $\overline{A.0}$ 2) $A.\overline{A}$ 3) A.0 4) $\overline{A+\overline{A}}$

157. An AND gate is followed by a NOT gate in series. With two inputs A & B, the Boolean expression for the out put Y will be :

1) A.B 2) A + B 3) $\overline{A+B}$ 4) $\overline{A.B}$

158. NOR gate is the series combination of

1) NOT gate followed by OR gate
2) OR gate followed by NOT gate
3) AND gate followed by OR gate
4) OR gate followed by AND gate

159. The gate that has only one input terminal

1) NOT 2) NOR 3) NAND 4) XOR

160. AND gate:

1) It has no equivalence to switching circuit.
2) It is equivalent to series switching circuit.
3) It is equivalent to parallel switching circuit
4) It is a mixture of series and parallel switching circuit

161. The gate that can act as a building block for the digital circuits is

1) OR 2) NOT 3) AND 4) NAND

162. Statement -1: NAND or NOR gates are called digital building blocks.
Statement-2: The repeated use of NAND (or NOR) gates can produce all the basis or complicated gates.

1) Statement -1 is false, statement -2 is true
2) statement 1-is true statement -2 is true statement -2 is correct explanation of statement-1.
3) statement 1-is true statement -2 is true statement -2 is not correct explanation of statement-1.
4) statement 1-is true statement -2 is false

163. Statement-1: NOT gate is also called invertor circuit
Statement-2: NOT gate inverts the input order.

1) Statement -1 is false, statement -2 is true
2) statement 1-is true statement -2 is true statement -2 is correct explanation of statement-1.
3) statement 1-is true statement -2 is true statement -2 is not correct explanation of statement-1.
4) statement 1-is true statement -2 is false

164. Match the quantities of column I with the column II when A=1 and B=0

Column - I	**Column - II**
a) $(A.B)+\overline{A}$	**(p) A**
b) $(A.\overline{B})+A$	**(q) B**
c) $(\overline{A}+B).A$	**(r) A + B**
d) $(A+\overline{B}).B$	**(s) A.B**

1) a-q,s b-p,r, c-q,s, d-q,s
2) a-p,s, b-p,s, c-q,r, d-r,q
3) a-q,r, b-p,s, c-q,r, d-s,p
4) a-r,s, b-p,r, c-q,s, d-p,s

KEY - CONCEPTUAL QUESTIONS

01)1	**02)3**	**03)3**	**04)2**	**05)2**	06)2	07)2	08)1	09)3	10)3
11)4	12)1	13)4	14)4	15)1	16)1	17)1	18)2	19)4	20)2
21)3	22)3	23)1	24)3	25)3	26)3	27)1	28)1	29)3	30)3
31)3	32)3	33)4	34)1	35)4	36)3	37)1	38)1	39)1	40)2
41)1	42)2	43)2	44)3	45)1	46)2	47)3	48)1	49)1	50)4
51)3	52)3	53)3	54)2	55)4	56)2	57)2	58)4	59)1	60)4
61)1	62)1	63)1	64)4	65)1	66)1	67)3	68)3	69)2	70)4
71)1	72)2	73)2	74)3	75)3	76)3	77)1	78)1	79)2	80)3
81)3	82)3	83)2	84)2	85)1	86)1	87)3	88)1	89)2	90)2
91)3	92)1	93)4	94)1	95)1	96)1	97)2	98)1	99)1	100)4
101)1	102)1	103)4	104)2	105)1	106)2	107)3	108)4	109)1	110)3
111)2	112)2	113)1	114)3	115)3	116)1	117)1	118)4	119)2	120)3
121)3	122)1	123)1	124)2	125)4	126)1	127)1	128)4	129)2	130)3
131)3	132)3	133)2	134)3	135)3	136)3	137)1	138)1	139)1	140)3
141)1	142)4	143)2	144)2	145)2	146)4	147)2	148)4	149)3	150)3
151)1	152)1	153)2	154)4	155)1	156)1	157)4	158)2	159)1	160)2
161)4	162)2	163)2	164)1						

LEVEL - I -1

INTRINSIC, EXTRINSIC SEMICONDUCTORS AND DIODES

1. **The electrical conductivity of a semiconductor increases when electomagnetic radiation of wavelength shorter than 2480nm is incident on it. The band gap in (eV) for the semiconductor is**

   1) $0.7eV$ 2) $0.5eV$ 3) $2.5eV$ 4) $1.2eV$

2. **Pure Si at 300 K has equal electron (n_i) concentrations of $1.5\times10^{16}m^{-3}$. Doping by indium increases n_h $4.5\times10^{22}m^{-3}$. n_e in the doped Si is**

   1) 5×10^9 2) 7×10^9 3) 9×10^9 4) 8×10^9

3. **In a p-n junction the depletion region is 400nm wide and electric field of $5\times10^5 Vm^{-1}$ exists in it. The minimum energy of a conduction electron, which can diffuse from n-side to the p-side is**

   1) 4eV 2) 5eV 3) 0.4eV 4) 0.2eV

4. **The reverse bias in a junction diode is changed from 5V to 15V then the value of current changes from 38 μ A to 88 μ A. The resistance of junction diode will be**

   1) $4\times10^5\Omega$ 2) $3\times10^5\Omega$ 3) $2\times10^5\Omega$ 4) $10^6\Omega$

5. **A diode made of silicon has a barrier potential of 0.7V and a currrent of 20mA passes through the diode when a battery of emf 3V and a resistor is connected to it. The wattage of the resistor and diode are**

   1) 0.046W, 0.014W 2) 4.6W, 0.14W 3) 0.46W, 0.14W 4) 46W, 14W

6. **In a half wave rectifier output is taken across a 90 ohm load resistor. If the resistance of diode in forward biased condition is 10ohm, the efficiency of rectification of ac power into dc power is**

1) 40.6% 2) 81.2% 3) 73.08 % 4) 36.54%

7. **In a full wave rectifier output is taken across a load resistor of 800 ohm. If the resistance of diode in forward biased condition is 200 ohm, the efficiency of rectification of ac power into dc power is**

1) 64.96% 2) 40.6% 3) 81.2% 4) 80%

TRANSISTORS

8. **In a P-N-P transistor, the collector current is 10 mA. If 90% of the holes reach the collector, then emitter current will be:**

1) 13 mA 2) 12 mA 3) 11 mA 4) 10 mA

9. **A transistor has a base current of 1mA and emitter current 100mA. The current transfer ratio will be**

1) 0.9 2) 0.99 3) 1.1 4) 10.1

10. **When the base -emitter voltage of a transistor connected in the common -emitter mode is changed by 20 mV the collector current is changed by 25 mA. Find the transconductance.**

1) $1.25\Omega^{-1}$ 2) $2.50\Omega^{-1}$ 3) $0.5\Omega^{-1}$ 4) $5.5\Omega^{-1}$

11. **In a transistor circuit the base current changes from 30 μA to 90 μA. If the current gain of the transistor is 30, the change in the collector current is**

1) 4 mA 2) 2 mA 3) 3.6 mA 4) 1.8 mA

12. **The current gain of transistor in a common emitter circuit is 40. The ratio of emitter current to base current is**

1) 40 2) 41 3) 42 4) 43

13. **In a common base configuration the emitter current changes by 5mA when emitter voltage is changed by 200 mV at a fixed collector to base voltage. The input resistance is**

1) $40\,\Omega$ 2) $1000\,\Omega$ 3) $2.5\,\Omega$ 4) $4\,\Omega$

14. **For a common base amplifier, the values of resistance gain and voltage gain are 3000 and 2800 respectively. The current gain will be**

1) 0.93 2) 0.83 3) 0.73 4) 0.63

15. **In a transistor amplifier $\beta = 62$, $R_L = 5000\,\Omega$ and internal resistance of the transistor is $500\,\Omega$. Its power amplification will be**

1) 25580 2) 33760 3) 38440 4) 55280

LOGIC GATES

16. **Decimal number 15 is equivalent to the binary number:**

1) 110001 2) 000101 3) 101101 4) 001111

17. **Binary number 1001001 is equivalent to the decimal number:**

1) 37 2) 73 3) 41 4) 32

18. **In the Binary number system 1+1=**

1)2 2)1 3)10 4)100

19. **If A = B = 1, then in terms of Boolean algebra the value of A.B + A is not equal to**

1) B.A+B 2) B+A 3) B 4) $\overline{A}.B$

20. In the Boolean algebra, the following one which is not equal to A is

1) A.A 2) A+A 3) $\overline{A}.A$ 4) $\overline{\overline{A}+\overline{A}}$

21. The logic expression which is NOT true in Boolean algebra is

1) $[\bar{1}+\bar{1}].1=0$ 2) $[\bar{1}+0].1=0$ 3) $[\bar{1}+0].\bar{1}=0$ 4) $[1+1].1=0$

KEY LEVEL - I -1

1)2	2)1	3)4	4)3	5)1	6)4	7)1	8)3	9)2	10)1
11)4	12)2	13)1	14)1	15)3	16)4	17)2	18)3	19)4	20)3
21)4									

HINTS LEVEL - I -1

1. $$E_g=\frac{hc}{\lambda}=\frac{6.6\times10^{-34}\times3\times10^{8}}{2480\times10^{-9}\times(1.6\times10^{-19})}=0.5eV$$

2. We know that ofr a doped semiconductor in thermal equilibrium, we have

$n_en_h=n_i^2$

As per given data, $n_i=1.5\times10^{16}m^{-3}$

$n_h=4.5\times10^{22}m^{-3}$

Thus $n_e=\frac{n_i^2}{n_h}=\frac{(1.5\times10^{16})^2m^{-6}}{4.5\times10^{22}m^{-3}}=5.0\times10^{9}m^{-3}$

3. $W=Vq$ =Edq

$=5\times10^5\times400\times10^{-9}\times1.6\times10^{-19}J$

$=\frac{5\times10^5\times400\times10^{-9}\times1.6\times10^{-19}J}{1.6\times10^{-19}}eV$

= 2000 X 10^{-4} eV = 0.2 eV

4. $$R=\frac{\Delta V}{\Delta I}=\frac{15-5}{(88-38)\times10^{-6}}=\frac{10}{50\times10^{-6}}=\frac{10^6}{5}=\frac{10}{5}\times10^5=2\times10^5\Omega$$

5. P = VI (for resistor)

$=(3-0.7)20\times10^{-3}=2.3\times20\times10^{-3}=46\times10^{-3}=0.046W$

for diode, $P=0.7\times20\times10^{-3}=0.014W$

6. $$\eta=\frac{0.406R_L}{r_f+R_L}$$

7. $$\eta=\frac{0.812R_L}{r_f+R_L}$$

8. $I_C = \frac{90}{100} I_E \Rightarrow I_E = \frac{10}{9} I_C$

9. $\alpha = \frac{I_c}{I_e}, I_e = I_b + I_c$

10. $g_m = \frac{\Delta I_c}{\Delta V_{BE}} = \frac{25\times10^{-3}}{20\times10^{-3}} = 1.25\Omega^{-1}$

11. $\Delta I_B = \Delta I_E - \Delta I_C$ and $\beta = \frac{\Delta I_C}{\Delta I_B}$

12. $\beta = \frac{I_C}{I_B}$ and $\frac{I_E}{I_B} = \frac{I_C + I_B}{I_B} = \frac{I_C}{I_B} + 1 = \beta + 1$

13. $V_{BE} = I_e R_i$

14. $A_v = A_R \beta$

15. $A_P = \beta^2 \frac{R_L}{R_i}$

16.
$$\begin{array}{r|l} 2 & 15 \\ \hline 2 & 7-1 \\ \hline 2 & 3-1 \\ \hline 2 & 1-1 \\ \hline & 0-1 \end{array}$$

$\therefore 15_{(10)} = 1111_{(2)}$

17. $\overset{6}{1}\overset{5}{0}\overset{4}{0}\overset{3}{1}\overset{2}{0}\overset{1}{0}\overset{0}{1}_{(2)} = 1\times2^6 + 0\times2^5 + 0\times2^4 + 1\times2^3$

$+0\times2^2 + 0\times2^1 + 1\times2^0 = 73$

18. 1+1=10 in binary number system

19. If A=B=1; then A.B+A=1.1+1=1

then $\overline{A}.B = \overline{1}.1 = 0.1 = 0$

20. $\overline{A} \bullet A = 0$

21. logic, OR and AND operation

LEVEL - I -2

INTRINSIC, EXTRINSIC SEMICONDUCTORS AND DIODES

1. **The electrical conductivity of a semi conductor increases when electromagnetic radiation of wavelength shorter than 1240 nm is incident on it. The forbidden band energy for the semi conductor is (in eV)**

   1) 0.5 2) 0.97 3) 0.7 4) 1.1

2. **A semiconductor is known to have an electron concentration of $5\times10^{13}/cm^3$ and hole concentration of $8\times10^{12}/cm^3$. The semiconductor is**

   1) n-type 2) p-type 3) intrinsic 4) insulator

3. **A potential barrier of 0.5V exists across a p-n junction. If the width of depletion layer is 10^{-6}m, then intensity of electric field in this region will be**

   1) 1×10^6 V/m 2) 5×10^5 V/m 3) 4×10^4 V/m 4) 2×10^6 V/m

4. **A p - n junction diode has breakdown voltage of 28V. If applied external voltage in reverse bias is 40V the current through it is**

   1) Zero 2) infinite 3) 10 A 4) 15A

5. **The value of current in the following diagrams is (diode assumed to be ideal one)**

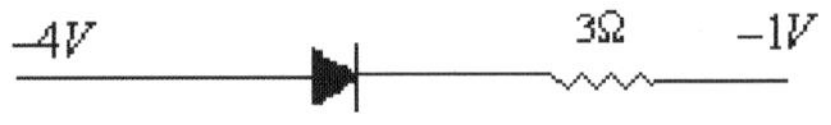

   1) 0.1 amp 2) 0.01amp 3) 1 amp 4) zero

6. **A half -wave rectifier is used to convert 50Hz A.C. to D.C. voltage. The number of pulses per second in the rectified voltage are**

   1) 50 2) 25 3) 100 4) 75

7. **If a full wave rectifier circuit is operating from 50Hz mains, the fundamental frequency in the ripple will be**

   1) 25 Hz 2) 50 Hz 3) 70.7 Hz 4) 100 Hz

TRANSISTORS

8. **In an npn transistor the base and collector currents are 100 μ A and 9mA respectively. Then the emitter current will be**

   1) 9.1mA 2) 18.2mA 3) 3.91 μ A 4) 18.2 μ A

9. **A change of 8mA in the emitter current brings a change of 7.9mA in the collector current. The change in base current required to have the same change in the collector current is**

   1) 0.01mA 2) 1A 3) 10mA 4) 0.1mA

10. For a p-n-p transistor in CB configuration, the emitter current I_E **is 1mA and α = 0.95. The base current and collector current are**

    1) 0.95 mA, 0.05mA 2) 0.05mA, 0.95mA 3) 9.5mA, 0.5mA 4) 0.5mA, 9.5mA

11. **If a change of 100 μ A in the base current of an n-p-n transistor in CE causes a change of 10mA in the collector current, the ac current gain of the transistor is**

    1) 10 2) 100 3) 1000 4) 10000

12. **For a common emitter amplifier, current gain is 70. If the emitter current is 8.4mA, then the base current is**

    1) 0.236mA 2) 0.118mA 3) 0.59mA 4) 8.3mA

13. **The base current of a transistor is 105 μ A and the collector current is 2.05mA. Then β of the transistor is**

1) 1.952 2) 19.52 3) 195.2 4) 1952

14. **For a transistor the value of α is 0.9. Its β value is**

1) 9 2) 0.9 3) 0.09 4) 90

15. **For a transistor the current amplification factor is 0.8. The transistor is connected in common emitter configuration, the change in collector current when the base current changes by 6mA is**

1) 6mA 2) 4.8mA 3) 24mA 4) 8mA

16. **A change of 400mV in base-emitter voltage causes a change of 200μ A in the base current. The input resistance of the transistor is**

1) $1K\Omega$ 2) $6K\Omega$ 3) $2K\Omega$ 4) $8K\Omega$

17. **In a common base circuit, if the collector base voltage is changed by 0.6V, collector current changes by 0.02mA. The output resistance will be**

1) $10^4\Omega$ 2) $2 \times 10^4\Omega$ 3) $3 \times 10^4\Omega$ 4) $4 \times 10^4\Omega$

18. **A common emitter transistor amplifier has a current gain of 50. If the load resistance is 4k Ω, and input resistance is 500 Ω, the voltage gain of amplifier is**

1) 100 2) 200 3) 300 4) 400

LOGIC GATES

19. **Equivalent of decimal number 8 in the binary number is**

1)10 2)101 3)1000 4)1011

20. **The equivalent of 110in the decimal number is**

1)2 2)4 3)8 4)6

21. **If A = 1, B = 0 then the value of $\overline{A}+B$ in terms of Boolean algebra is**

1) A 2) B 3) B + A 4) $A.\overline{B}$

22. **In the Boolean algebra : A + B =**

1) $\overline{A}+\overline{B}$ 2) A.B 3) $\overline{\overline{A}}+\overline{\overline{B}}$ 4) $\overline{\overline{A}}+\overline{B}$

23. **The following one represents logic addition is**

1) 1 + 1 = 2 2) 1 + 1 = 10 3) 1 + 1 = 1 4) 1 + 1 = 11

24. **In the Boolean algebra : $\overline{A}.\overline{B}$ = ...**

1) $\overline{A+B}$ 2) A.B 3) $\overline{\overline{A}+\overline{B}}$ 4) A + B

25. **In the Boolean algebra, which gate is expressed as $Y=\overline{A+B}$**

1) OR 2) NAND 3) AND 4) NOR

26. **The truth table for NOT gate is**

1) $\begin{bmatrix} 1 & 1 \\ 0 & 0 \end{bmatrix}$ 2) $\begin{bmatrix} 1 & 0 \\ 0 & 0 \end{bmatrix}$ 3) $\begin{bmatrix} 1 & 0 \\ 0 & 1 \end{bmatrix}$ 4) $\begin{bmatrix} 0 & 1 \\ 1 & 1 \end{bmatrix}$

KEY LEVEL - I -2

1) 2	2)1	3)2	4)2	5)4	6)1	7)4	8)1	9)4	10)2
11)2	12)2	13)2	14)1	15)3	16)3	17)3	18)4	19)3	20)4
21)2	22)3	23)3	24)1	25)4	26)3				

HINTS LEVEL - I -2

1. $E = \frac{hc}{\lambda} = \frac{12000}{12400 A^{0}}(e_{v})$

2. $n_e > n_h$

3. $E = \frac{V}{d}$

$\frac{10^{24}}{10^{18}} = 10^{6}$

4. Since $V_a > V_b$

5. diode does not conduct in reverse bias as $V_p < V_n$

6. n Hz

7. 2n Hz

8. $I_e = I_b + I_c$

9. $\Delta I_B = \Delta I_E - \Delta I_C$

10. $\alpha = \frac{I_c}{I_e}, I_e = I_b + I_c$

11. $\beta = \frac{\Delta I_c}{\Delta I_b}$

12. $\beta = \frac{I_c}{I_b}$ and $I_e = I_b + I_c$

13. $\beta = \frac{I_c}{I_b}$

14. $\beta = \frac{\alpha}{1-\alpha}$

15. $\beta = \frac{\alpha}{1-\alpha}$ and $\beta = \frac{\Delta I_c}{\Delta I_b}$

16. $\Delta V_{BE} = \Delta I_e \times R_i$

17. $V_{CB} = I_C \times R_{out}$

18. $A_V = \beta \frac{R_L}{R_i}$

19.
$$\begin{array}{r|l} 2 & 8 \\ \hline 2 & 4-0 \\ \hline 2 & 2-0 \\ \hline 2 & 1-0 \\ \hline & 0-1 \end{array}$$
$\therefore 8_{(10)} = 1000_{(2)}$

20. $\overset{2\,1\,0}{110}_{(2)} = 1\times 2^2 + 1\times 2^1 + 0\times 2^0 = 6$

21. If A=1, B=0 then $\overline{A} + B = \overline{1} + 0 = 0 + 0 = 0 = B$

22. $\overline{\overline{A}} = A$ and $\overline{\overline{B}} = B$

23. acording to Boolean algebra, 1+1=1

24. According to Demorgon's theorem $\overline{A} \bullet \overline{B} = \overline{A+B}$

25. $Y = \overline{A+B}$ is the output of NOR gate

26. conceptual

LEVEL - II -1

INTRINSIC,EXTRINSIC SEMI CONDUCTORS AND DIODES

1. Mobilities of electrons and holes in a sample of intrinsic germanium at room temperature are $0.36 m^2/Vs$ and $0.17 m^2/Vs$. The electron and hole densities are each equal to 2.5 $\times 10^{19} m^{-3}$. The electrical conductivity of germanium is

1) $0.47\, S/m$ 2) $5.18\, S/m$ 3) $2.12\, S/m$ 4) $1.09\, S/m$

2. In a p-n junction diode the thickness of deplection layer is $2\times 10^{-6} m$ and barrier potential is 0.3V. The intensity of the electrical field at the junction is

1) $0.6\times 10^{-6} Vm^{-1}$ from n to p side
2) $0.6\times 10^{-6} Vm^{-1}$ from p to n side
3) $1.5\times 10^{5} Vm^{-1}$ from n to p side
4) $1.5\times 10^{5} Vm^{-1}$ from p to n side

3. A potential barrier V volts exists across a P-N junction. The thickness of the depletion region is 'd'. An electron with velocity 'v' approches P-N junction from N-side. The velocity of the electron acrossing the junction is

1) $\sqrt{v^2 + \frac{2Ve}{m}}$ 2) $\sqrt{v^2 - \frac{2Ve}{m}}$ 3) v 4) $\sqrt{\frac{2Ve}{m}}$

4. In the following , reverse biased diode is

1) 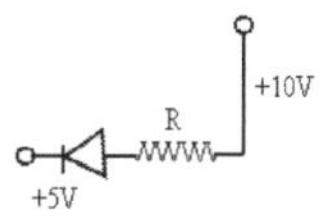

2) 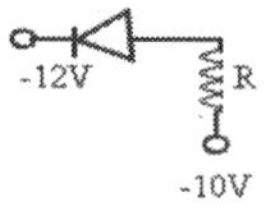

3) 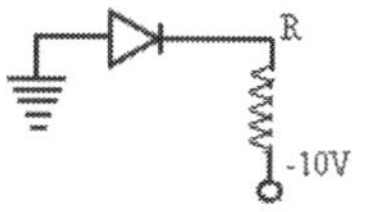

4) 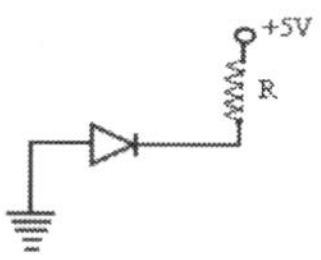

5. Two similar p-n junctions can be connected in three different ways as shown in the figures. The two connections across which the potential difference is same are

a) 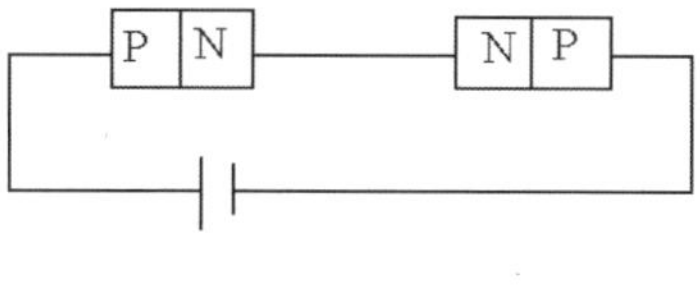

b) 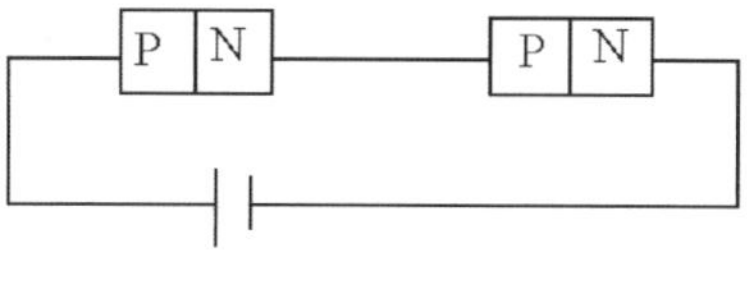

c) 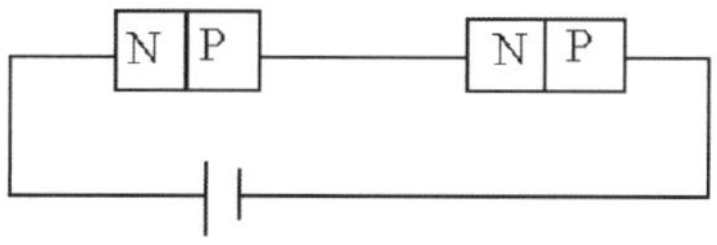

1) circuits a and b 2) circuits b and c 3) circuits a and c 4) all the circuits

6. If $V_1 > V_2$, r is resistance offered by diode in forward bias then current through the diode is

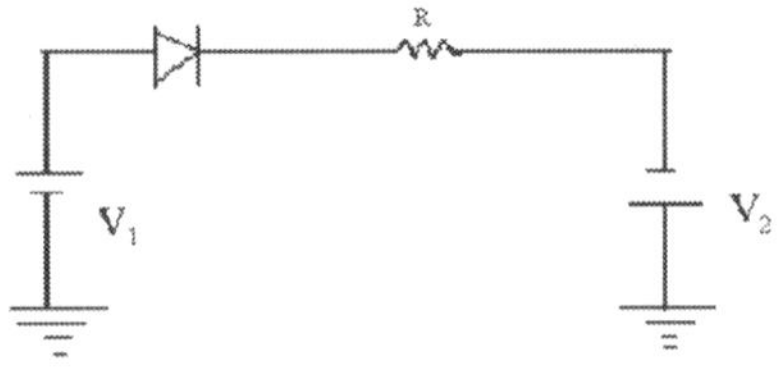

1) 0

2) $\frac{V_1 + V_2}{R + r}$

3) $\frac{V_1 - V_2}{R + r}$

4) None

7. A PN junction diode when forward biased has a drop of 0.5V which is assumed to be independent of current. The current in excess of 10mA through the diode produces large joule heating which damages the diode. If we want to use a 1.5V battery to forward bias the diode, the resistor used in series with the diode so that the maximum curent does not exceed 5mA is

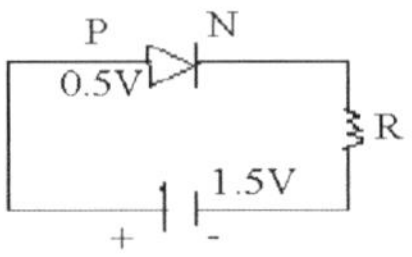

1) $2 \times 10^2 \Omega$

2) $2 \times 10^5 \Omega$

3) $2 \times 10^3 \Omega$

4) $2 \times 10^4 \Omega$

8. The circuit shown in figure (1) Contains two diodes each with a forward resistance of 50 ohm and with infinite reverse resistance. If the battery voltage is 6 V, the current through the 100 ohm resistance is.

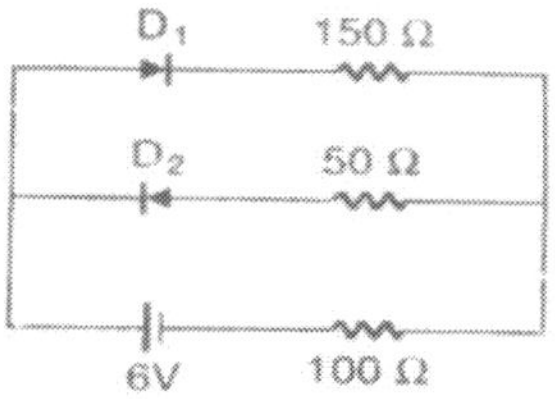

1) 0.01A

2) 0.02A

3) 0.03A

4) 0.04A

9. 4 ideal diodes are connected as shown in the circuit the current through 50 Ω is

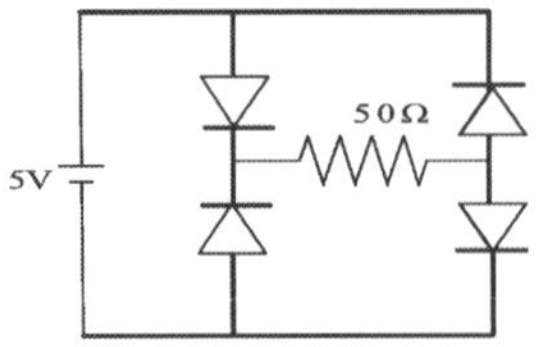

1) 0.1 A

2) 0.5 A

3) 0.6 A

4) 1 A

10.. Find the effective resistance between A & B

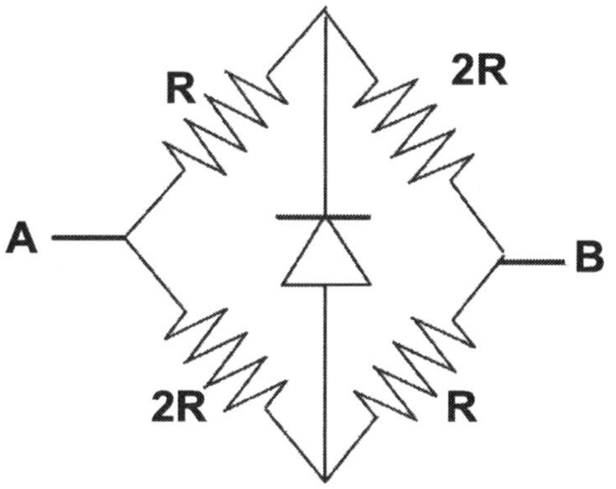

1) $\frac{2}{3R}$ 2) $\frac{3R}{2}$ 3) $\frac{2R}{3}$ 4) R

11 The equivalent resistance of the circuit across AB is given by

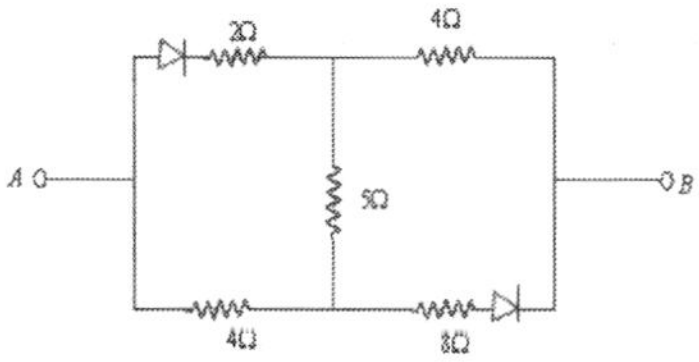

1) 4Ω 2) 13Ω 3) 4Ω *or* 13Ω 4) 4Ω zero

12. The equivalent resistance between A and B is

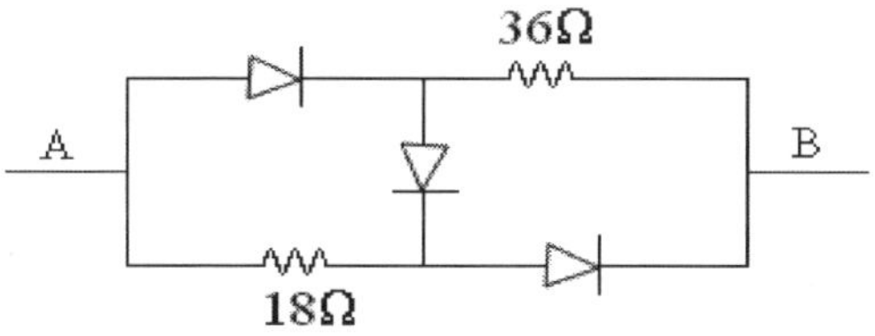

1) $36\Omega\, if\ V_A > V_B$ 2) $18\Omega\ if\ V_A < V_B$

3) Zero if $V_A < V_B$ and $54\Omega\, if\ V_A > V_B$ 4) Zero if $V_A > V_B$ and $54\Omega\, if\ V_A < V_B$

13. A junction diode is connected to a 10 V source and $10^3\ \Omega$ rheostat. The slope of load line on the characteristic curve of diode by (in A/V)

1) 10^{-1} 2) 10^{-2} 3) 10^{-3} 4) 10^{-4}

14. A p–n junction (D) shown in the figure can act as a rectifier. An alternating curent source (V) is connected in the circuit

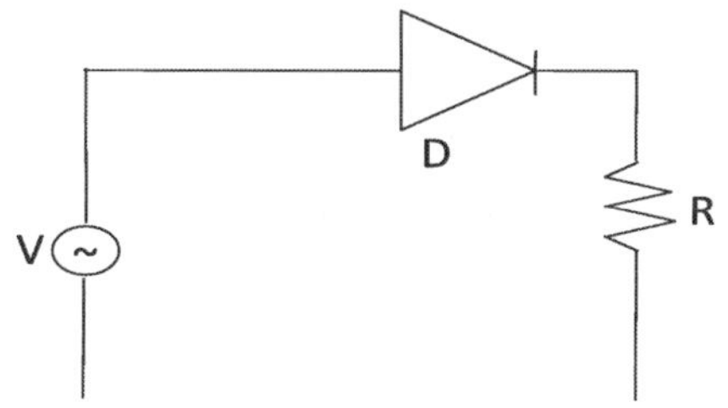

1) 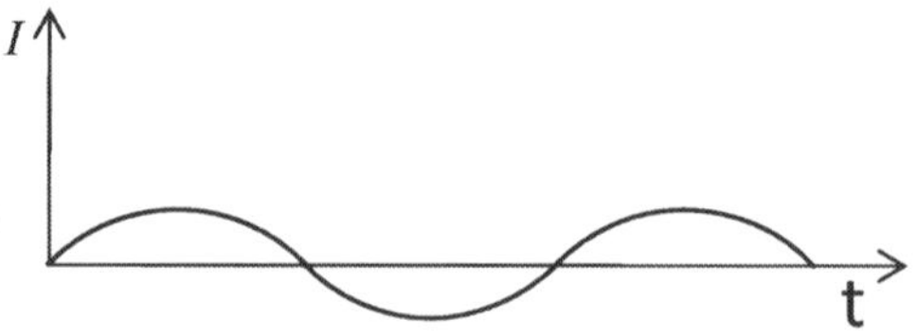

2) 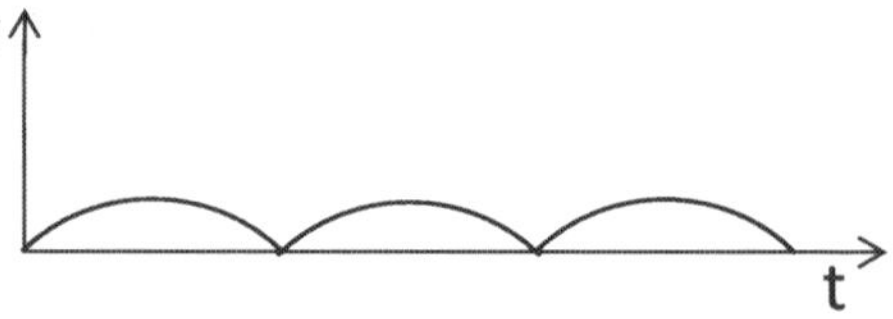

3) 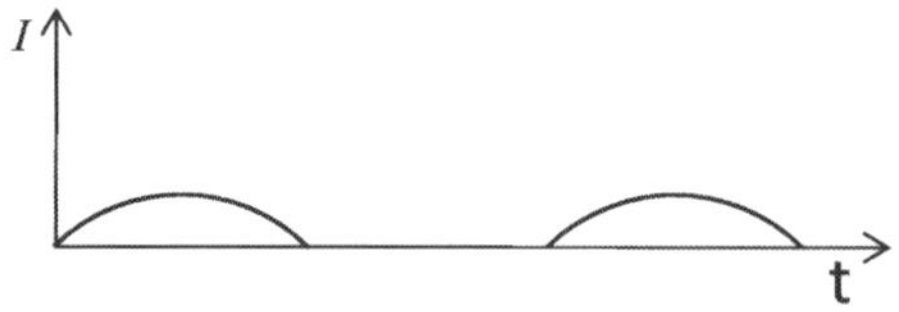

4) 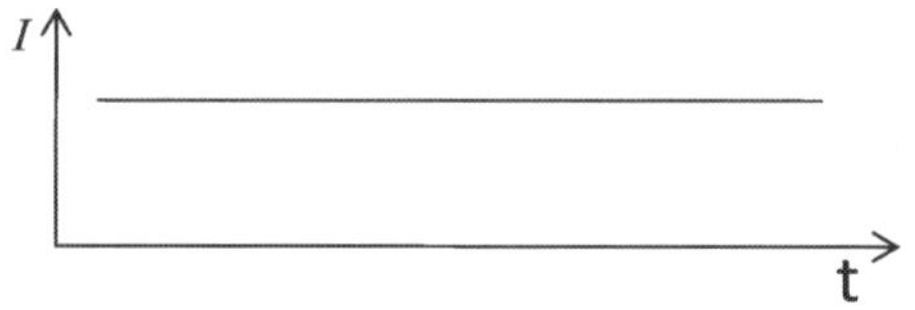

15. In the figure, the input is across terminals A and C and the output is across B and D. Then the output is

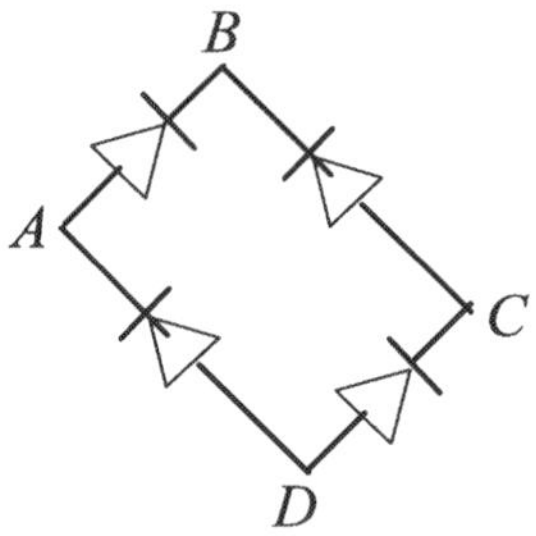

1) Zero
2) Same as the input
3) Full wave rectified
4) Half-wave rectified

16. The I-V characteristic of an LED is

1) 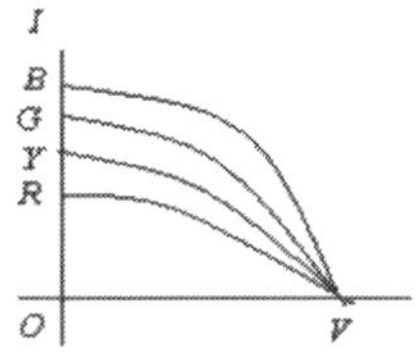

2) 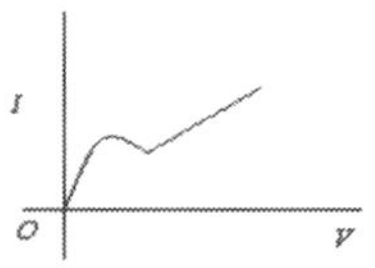

3) 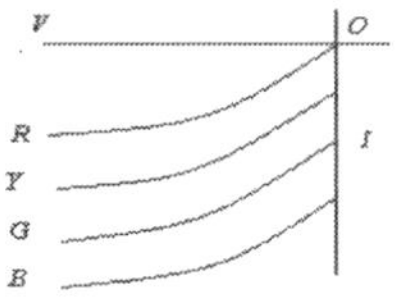

4) 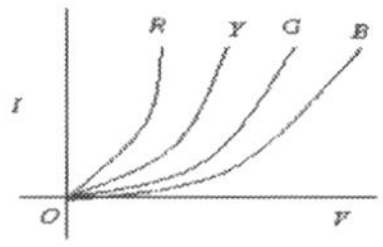

17. In the figure shown, the potential drop across each capacitor is (assume diodes to be ideal)

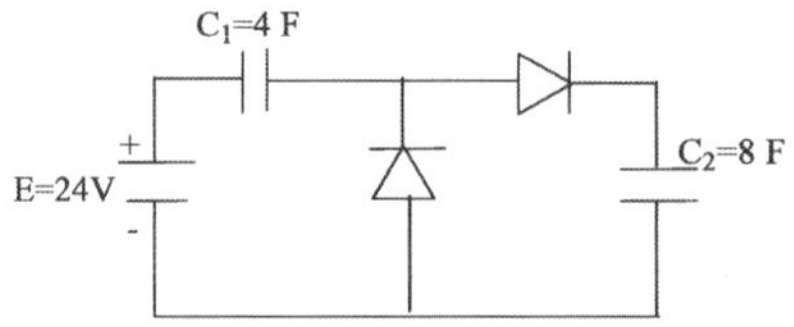

1) 12V, 12V
2) 16V, 8V
3) zero, 24V
4) 8V, zero

18. From the circuit shown below, the maximum and minimum values of zener diode current are

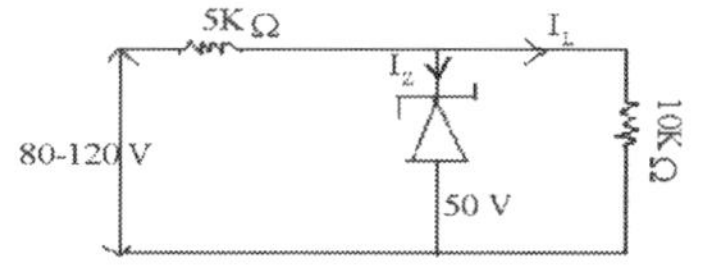

1) 6mA, 5mA 2) 14mA, 5mA 3) 9 mA, 1mA 4) 3mA, 2mA

TRANSISTORS

19. In an n-p-n transistor circuit, the collector current is 10mA. If 90% of the electrons emitted reach the collector.

1) the emitter current , the will be 9mA
2) the base current will be 1 mA
3) the emitter current will be 11mA
4) both 2 & 3

20. The constant α of a transistor is 0.9. What would be the change in the collector current corresponding to a change of 4 mA in the base current in a common emitter arrangement ?

1) 30mA 2) 63mA 3)36 mA. 4) 3.6 mA

21. A voltage amplifier operated from a 12 volt battery has a collector load $6k\Omega$. Calculate the maximum collector current in the circuit.

1) 0.5mA 2) 1 mA 3) 3 mA 4) 2mA

22. In a transistor circuit shown here the base curent is $35\mu A$ The value of the resistor R_b is

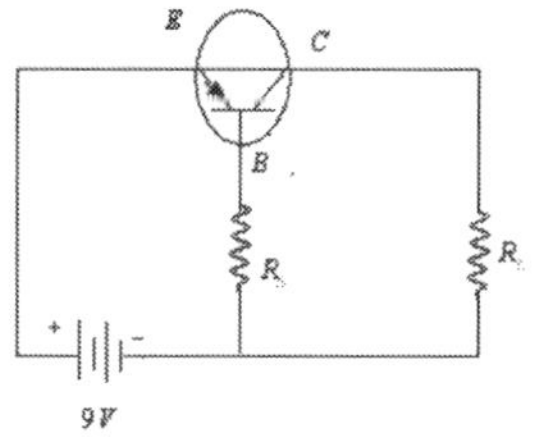

1) $124k\Omega$ 2) $257k\Omega$ 3) $352k\Omega$ 4) None of these

23. In a single state transistor amplifier, when the signal changes by 0.02 V, the base current change by $10\mu A$ and collector current by 1mA. If collector load $R_C = 2k\Omega$ and $R_L = 10k\Omega$, Calculate Current Gain, Input impedance, Effective a.c.load, Voltage gain and Power gain.

1)50, 2kΩ,1.66kΩ, 83, 8300
2) 100, 1kΩ,1.66kΩ, 83, 8300
3) 100, 2kΩ,1.66kΩ, 83, 830
4) 100, 2kΩ,1.66kΩ, 83, 8300

LOGIC GATES

24. On subtracting 010101 from 101010, we get:

1) 001011 2) 001100 3) 010101 4) 011111

25. The minimum number of gates required to realise this expression $Z = DABC + D\overline{ABC}$ is

1) One 2) Two 3) Eight 4) Five

26. In the circuit below, A and B represent two inputs and C represents the output, the circuit reperesents

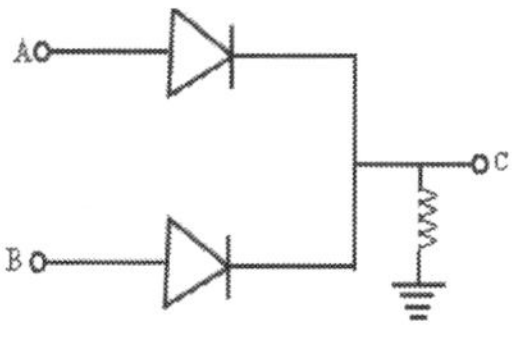

1) NOR gate 2) AND gate 3) gate 4) OR gate

27. **In the following circuit the output Y becomes zero for the inputs**

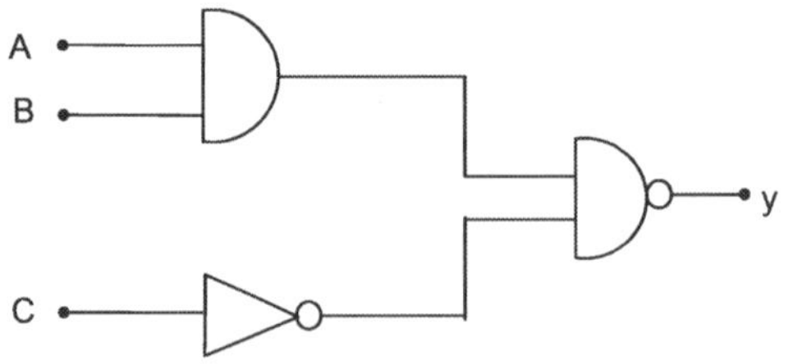

1) $A=1, B=0, C=0$
2) $A=0, B=1, C=1$
3) $A=0, B=0, C=0$
4) $A=1, B=1, C=0$

28. **The combinatin of gates shown below yields**

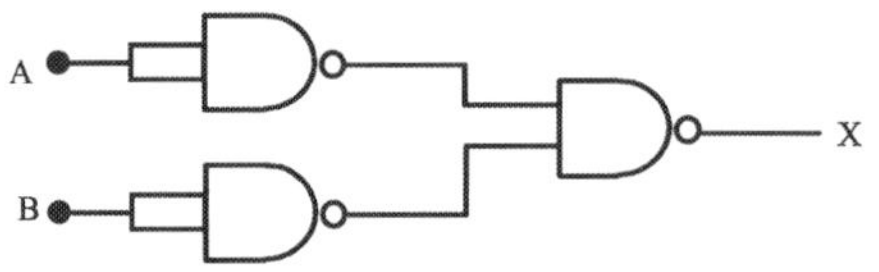

1) NAND gate
2) OR gate
3) NOT gate
4) XOR gate

29. **The logic gate having an output of 1 is**

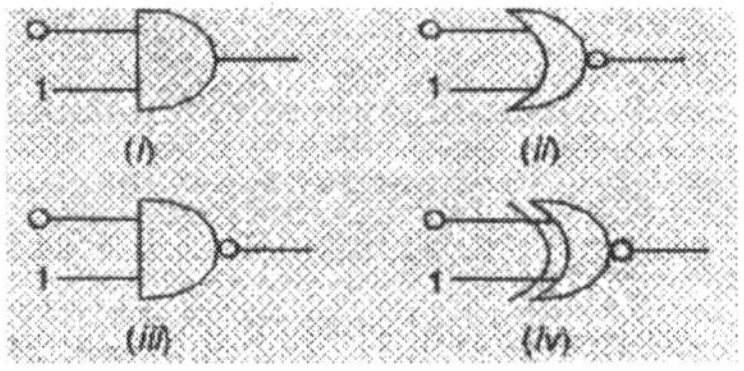

1)(iv)
2)(i)
3)(ii)
4)(iii)

30. **In a given circuit as shown the two input wave forms A and B are applied simultaneously**

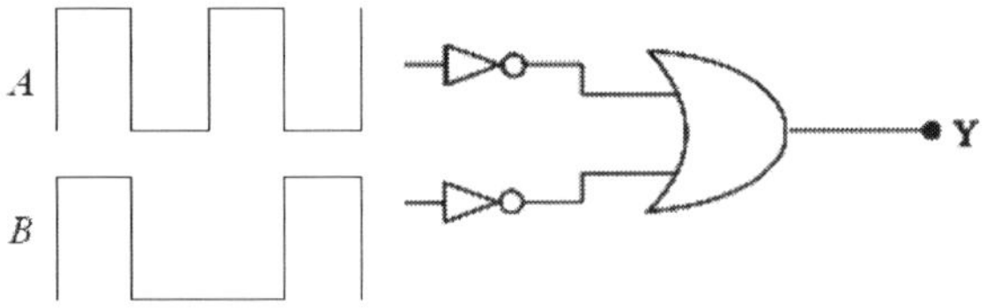

The resultant wave form at Y is

1) 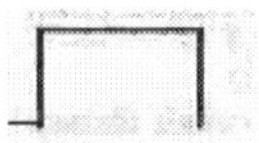
2) 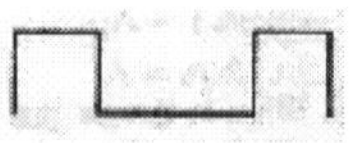
3) 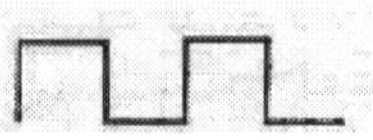
4) 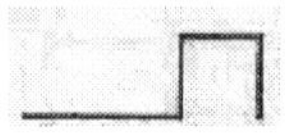

31. **The output of the combination of the gates shown in the figure below is**

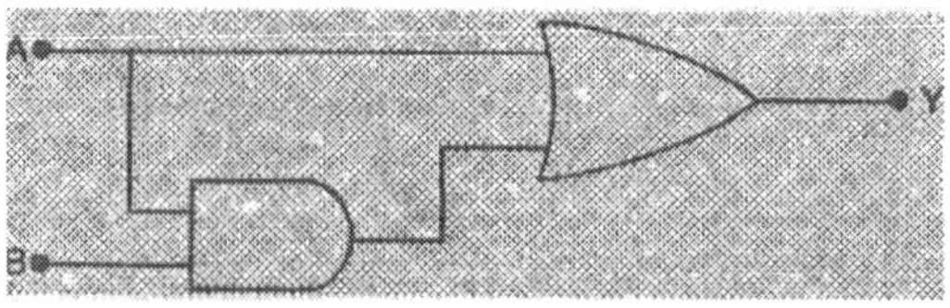

1) A + A. B
2) $(A+B)A+\overline{B}$
3) $(A.B)+(\overline{A.B})$
4) $(A+B)(\overline{A.B})$

32. The expression of Y in following circuit is

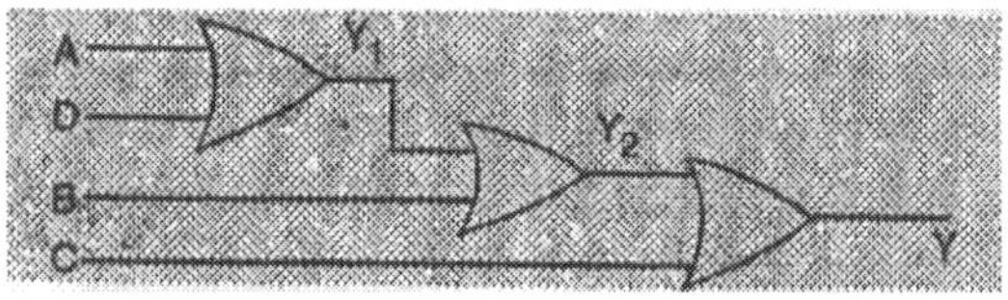

1) ABCD   2) A+BCD   3) A+B+C+D   4) AB+CD

33. How many NAND gates are used in an OR gate ?

1) four   2) two   3) three   4) Five

KEY LEVEL - II

1)3	2)3	3)2	4)4	5)2	6)2	7)1	8)2	9)1	10)2
11)3	12)4	13)3	14)3	15)3	16)4	17)2	18)3	19)4	20)3
21)4	22)2	23)4	24)3	25)1	26)4	27)4	28)2	29)4	30)1
31)1	32)3	33)3							

HINTS LEVEL - II (CW)

1. $\sigma = \frac{1}{\rho} = e(\mu_e n_e + \mu_h n_h)$

2.. $E = \frac{v}{d} = \frac{0.3}{2\times10^{-6}} = 1.5\times10^{5} Vm^{-1}$

Its direction from n to p side.

3. $v^2 - u^2 = 2as$

4. For reverse biasing of an ideal diode, the potential of n-side should be higher then potential of p-side. Only option (4) is satisfying the criterion for reverse biasing.

5. In (1) one is forward biased and the other one is reverse biased.In (2) both are forward biased In (3) both are reverse biased

6. $i = \frac{total\ P.D}{total\ resistence}$

7. $R = \frac{E - V_b}{i} = \frac{1.5 - 0.5}{5\times10^{-3}} = \frac{1}{5\times10^{-3}}$

=0.2′ 10^3 =2′ 10^2W

8. As per given circuit, diode D_1 is forward biased and offers a resistance of 50 ohm. Diode D_2 is reverse biased and as its corresponding resistance is infinite, no current flows through it. Thus the equivalent circuit is as shown in Figure(2). As all the three resistances are in series, the current through then is

$I = \frac{6V}{(50+150+100)\Omega} \quad = \frac{6}{300} A = 0.02A$

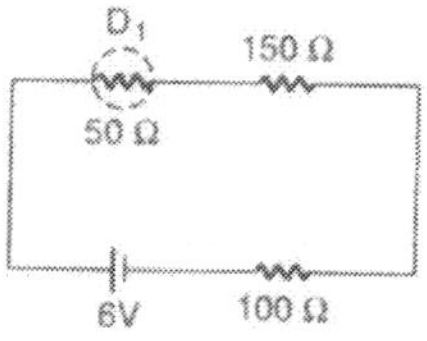

9. $i = \frac{V}{R} = \frac{5}{50}$

10. current does not flow through the diode.

11 If $V_A > V_B$ both the junction diodes are forward biased and the given circuit diagram becomes a balanced Wheatestone bridge The equivalent resistance in this case becomes 4Ω ,if $V_A < V_B$ the diodes are reveres biased in that case $4\Omega, 5\Omega$ and 4Ω are in series

12. During FB, diode conducts and during RB, diode does not conduct.

13. Slope $= \frac{\Delta I}{\Delta V}$

14 Given figure is halfwave rectifier

15. The given question is based on Bridge Rectifier. This is the most widely used full-wave rectifier. It makes use of four diodes D_1, D_2, D_3, D_4 connected in the four arms of a bridge. Bridge rectifier does not require a centre-tapped transformer.

16. Conceptual

17. The diode connected parallel to the battery is reverse bias. so current will not pass through it. so total emf divided among C_1 and C_2 and in the inverse ratio of their capacities

$$V = \frac{q}{C}; \frac{V_1}{V_2} = \frac{C_2}{C_1}$$

$$V_1 = \frac{VC_2}{C_1 + C_2}, V_2 = \frac{VC_1}{C_1 + C_2}$$

18. $V = iR$

19. $I_C = \frac{90}{100} I_E \Rightarrow I_E = \frac{10}{9} I_C$ and $I_e = I_b + I_c$

20. $\beta = \frac{\alpha}{1-\alpha}$ But $\beta = \left(\frac{\delta I_C}{\delta I_B}\right)_{V_{CE}}$

$$\Rightarrow \delta I_C = \beta . \delta I_B = 9 \times 4\text{mA}$$

21. $I_c = \frac{V_{cc}}{R_L}$

22 $V = i_b R_b$

$$\therefore \; R_B = \frac{V}{i_b} = \frac{9}{35 \times 10^{-6}} = 257 k\Omega$$

23. (i) Current Gain $\beta = \frac{\delta i_c}{\delta i_b}$

(ii) Input impedance $R_i = \frac{\delta V_{BE}}{\delta i_b}$

(iii) Effective (ac) load $R_{AC} = \frac{R_C R_L}{R_C + R_L}$

(iv) Voltage gain $A_v = \beta \times \frac{R_{AC}}{R_{in}}$

(v) Power gain, $A_p = A_v \times \beta$

24.
$$\begin{array}{r} \overset{1}{1'}\,{}^{1}0\overset{1}{1'}\,{}^{1}0\overset{1}{1'}\,{}^{1}0 \\ -0\;1\,0\;\;1\,0\;\;1 \\ \hline 0\;1\,0\;\;1\,0\;\;1 \end{array}$$

25. $Z = DABC + DA\overline{BC}$

$= DA\left(BC + \overline{BC}\right) = DA(1)$

$= DA$ i.e One AND gate.

26. It is diode-diode logic ckt of OR gate

27. Here $Y = \overline{(A.B).\overline{C}} = \left(\overline{A.B}\right) + C = \overline{A} + \overline{B} + C$

$\therefore Y = 0$ if $A = 1,\ B = 1$ and $C = 0$

28. $X = \overline{\overline{A}.\overline{B}} = \overline{\overline{A}} + \overline{\overline{B}} = A + B$. It is OR gate

29. $Y = \overline{A.B} = \overline{1.0} = 1 = NAND$

30. Logic of A is 1010, Logic of B is 1001
Logic of Y is 0111

31. The input to OR gate is A and A.B. Hence Y = A + BA

32. $Y_1 = A + D, Y_2 = A + D + B$,

$Y = A + D + B + C$

33. For this purpose we use NAND gate in manner as shown. The first two NAND gates are operated as NOT gates adn their output are fed to the third The resulting circuit is OR gate. Here, there gates are used

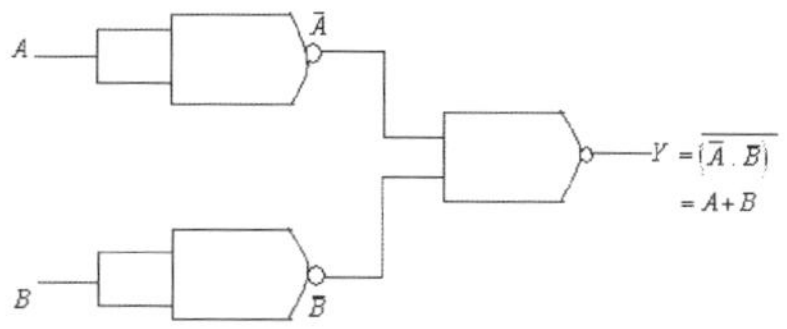

LEVEL - II -2

INTRINSIC,EXTRINSIC SEMI CONDUCTORS AND DIODES

1. If the lattice constant of this semiconductor is decreased, then which of the following is correct?

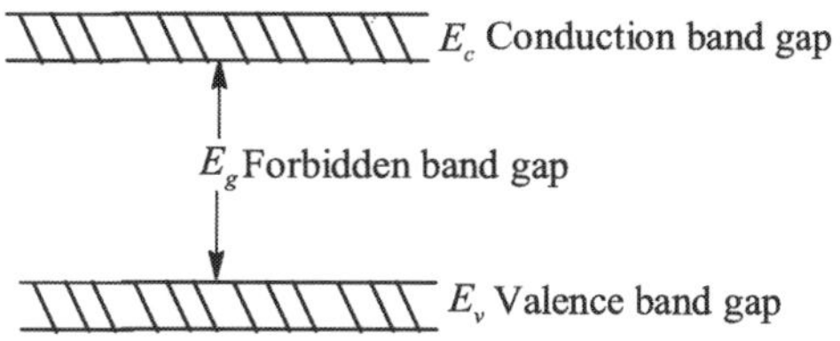

1) All E_c, E_g & E_v decrease

2) All E_c, E_g & E_v increase

3) E_c and E_v increase, but E_g decrease

4) E_c and E_v decrease but E_g increase

2. A Ge specimen is doped with Al. The concentration of acceptor atoms is $10^{21}\ atoms/m^3$. Given that the intrinsic concentration of electron -hole pairs is $10^{19}/m^3$, the concentration of electrons in the specimen is

1) $10^{17}/m^3$ 2) $10^{15}/m^3$ 3) $10^4/m^3$ 4) $10^2/m^3$

3. The following data are for intrinsic germanium at 300 K. $n_i = 2.4\times10^{19}/m^3$,

$\mu_e = 0.39m^2/Vs, \mu_h = 0.19m^2/Vs$. Calculate the conductivity of intrinsic germanium.

1) $4.3\,Sm^{-1}$ 2) $1.21\,Sm^{-1}$ 3) $2.22\,Sm^{-1}$ 4) $4.22\,Sm^{-1}$

4. The diagram correctly represent the direction of flow of charge carriers in the forward bias of p-n junction is

1) 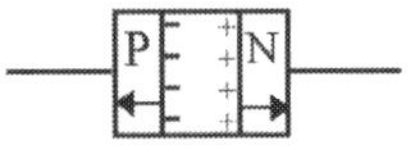
2) 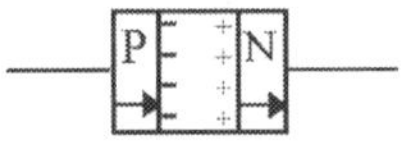
3) 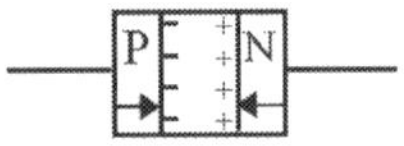
4) 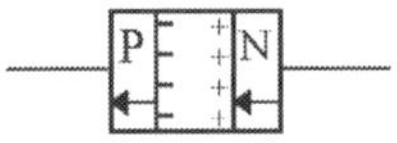

5. In the figurs shown below

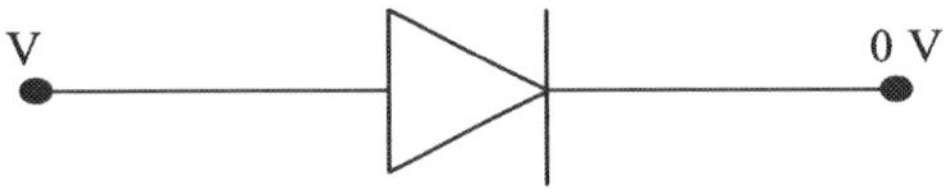

Fig a

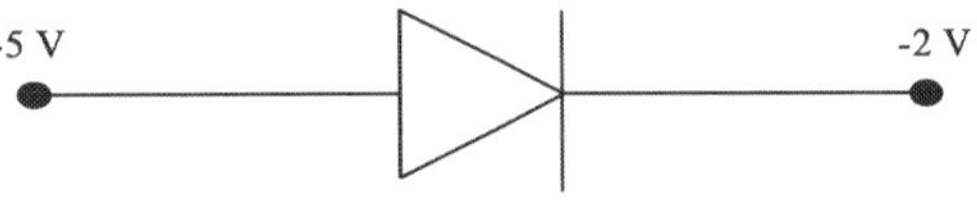

Fig b

1) In both Fig a and Fig b the diodes are forward biased

2) In both Fig, a and Fig b the diodes are reverse biased

3) In Fig a the diode is foward biased and in Fig b, the diode is reverse biased

4) In Fig a the diode is reverse biased and in fig b, it is forward biased

6. A P-N junction diode can withstand currents up to 10 mA. Under forward bias, The diode has a potential drop of 0.5 V across it which is assumed to be independent of current.The maximum voltage of the battery used to forward bias the diode when a resistance of 200Ω is connected in series with it is

1) 2.5V 2) 2.6V 3) 2.7V 4) 2.8V

7. A cell of emf 4.5V is connected to a junction diode whose barrier potential is 0.7V. If the external resistance in the circuit is $190\,\Omega$, the current in the circuit is

1) 20 mA 2) 2mA 3) 23mA 4) 200mA

8. V_A *and* V_B **denote potentials of A and B, then the equivalent resistance between A and B in the adjoining circuit is(ideal diode)**

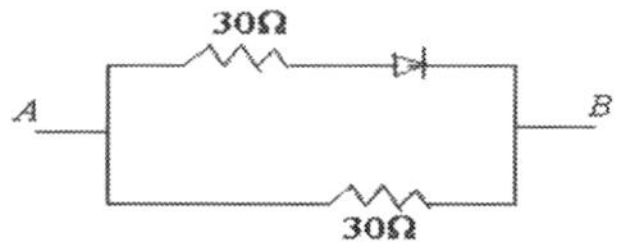

1) 15Ω *if* $V_A > V_B$ 2) 30Ω *if* $V_A < V_B$ 3) Both 1 and 2 4) neither 1 nor 2

9. **Two ideal junction diodes D_1, D_2 are connected as shown in the figure. A 3V battery is connected between A and B. The current supplied by the battery if its positive terminal is connected to A is**

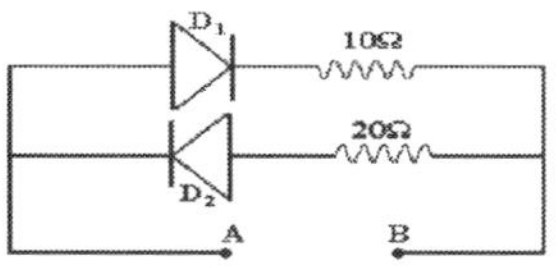

1) 0.1A 2) 0.3 A 3) 0.9 A 4) 90 A

10. **Find the effective resistance between A and B**

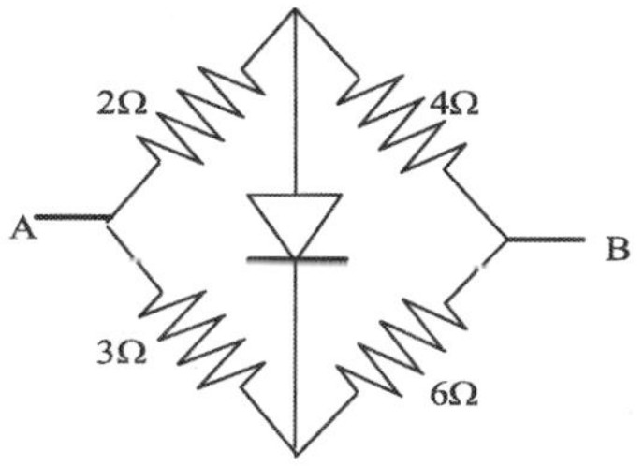

1) 5/18 Ω 2) 9/5 Ω 3) 18/5 Ω 4) 5/9 Ω

11. **The peak voltage in the output of a half-wave diode rectifier fed with a sinusoidal signal without filter is 10V. The d.c. component of the output voltage is**

1) $10/\sqrt{2}\ V$ 2) $10/\pi\ V$ 3) 10 V 4) $20/\pi\ V$

12. **In the figure shown the potential drop across the series resistor is**

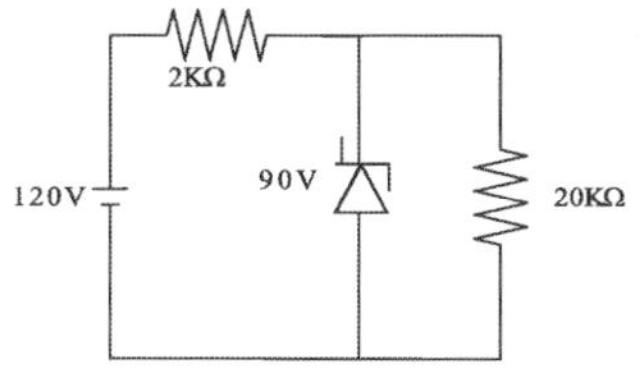

1) 30 V 2) 60 V 3) 90 V 4) 120 V

13. **A 220V ac supply is connected between points A and B (Fig). What will be the potential difference V across the capacitor?**

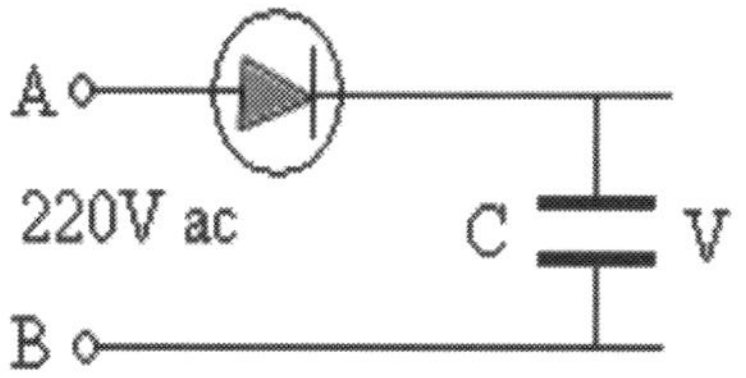

1) 220V 2) 110V 3) 0V 4) $220\sqrt{2}V$

14 **In the circuit shown (Fig). if the diode forward voltage drop is 0.3V, the voltage difference between A and B is :**

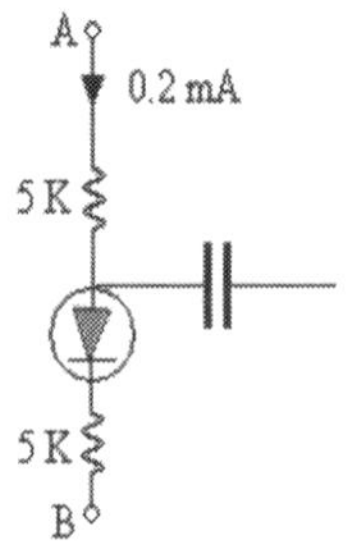

1) 1.3V 2) 2.3V 3) 0 4) 0.5V

TRANSISTORS

15. **In a n-p-n transistor 10^{10} electron enter the emitter in 10^{-6}s. If 2% of the electrons are lost in the base, the current transfer ratio and the current amplification factor are**

1) 0.98, 49 2) 0.49, 49 3) 0.98, 98 4) 0.49, 98

16. **In a common base mode of transistor, collector current is 5.488 mA for an emitter current of 5.60mA. The value of the base current amplification factor (β) will be:**

1) 48 2) 49 3) 50 4) 51

17. **Current amplification factor of a common base configuration is 0.88. Find the value of base current when the emitter current is 1 mA.**

1) 0.12 mA. 2)0.1 mA 3) 0.5 mA 4) 0.2 mA

18. **For a transistor $\beta = 40$ and $I_B = 25\mu A$. Find the value of I_E.**

1) 1mA 2) 1.025mA 3) 2mA 4) 1.2 mA

19. **In a transistor if $\frac{I_C}{I_E} = \alpha$ and $\frac{I_C}{I_B} = \beta$, If α varies between $\frac{20}{21}$ and $\frac{100}{101}$, then the value of β lied between**

1)1-10 2)0.95-0.99 3)20-100 4)200-300

20. **For a transistor $x = \frac{1}{\alpha}$ & $y = \frac{1}{\beta}$ where α & β are current gains in common base and common emitter configuration. Then**

1) $x + y = 1$ 2) $x - y = 1$ 3) $2x = 1 - y$ 4) $x + y = 0$

21. **A voltage amplifier operated from a 12 volt battery has a collector load $6k\Omega$. Calculate the maximum collector current in the circuit.**

1) 0.5mA 2) 1 mA 3) 3 mA 4) 2mA

22. **A CE amplifier is designed with a transistor having $\alpha = 0.99$. Input impedance is 1 $k\Omega$ and load is 10 $k\Omega$. Voltage gain will be:**

1) 9900 2) 99000 3) 99 4) 990

23. **In a common emitter amplifier the load resistance of the output circuit is 792 times the resistance of the input circuit. If α = 0.99, the voltage gain is**

1) 79200 2) 39600 3) 7920 4) 3960

24. **In a transistor amplifier β = 62, R_L = 5000 Ω and internal resistance of the transistor is 500 Ω. Its power amplification will be**

1) 25580 2) 33760 3) 38440 4) 55760

25. **The tuned collector oscillator circuit used in the local oscillator of a radio receiver makes use of a tuned circuit with L = 60 μ H and C = 400 pF. Calculate the frequency of oscillations.**

1) 1.03 KHz 2) 1.03 Hz 3) 1.03 GHz 4)1.03 MHz

LOGIC GATES

26. **When we add binary numbers 111 and 111 we get the binary number:**

1) 222 2) 1000 3) 1110 4) 000

27. **If A= B=C=1 and** $X = A\overline{BC} + B\overline{CA} + C\overline{AB}$**, then X=**

1) 0 2) 1 3) 100 4) 110

28. **The input of A and B for the Boolean expression** $\left(\overline{A+B}\right).\left(\overline{A.B}\right)=1$ **is**

1) 0,0 2) 0, 1 3) 1,0 4) 1, 1

29. Write the name of the follwoing gate that the circuit shown in figure.

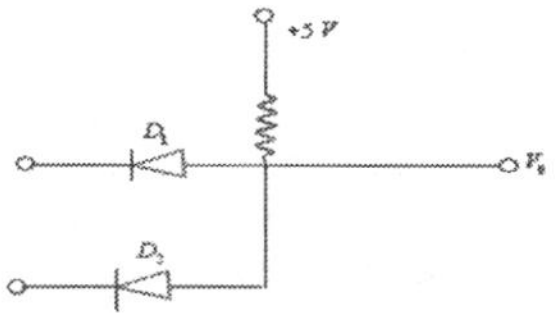

1)AND gate 2)OR gate 3)NOR gate 4)XOR gate

30. **Consider a two-input AND gate of figure below. Out of the four entries for the Truth Table given here, the correct ones are**

	Input A	Input B	Output Y
1	0	1	0
2	1	0	0
3	1	1	1
4	0	0	1

1) All 2) 1 and 2 only 3) 1, 2 and 3 only 4) 1, 3 and 4 only

31. **A Truth table is given below. The logic gate having following truth table is**

A	B	Y
0	0	1
1	0	0
0	1	0
1	1	0

1)NAND gate 2)NOR gate 3)AND gate 4)OR gate

32. **For a logic 0101 the waveform is**

1) 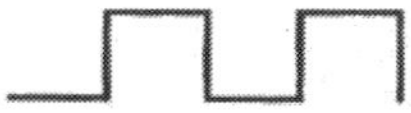

2) 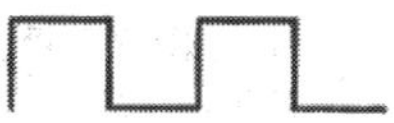

3) 

4) 

33. For the given combination of gates, if the logic states of inputs A, B, C are as follows A=B=C=0 and A=B=1, C=0 then the logic states of output D are

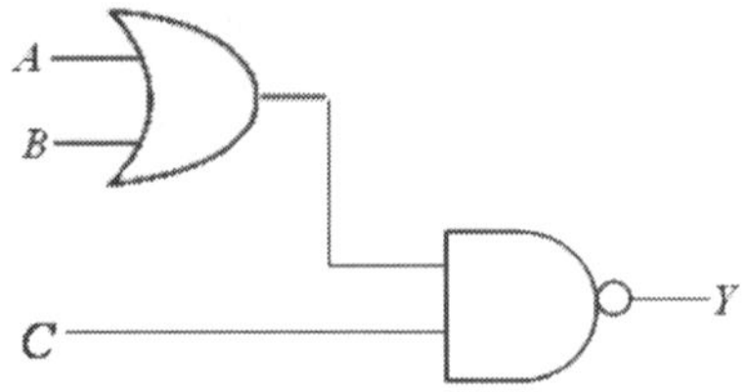

1) 0,0 2) 0,1 3) 1, 0 4) 1,1

34. Identify the gate represented by the block diagram as shown in fig.

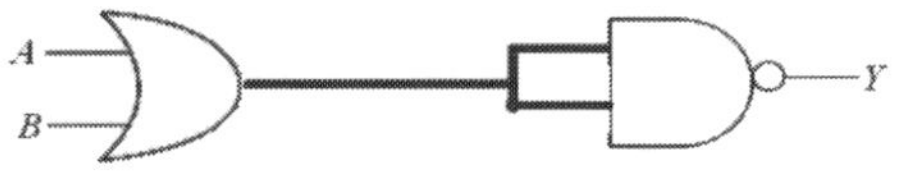

1) AND gate 2) NOT gate 3) NAND gate 4) NOR gate

35. The Boolean expression for the gate circuit shown below is

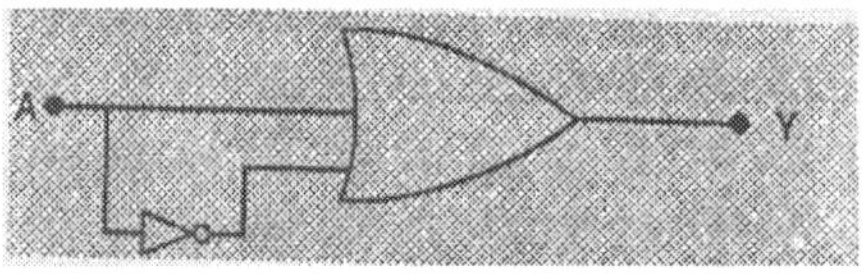

1) $A+\overline{A}=1$ 2) $A+1=1$ 3) $A+A=A$ 4) $A+0=A$

36. The output Y of the gate circuit shown in the figure below is

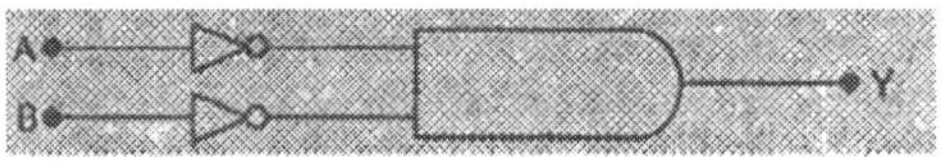

1) $\overline{A.B}$ 2) $\overline{A}.\overline{B}$ 3) $\overline{\overline{A.B}}$ 4) $\overline{A}+\overline{B}$

KEY LEVEL - II -2

1) 4	2)1	3)3	4)3	5)3	6)1	7)1	8)3	9)2	10)3
11)2	12)1	13)4	14)2	15)1	16)2	17)1	18)2	19)3	20)2
21)4	22)4	23)1	24)3	25)4	26)3	27)1	28)1	29)1	30)3
31)2	32)1	33)4	34)4	35)1	36)2				

HINTS - LEVEL - II -2

1. Conceptual
2. When Ge specimen is doped with Al, then concentration of acceptor atoms is also called concentration of holes. Using formula, $n_i^2 = n_0 p_0$ where

n_1 = concentration of electron hole pair = $10^{19} m^{-3}$

n_0 = concentration of electron

p_0 = concentration of holes = $10^{21}\, atom\; m^{-3}$

$\Rightarrow (10^{19})^2 = 10^{21} \times n_0 \qquad \Rightarrow n_0 = 10^{17} m^{-3}$

3. $\sigma_i = n_i e(\mu_e + \mu_h)$

4. In forward bias holes move from P to n, side electrons move from n to P

5. for forward bias, $V_p > V_n$, and for reverse bias $V_p < V_n$

6. $i = \frac{E - V_b}{R}$

7. V=4.5- 0.7=3.8V ; $R = 190\Omega$, $i = \frac{V}{R}$

8. $R = \infty$ in the case of reversebias

9. If positive terminal is connected to A, D_1 is in forward bias and D_2 is in reverse bias. So 20W is ineffective.

$\therefore i = \frac{V}{R} = \frac{3}{10} = 0.3A$

10. Use Wheat stones bridge principle

11. $V_{dc} = \frac{V_m}{\pi} = \frac{10}{\pi} volt$

12. $V_{in} = V + V_Z$

13. Capacitor once gets charged upto maximum potential .After that for any other lesser value of p.d across A and B diode is reverse biased and it does not allow charge to flow in opposite direction.

$From\, V_{rms} = \frac{V_{max}}{\sqrt{2}}$

$\Rightarrow V_{max} = V_{rms}\sqrt{2}$

$= 220\sqrt{2}\, volts$

14. $V_A - 0.2\times10^{-3}\times5\times10^3 - 0.3$

$-0.2\times10^{-3}\times5\times10^3 - V_B = 0$

$\Rightarrow V_A - V_B = 2.3\, volt$

15. $\alpha = \frac{I_c}{I_e}, I_c = \frac{98}{100} I_e, \beta = \frac{\alpha}{1-\alpha}$

16. $I_e = I_b + I_c$ and $\beta = \frac{I_C}{I_B}$

17. In a common-base arrangement, the current amplifica-tion factor.

$\alpha = \frac{I_C}{I_E} \Rightarrow I_C = \alpha I_E = 0.88\times1 = 0.88mA.$ and $I_B = I_E - I_C$

18. $\beta = \frac{I_C}{I_B} \Rightarrow I_C = \beta I_B = 40\times25\times10^{-6}$ and $I_E = I_B + I_C$

19. $\beta = \frac{\alpha}{1-\alpha}$

20. $\alpha = \frac{\Delta I_c}{\Delta I_e}, \beta = \frac{\Delta I_c}{\Delta I_b}$

21. $I = \frac{\text{Battery voltage}}{\text{Collector load}} = \frac{12\,\text{volt}}{6\text{k}\Omega} =$

22. $\beta = \frac{\alpha}{1-\alpha}$ and $A_v = \beta \times \frac{R_{AC}}{R_{in}}$

23. $\beta = \frac{\alpha}{1-\alpha}, \quad A_V = \beta \frac{R_L}{R_i}$

24. $A_P = \beta^2 \frac{R_L}{R_i}$

25. $v = \frac{1}{2\pi\sqrt{LC}}$

26.
$$\begin{array}{r} 111 \\ +111 \\ \hline 1110 \end{array}$$
(carries: 1 1 1)

27. Check the value of x by substituing A=B=C=1

28. $\overline{A+B} = 1$, $\overline{A.B} = 1$

$A+B = 0$, $A.B = 0$

29. conceptual

30. Check the truth table

31. $Y = \overline{A+B}$

32. concpetual

33.

A	B	Y
1	1	1
1	0	0
0	1	0
0	0	0

34. The boolean expression of this gate is

$Y = \overline{(A+B).(A+B)} = \overline{(A+B)} + \overline{(A+B)} = \overline{(A+B)}$

which is for NOR gate.

35. The input of OR gate is A & $\overline{A}$

$Y = A + \overline{A} = 1$

36. The input of AND gate is $\overline{A}$ and $\overline{B} \Rightarrow Y = \overline{A}.\overline{B}$

LEVEL-III

1. If the ratio of the concentration of electrons to that of holes in a semi-conductors is 7/5 and the ratio of currents is 7/4, then the ratio of drift velocities is

1) 5/8 2) 4/5 3) 5/4 4) 4/7

2. If the resistivity of copper is $1.7\times10^{-6}\Omega$ cm, then the mobility of electrons in copper, if each atom of copper contributes one free electron for conduction, is [The atomic weight of copper is 63.54 and its density is 8.96 g/cc]:

1) $23.36cm^2/Vs$ 2) $503.03cm^2/Vs$ 3) $43.25cm^2/Vs$ 4) $88.0cm^2/Vs$

3. A pure silicon crystal of length $l(0.1m)$ and area $A(10^{-4}m^2)$ has the mobility of electron (μ_e) and holes (μ_h) as $0.135m^2/Vs$ and $0.048m^2/Vs$, respectively, If the voltage applied across it is 2V and the intrinsic charge concen-tration is $n_i=1.5\times10^6m^{-3}$, then the total current flowing through the crystal is.

1) $8.78\times10^{-17}A$ 2) $6.25\times10^{-17}A$ 3) $7.89\times10^{-17}A$ 4) $2.456\times10^{-17}A$

4. Find the current produced at room temperature in a pure germanium plate of area $2\times10^{-4}m^2$ and of thickness 1.2×10^{-3} m when a potential of 5 V is applied across the faces. Concentration of carriers in germanium at room temperature is 1.6×10^6 per cubic metre. The mobilities of electrons and holes are $0.4m^2V^{-1}s^{-1}$ and $0.2m^2V^{-1}s^{-1}$ respectively. The heat energy generated in the plate in 100 second is

1) $2.4\times10^{-11}J$ 2) $3.4\times10^{-11}J$ 3) $5.4\times10^{-11}J$ 4) $6.4\times10^{-11}J$

5. An n-type semiconductor has impurity level 20meV below the conduction band. In a thermal collision, transferble energy is KT. The value of T for which electrons start to jump in conduction band is :

1) 232 K 2) 348 K 3) 400 K 4) 600 K

6. Assume that the number of hole-electron pair in an intrinsic semiconductor is proportional to $e^{-\Delta E/2KT}$. Here ΔE = energy gap and $k=8.62\times10^{-5}eV/kelvin$. The energy gap for silicon is $1.1eV$. The ratio of electron hole pairs at 300K and 400 K is :

1) $e^{-5.31}$ 2) e^{-5} 3) e 4) e^2

7. In the circuit shown in figure (1), the V_0, I_1, I_{D_1}, and I_{D_3} are respectively

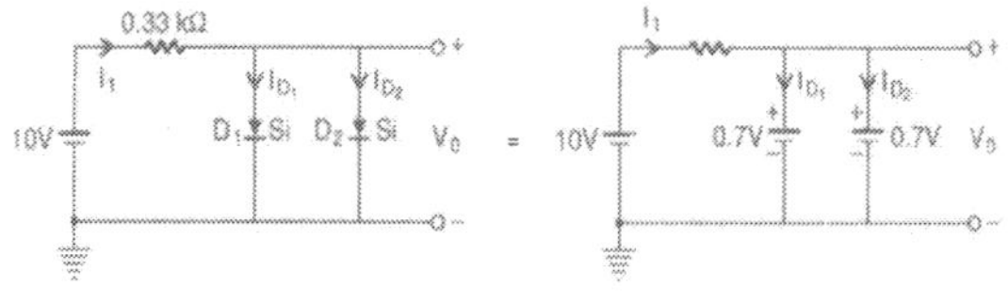

(A) (B)

1) 0.5V, 25 mA, 15 mA 2) 0.7V, 28.18 mA, 14.09 mA

3) 0.4V, 15 mA, 20 mA 4) 0.3V, 15.06 mA, 20.18 mA

8. For a junction diode, the ratio of forward current (I_f) and reverse current is

[I_e = electronic charge, V = voltage applied across junction,

k = Boltzmann constant T = temperature in kelvin]

1) $e^{-v/kT}$ 2) $e^{V/kT}$ 3) $(e^{eV/kT}-1)$ 4) $(e^{V/kT}-1)$

9. **In the diagram D an ideal diode and an alternating voltage of peak value 10V is connected as input V_1. Which of the following diagram represents the correct waveform of output voltage V_0?**

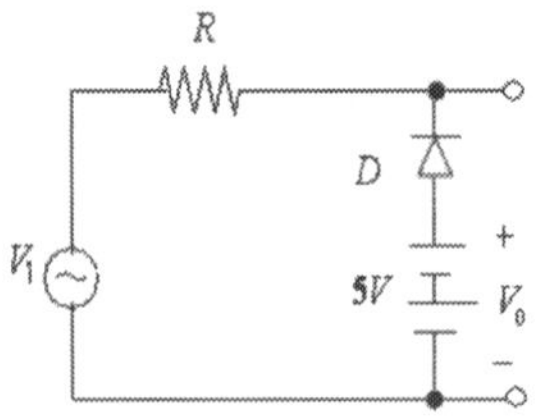

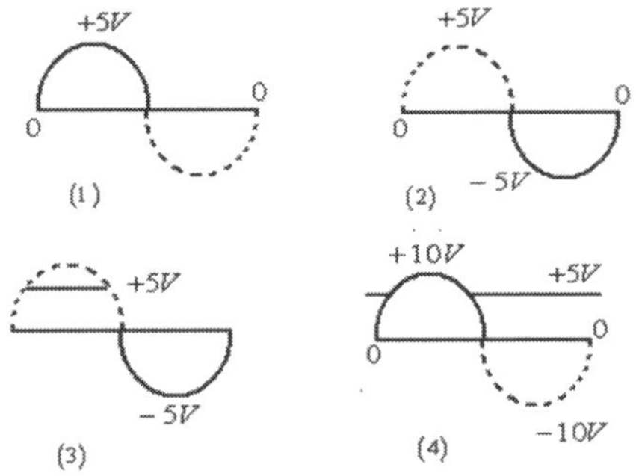

10. **For a CE-transistor amplifier, the audio signal voltage across the collector resistance of $2k\Omega$ is 2V. Suppose the current amplification factor of the transistor is 100. Find the input signal voltage and base current, if the base resistance is $1k\Omega$.**

1) 0.02V 2) 0.01V 3) 0.03V 4) 0.04V

11. **In the circuit, transistor has a current**

$(\beta) = 100$. What should be the base resistor R_B neglect V_{BE}, so that $V_{CE} = 5V$:

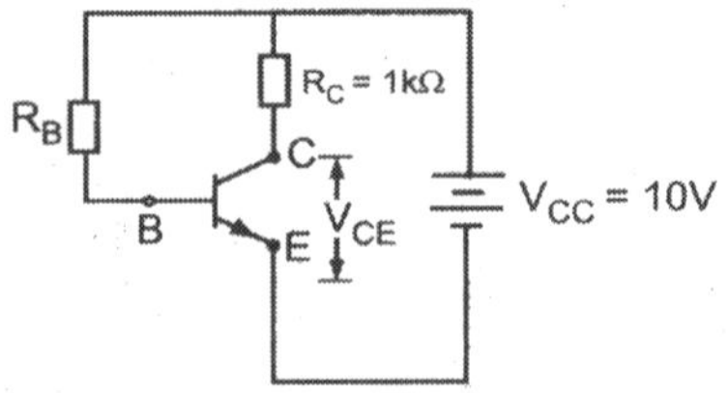

1) 200 kΩ 2) 1 kΩ 3) 500 kΩ 4) 2 kΩ

12. **An $N-P-N$ transistor is connected in common - emitter configuration in which collector supply is 8V and the voltage drop across the load resistance of 800 Ω connected in the collector circuit is 0.8V. If current amplification factor is 25/26 (If the internal resistance of the transistor is 200 Ω), the collector-emitter voltage, voltage gain and power gain are respectively.**

1) 5.2V, 1.86, 3 2) 6.2V, 186, 5.5 3) 7.2V, 3.86, 3.698 4) 8.2V, 4.91, 3.15

13 **For a CE transistor amplifier, the audio signal voltage across the collector resistance of $2\,k\Omega$ is 2V. Suppose the current amplification factor of the transistor is 100. The value of R_B in series with V_{BB} supply of 2V, if the DC base current has to be10 times the signal current is**

1) $4\,k\Omega$ 2) $14\,k\Omega$ 3) $28\,k\Omega$ 4) $54\,k\Omega$

14 **Figure shows the transfer characteristics of a base biased CE transistor. Which of the following statements are true ?**

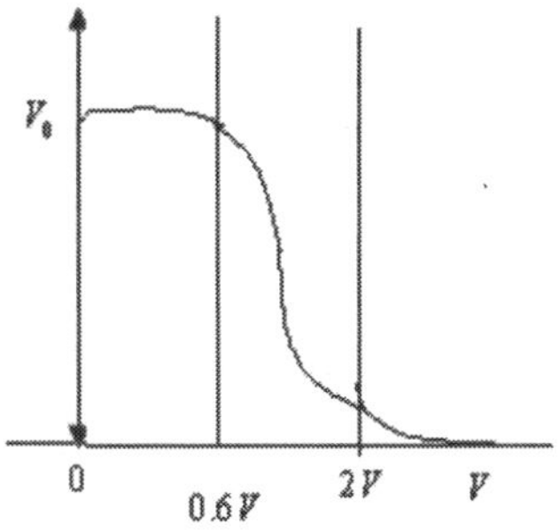

A) At $V_1 = 0.4\ V$ transistor is in active state

B) At $V_1 = 1\ V$ it can be used as an amplifier

C) At $V_1 = 0.5\ V$, It can be used as a switch turned off

D) At $V_1 = 2.5\ V$, it can be used as a switch turned on

1)A,B,C 2)B,C,D 3)A,C,D 4)A,B,D

15. **The following configuration of gates is equivalent to**

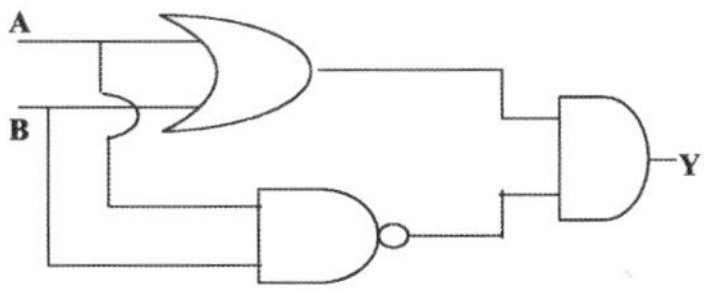

1) NAND gate 2) XOR gate 3) OR gate 4) NOR gate

16. **The combination of the gates shown below produces**

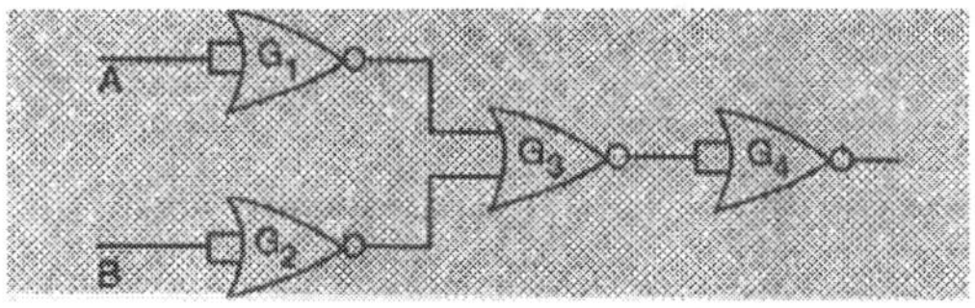

1) AND gate 2) XOR gate 3) NOR gate 4) NAND gate

17. Which of the following truth tables is true?

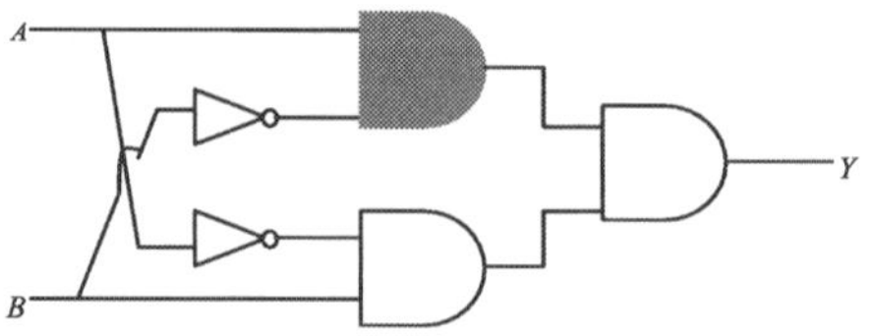

1)

A	B	Y
0	0	0
1	0	0
0	1	0
1	1	0

2)

A	B	Y
0	0	0
1	0	0
0	1	1
1	1	1

3)

A	B	Y
0	0	0
1	0	1
0	1	1
1	1	1

4)

A	B	Y
0	0	0
1	0	1
0	1	1
1	1	0

18. Truth table for system of four NAND gates as shown in figure is :

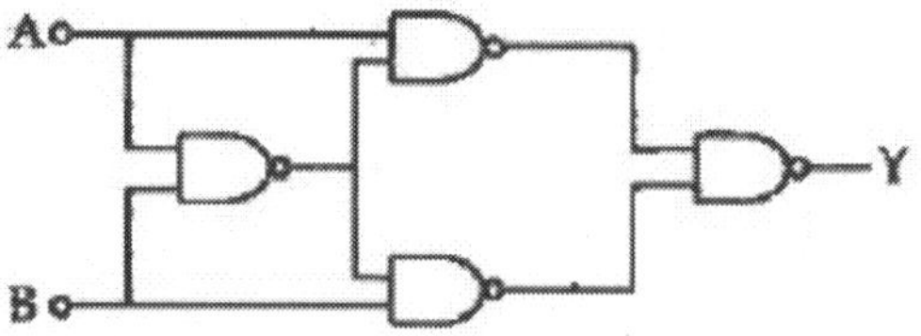

1)

A	B	Y
0	0	0
0	1	1
1	0	1
1	1	0

2)

A	B	Y
0	0	0
0	1	0
1	0	1
1	1	1

3)

A	B	Y
0	0	1
0	1	1
1	0	0
1	1	0

4)

A	B	Y
0	0	1
0	1	0
1	0	0
1	1	1

19. The logic circuit shown below has the input waveforms 'A' and 'B' as shown. Pick out the correct output waveform

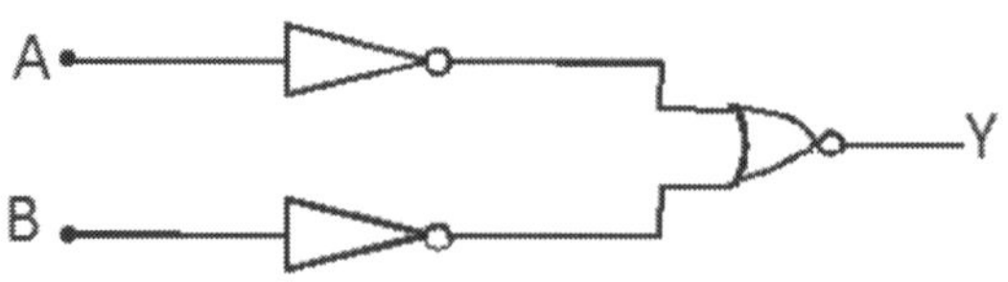

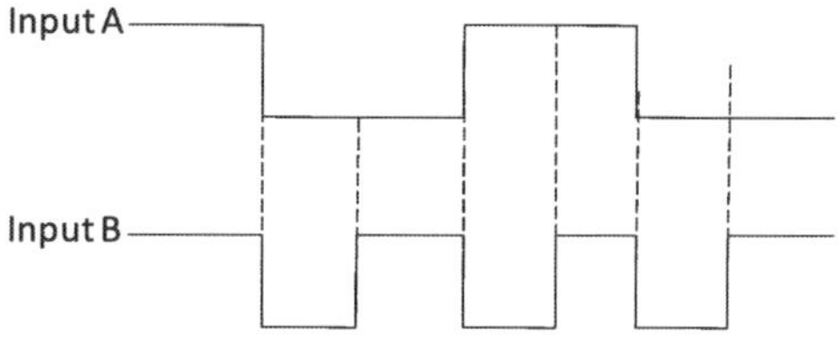

1) 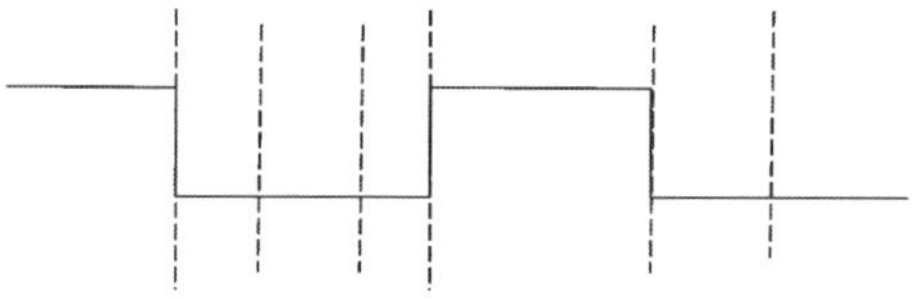

2) 

3) 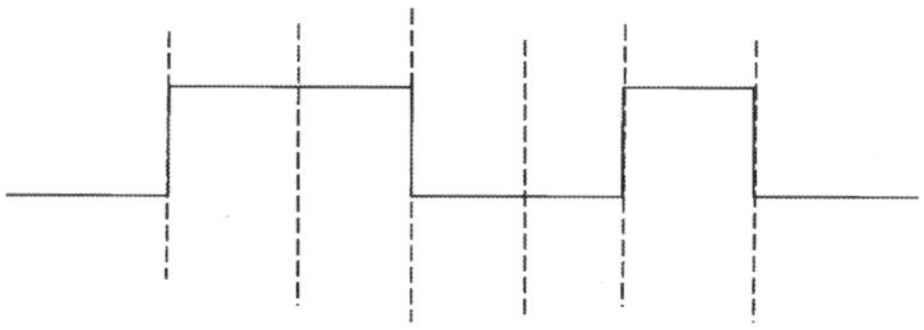

4) 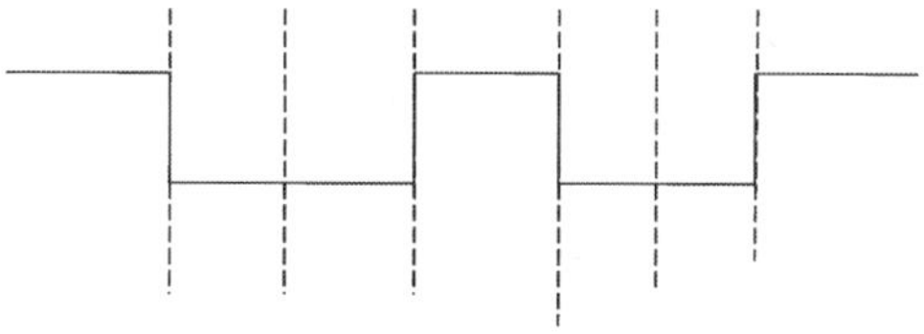

20. Logic gates X and Y have the truth tables shwon below

P	Q	R	P	R
0	0	0	0	1
1	0	0	1	0
0	1	0		
1	1	1		

When the output of X is connected to the input of Y, the resulting combination is equivalent to a single .

1) NOT gate 2) OR gate 3) NOR gate 4) NAND gate

KEY LEVEL - III

1) 3	2)3	3)1	4)4	5)1	6) 1	7)2	8)3	9)4	10)2
11)1	12)3	13)2	14)2	15)2	16)4	17)1	18)1	19)1	20)4

HINTS LEVEL - III

1. $I = n\ e\ Av_d$

$$\frac{I_e}{I_h} = \frac{n_e \times (v_d)_e}{n_h \times (v_d)_h}$$

Here $\frac{n_e}{n} = \frac{7}{5}, \frac{I_e}{I} = \frac{7}{4}$

$\therefore \frac{7}{4}=\frac{7}{5}\times\frac{(v_d)_e}{(v_d)_h} \qquad \Rightarrow \frac{(v_d)_e}{(v_d)_h}=\frac{5}{7}\times\frac{7}{4}=\frac{5}{4}$

2. Mobility of electron $\mu=\frac{\sigma}{ne}$ - (1)

also resistivity $\rho=\frac{1}{\sigma}$

From Eqs. (i) and (ii) , we get $\mu=\frac{1}{ne\rho}$

Here, n = number of free electrons per unit volume

$n=\frac{N_0\times d}{atomicweight}$

$n=\frac{6.023\times10^{23}\times8.96}{63.54}=8.5\times10^{22}$

From Eqs. (iii) and (iv) , we get

$\mu=\frac{1}{8.5\times10^{22}\times1.6\times10^{-19}\times1.7\times10^{-6}}$

$\therefore \mu=43.25cm^2/Vs$

3. $I=e(n_e\mu_e+n_h\mu_h)\frac{AV}{l}$

Putting the values, we get

$I=8.78\times10^{-17}A$

4. $n_e=n_h=n_i$

$\therefore \sigma=n_ie(\mu_e+\mu_h)=1.6\times1.6\times0.6\times10^{-13}$

Current produced in germanium plate.

$I=JA=\sigma E.A=\sigma\left(\frac{V}{d}\right)A$

Heat generated in the plate,

$H=V\times I\times t$

$=5\times1.28\times10^{-13}\times100=6.4\times10^{-11}$ joule

5. $KT=20\times10^{-3}eV$

$\therefore T=\frac{20\times10^{-3}}{k}eV=\frac{20\times10^{-3}eV}{8.62\times10^{-5}eV/K}=232$ K

6. $\frac{N_1}{N_2}=\frac{e^{-\Delta E/2KT_1}}{e^{-\Delta E/2KT_2}}$

$=e^{\frac{\Delta E}{2K}\left(\frac{1}{T_2}-\frac{1}{T_1}\right)}=\frac{1.1}{e^{2\times8.62\times10^{-5}}}\left(\frac{1}{400}-\frac{1}{300}\right)$

7. Since applied voltage is greater than 0.7 V and direction of current is same as direction of arrow head, both the diodes are in "on" state (see above figure(2)). Now V_0 is parallel to D_1 and D_2.

Hence V_0=0.7 V

Current $I_1 = \frac{V_R}{R} = \frac{E - V_D}{R} = \frac{10V - 0.7V}{0.33k\Omega} = 28.18$ mA

Bothe the silecon diodes have similar characteristics.

Hence $I_{D_1} = I_{D_2} = \frac{I_1}{2} = 14.09mA$

8. Here; $v_0 = 2V, \beta = 100,$

R_C (collector resistance) $= 2k\Omega, R_B = 1k\Omega$

As $A_v = \frac{v_0}{v_1} = \beta\left(\frac{R_C}{R_B}\right), v_i = v_0\left(\frac{R_B}{R_C}\right)\left(\frac{1}{\beta}\right)$

$=(2V)\left(\frac{1k\Omega}{2k\Omega}\right)\left(\frac{1}{100}\right)$=0.01V

9. Current in junction diode , $I_f = I_0\left(e^{eV/kT} - 1\right)$ In forward baising, V is positive and in reverse bias V is

$I_r = I_0$

negative Then, $\frac{I_f}{I_r} = \frac{I_0(e^{eV/KT} - 1)}{I_0} =$

$=\left(e^{eV/kT} - 1\right)$

10. For $V_1 < 5V$, the diode is forward-biased, output will be fixed at 5V. For $V_1 < 5V$, the diode is reverse biased. The output will follow V

11. Applying KVL to QCETSRQ, we get

$V_{CC} = I_C R_C + V_{CE}$ and calculate I_C

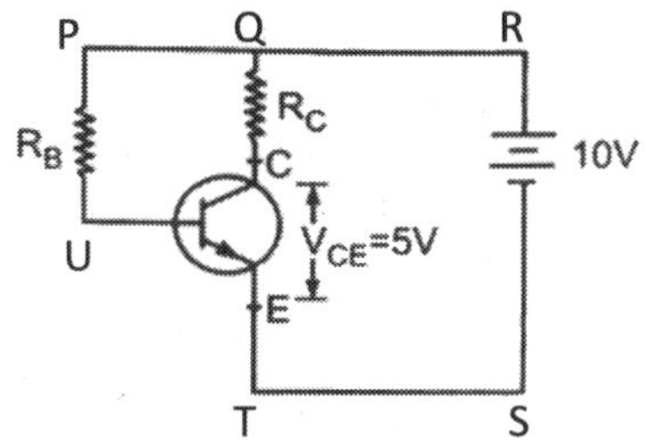

$\beta = \frac{I_C}{I_B} \Rightarrow I_B = \frac{I_C}{\beta}$

Applying KVL to PUTSRQP, we get

$V_{CC} = I_B R_B + V_{BE}$ here V_{BE} is negligible

$\therefore V_{CC} = I_B R_B \Rightarrow I_B = \frac{V_{CC}}{R_B}$

12. The circuit arrangement is shown in figure

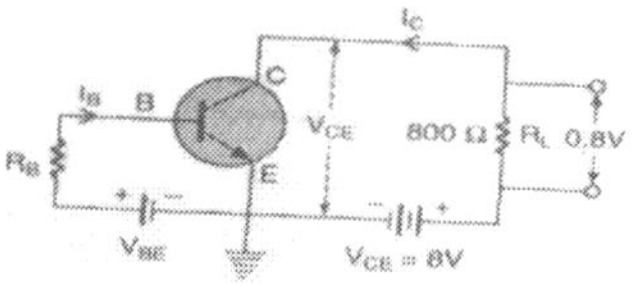

collector current $I_C = \dfrac{\text{Voltage drop across } R_L}{R_L}$

$= \dfrac{0.8}{800} = 10^{-3}\ amp.$

Now $V_{CE} = 8 - 0.8 = 7.2\,volt$

Current gain $\beta = \dfrac{I_C}{I_B}$ (or) $\dfrac{25}{26} = \dfrac{10^{-3}}{I_B}$

$\therefore\ I_B = \dfrac{26}{25} \times 10^{-3} = 1.04 \times 10^{-3}$ amp.

Voltage gain $= \beta \dfrac{R_L}{R_C} = \dfrac{25}{26} \times \dfrac{800}{200} = \dfrac{100}{26} = 3.846$

Power gain $= \beta^2 \dfrac{R_L}{R_C} = \left(\dfrac{25}{26}\right)^2 \times \dfrac{800}{200} = \left(\dfrac{25}{26}\right)^2 \times 4 = 3.698$

13 The output AC voltage is 2V. So the AC collector current $i_c = \dfrac{2}{2000} = 1mA$

The signal current through the base is therefore , given by $i_B = \dfrac{i_c}{\beta}$

The D.C base current has to be $10 \times i_B$

$\therefore\ R_B = \dfrac{(V_{BB} - V_{BE})}{I_B}, V_{BE} = 0.6V$

$\therefore\ R_B = \dfrac{(2 - 0.6)}{0.10} = 14\,k\Omega$

14. From the given transfer characteristics of a base biased common emitter transsistor, we note that

ii) When $V_1 = 1V$ (which is in between 0.6V to 2.V), the transistor circuit is in active state and it used as an amplifier

iii) When $V_1 = 0.5V$ there is no collector current , The transistor is in cut-off state. The transistor circuit can be used as a switch turned off

iv) When $V_1 = 2.5V$ the collector current becomes maximum and transistor is in saturation and can be used as switch turned on state

15. $Y = (A + B).(\overline{A.B}) = (A + B).(\overline{A} + \overline{B})$

$= A.\overline{A} + A.\overline{B} + \overline{A}.B + B.\overline{B}$ $Y = (A.\overline{B}) + (\overline{A}.B)$. it is XOR gate

16. The output of G_1 is $\overline{A}$ G_2 is $\overline{B}$. Hence output of G_3 is AB $\therefore$ $y = \overline{AB}$ i.e., NAND gate.

17. $Y = \left(A.\overline{B}\right).\left(\overline{A}.B\right)$

18. $Y = \overline{\left(\overline{A.\overline{A.B}}\right).\left(\overline{\overline{A.B}.B}\right)} = \left(\overline{\overline{A.\overline{A.B}}}\right) + \left(\overline{\overline{\overline{A.B}.B}}\right)$

$Y = A.\left(\overline{A} + \overline{B}\right) + \left(\overline{A} + \overline{B}\right).B$

$= A.\overline{A} + A.\overline{B} + \overline{A}.B + B.\overline{B}$ $Y = \left(A.\overline{B}\right) + \left(\overline{A}.B\right)$.

19. Truth table:

A	B	Y
1	1	1
1	0	0
0	1	0
0	0	0

20. Verify truth table

LEVEL-IV - COMPREHENSION TYPE

PASSAGE - I

A block of pure silicon at 300K has a length of 10cm and an area of 1.0cm^2. A battery of emf 2V is connected across it. The mobility of electron is $0.14\,m^2v^{-1}S^{-1}$ and their number density is 1.5×10^{16} m^{-3}. The mobility of holes is $0.05\,m^2v^{-1}S^{-1}$

1. The electron current is

1) $6.72\times10^{-4}A$ 2) $6.72\times10^{-5}A$ 3) $6.72\times10^{-6}A$ 4) $6.72\times10^{-7}A$

2. The hole current is

1) $2.0\times10^{-7}A$ 2) $2.2\times10^{-7}A$ 3) $2.4\times10^{-7}A$ 4) 2.6

3. The total current in the block is

1) $2.4\times10^{-7}A$ 2) $6.72\times10^{-7}A$ 3) $4.32\times10^{-7}A$ 4) $9.12\times10^{-7}A$

PASSAGE:2

The input and output resistances in a common base amplifier circuits are 400Ω and 400KΩ respectively. The emitter current is 2mA and current gain is 0.98.

4. The collector current is

1) 1.84mA 2)1.96mA 3)1.2mA 4)2.04mA

5. The base current is

1) 0.012mA 2) 0.022mA 3)0.032mA 4) 0.042mA

6. Voltage gain of transistor is

1)960 2)970 3)980 4)990

7. Power gain of transistor is

1) 950 2) 960 3)970 4)980

8. **The logic circuits are given in column I and the Boolean expressions in column II.**

Column - I	Column - II
A) 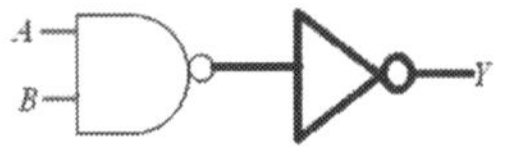	(P) $y = \overline{A} + \overline{B}$
B) 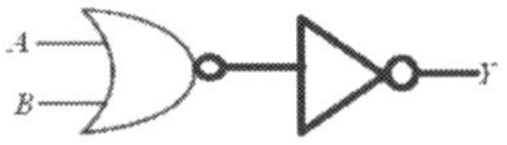	(Q) $y = \overline{A}.\overline{B}$
C) 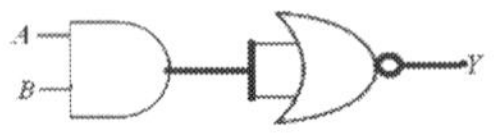	(R) $y = A.B$
D) 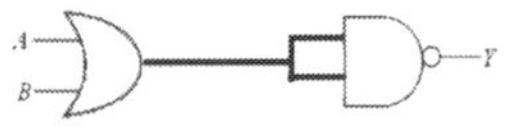	(S) y = A + B

1) $A \to R, B \to S, C \to p, D \to Q$ 2) $A \to S, B \to R, C \to P, D \to Q$

3) $A \to P, B \to Q, C \to R, D \to S$ 4) $A \to S, B \to P, C \to Q, D \to R$

KEY LEVEL - IV

1) 4 2)3 3)4 4)2 5)4 6)3 7)2 8)1

HINTS LEVEL - IV

1. $E = \frac{V}{l} = \frac{2}{0.1} = 20V/m$

$A = 1.0cm^2, v_e = \mu_e E = 0.14 \times 20 = 2.8ms^{-1}$

$I_e = n_e AeV_e = 6.72 \times 10^{-7} A$

2. In a pure semi conductor,

$n_e = n_h = 1.5 \times 10^{16} m^{-3}$

$V_h = \mu_h E = 0.05 \times 20 = 1.0m/s$

$I_h = n_h Aev_h = 2.4 \times 10^{-7} A$

3. Total current

$I = I_e + I_h$

$= 9.12 \times 10^{-7} A$

4. $I_c = \alpha I_e = 0.98 \times 2 = 1.96mA$

5. $I_b = I_e - I_c = 2 - 1.96 = 0.04mA$

6. $A_v = \alpha \frac{R_0}{R_i} = 0.98 \times \frac{400 \times 10^3}{400} = 980$

7. Power amplification, $A_P = \alpha A_v$
$= 0.98 \times 980 = 960$

Chapter - 5

COMMUNICATION SYSTEMS

COMMUNICATION SYSTEMS

SYNOPSIS

Introduction: Communication is an act of exchange of information between the sender and the receiver. Over decades, methods have been evolved to develop languages, codes, signals etc to make communication effective. Communication through electrical signals has made things much simpler because they can be transmitted over extremely large distances in extremely short time as their speed is $3\times10^{8}\,m/s$.

Modern communicaton has its roots in the 19th and 20th century in the work of scientists like J.C. Bose, F.B. Morse, G.Macroni and Alexander Graham Bell. The pace of development seems to have increased dramatically after the first half of the 20th century. We can hope to see many more accomplishments in the coming decades. The aim of this chapter is to introduce the concepts of communication, namely the mode of communication, the need for modulation, production and deduction of amplitude modulation.

➢ **Communication is basically of two types:**

a) Point to point :- This takes place between a transmitter and a receiver. Telephonic conversion between two persons is a good example of it.

b) Broad cast mode :- Here, a large number of receivers receive the information from a single transmitter. Radio and television are good examples of broadcast mode.

Elements of Communication System

Basic units of a communication system.

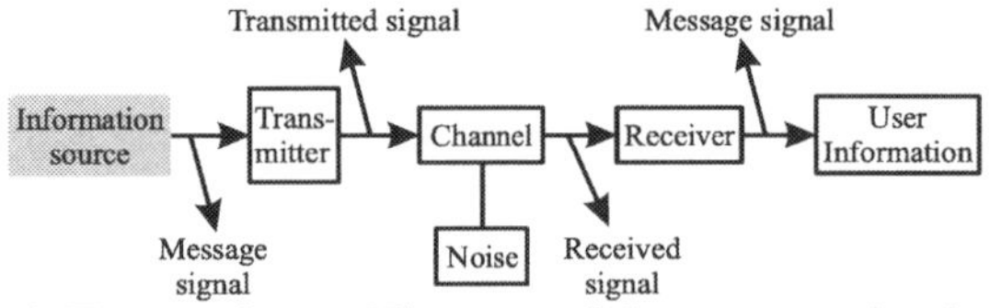

a) Transmitter: The part of the communication system, which sends out the information is called transmitter.

b) Transmission channel: The medium or the link, which transfers message signal from the transmitter to the receiver of a communication system is called channel.

c) Receiver: The part of the communication system, which picks up the information sent out by the transmitter is called **receiver.** The receiver consists of

➢ **Basic Terminology Used in Electronic Communication System**

Some important terms needed to understand the basic elements of communication

a) Information : It is nothing but, the message to be conveyed. The message may be a symbol, code, group of words etc. Amount of information in message is measured in "bits"

b) Communication Channel : Physical medium through which signals propagate between transmitting and receiving stations is called communication channel.

➢ **Transmitter:** Essential components of transmitter are as follows.

a)Transducer **:** Converts sound signals into electric signal. The device which converts a physical quantity (information) into electrical signal is known as ***transducer***.

b)Modulator **:** Mixing of audio electric signal with high frequency radio wave.

c)Amplifier: Boosting the power of modulated signal.

d)Antenna **:** Signal is radiated in the space with the aid of an antenna.

Receiver: Basic componenets of receiver.

a) ***Pickup antenna:*** To pick the signal

b) ***Demodulator:*** To separate out the audio signal from the modulated signal

c) ***Amplifier:*** To boost up the weak audio signal

d) ***Transducer:*** To convert back audio signal in the form of electrical pulses into sound waves.

Message Signal: Information converted in electrical form and suitable for transmission is called **signal**.

➢ A signal is defined as a single-valued function of time (that conveys the information) and which, at every instant of time has a unique value.

- ***Types of message signals***

  a) ***Analog signal:*** A signal, which is a continuous function of time (usually a sinusoidal function) is called *analog signal.*

  b) ***Digital signal:***A discrete signal (discontinuous function of time) which has only two levels is called digital signal.

- **Noise :** This refers to undesired signals which disturb the transmission and processing of signals.
- **Attenuation :** It is the loss of strength of a signal during propagation in a medium.
- **Amplification :** It is the process of increasing the strength of the signal (amplitude) using an amplifier.
- **Range :**It is the maximum distance from a source upto which the signal is received with sufficient strength.
- **Repeaters :** These are the devices used to increase the range of communication system
- **Band width of signals (speech, T.V and digital data) :** Band width is the frequency range over which an equipment operates.

(or)

It is the portion of the spectrum occupied by the signal.

Bandwidth of signals :

In a communication system, the message signal may be voice, music, picture or data etc. Each of these signals has a spread of different range of frequencies. Hence, the type of communication system needed depends upon the band of frequencies involved. Speech signal requires the band width of 2800 Hz (3100 Hz to 300 Hz). For music, a bandwidth of about 20KHz is required (due to high frequency produced by musical instruments). The audible range of frequencies extends from 20Hz to 20KHz. Video signals require band width of 4.2 MHz for picture transmission. However, a band width of 6MHz is needed for T.V signals. (as it contains both voice and picture)

Digital signals are in the form of rectangular waves as shown in Fig. One can show that this rectangular wave can be decomposed into a superposition of sinuosidal waves of frequencies $f_0 . 2f_0, 3f_0, 4f_0 .... nf_0$ where n is an integer extending to infinity and $f_0 = 1/T_0$. The fundamental (f_0), fundamental $(f_0)+$ second harmonic $(2f_0)$ and fundamental $(f_0)+$ second harmonic $(2f_0)+$ third harmonic $(3f_0)$, are shown in the same figure to illustrate this fact. It is clear that to reproduce the rectangular wave shape exactly we need to superimpose all the harmonics $f_0, 2f_0, 3f_0, 4f_0.....$ which implies an infinite bandwidth. However, for practical purposes, the contribution from higher harmonics can be neglected this limiting the bandwidth. As a result, received waves are a distorted version of the transmitted one. If the bandwidth is large enough to accommodate a few harmonics, the information is not lost and the rectangular signal is more or less recovered. This is so because the higher the harmonic. Less is its contribution to the wave form.

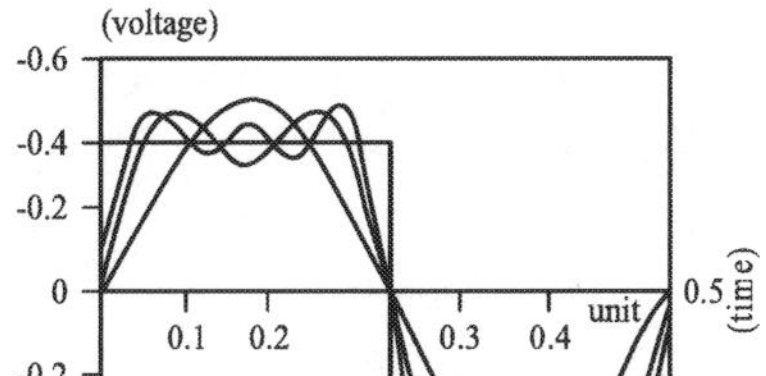

a) Rectangular wave

b) Fundamental (f_0)

c) Fundamental $(f_0)+$ second harmonic $(2f_0)$

d) Fundamental (f_0) Second harmonic $(2f_0)+$ third harmonic $(3f_0)$

- **Bandwidth of transmission medium :**

  The most used transmission media are wire, free space, and fibre optic cable. Different transmission media offer different band width. Coaxial cable offers a band width of about 750 MHz. Radio wave communication through free space takes place over a wide range of frequencies from 100kHz-GHz.

Service	Frequency bands	Comments
Standard AM broadcast	540-1600 kHz	
FM broadcast	88-108 MHz	
Television	54-72 MHz	VHF(Very high frequencies)
	76-88 MHz	TV
	174-216 MHz	UHF(ultra high frequencies)
	420-890 MHz	TV
Cellular Mobile	896-901 MHz	Mobile to base station
	840-935 MHz	Base station to mobile
Satellite	5.925-6.425 GHz	Uplink
Communication	3.7 - 4.2 GHz	Downlink

- Optical communication using fibres is performed in the frequency range of 1 THz to 1000 THz (microwaves to ultraviolet).
- An optical fibre can offer a transmission band width in excess of 100 GHz.
- **COMMUNICATION CHANNELS:**

  The medium or the link, which transfers message signal from the transmitter to the receiver of a communication system is called ***Communication channel***.

  (i) Space communication

  (i) Ground wave propagation

  (ii) Space wave propagation. (Tropospheric wave propagation. Surface wave propagation.)

  (iii) Sky wave propagation: A new dimension recently added to space communication is satellite communication.

  2) Line communication

  i) Two wire transmission line

  ii) Coaxial cable

  iii) Optical fibre cable

(1)SPACE COMMUNICATION

- **Propagation of EM waves in the atmosphere**

  The communication process utilizing the physical space around the earth is termed as space communication. Electromagnetic waves which are used in Radio, Television and other communication system are radio waves and microwaves.

  The velocity of electromagnetic waves of different frequency in a medium is different. It is more for red light and less for violet light. Electromagnetic waves are of transverse nautre.

- **Earth's atmosphere**

  (i) Earth's atmosphere is a gaseous envelope which surrounds the earth.

(ii) Earth atmosphere mainly consists of nitrogen 78% , oxygen 21% along with a little protion of argon, carbon dioxide, water vapour, hydrocarbons, sulphur compounds and dust particles.

(iii) The density of the atmospheric air goes on decreasing as we go up.

(iv) The electrical conductivity of the atmospheric air increases as we go up.

(v) The various regions of earth's atmosphere are:

- **Troposphere.** It extends upto a height of 12km
- **Stratosphere.** It extends from 12km to 50km. There is an ozone layer in stratosphere which mostly absorbs high energy radiations like ultraviolet radiations. etc. coming from outer space.
- **Mesosphere.** It extends from 50km to 80km.
- **Ionosphere.** (i) It extends from 80 km to 400km.

(ii) In this region, the temperature rises to some extent with height, hence it is called Thermosphere.

(iii) The ionosphere which is composed of ionised matter (i.e. electrons and positive ions) plays an important role in space communication.

(iv) The ionosphere is subdivided into four main layers as D, E, F_1 and F_2.

(v) D- layer is at a virtual height of 80km from surface of earth and having electron density $\approx 10^9 m^{-3}$.

(vi) The extent of ionisation of D layer depends upon the altitude of sun. This layer disappears at night. It reflects very low frequency (VLF) and low frequency (LF) electromagnetic waves, but absorbs medium frequency (MF) and high frquency (HF) electromagnetic waves to a certain degree,

(vii) E-layer is at a virtual height of 110km, from the surface of earth, having electron density $\approx 10^{11} m^{-3}$. The critical frequency* of this layer is about 4MHz. This layer helps to MF surface-wave propagation a little but reflects some high frequency waves in day time. It exists in day as well as in night time.

(viii) F_1 – layer is at a virtual height of 180km from the surface of earth, having electron density $\approx 5\times10^{11} m^{-3}$. The critical frequency for this layer is 5 MHz. It reflects some of the high frequency waves but most of the high frequencywaves pass through it and they get reflected from layer F_2 at night time.

(ix) F_2 layer is at a virtual height of about 300km in day time and about 350km in night time. The electron density of this layer is $\approx 8\times10^{11} m^{-3}$. The cirtical frequency of this layer is 8MHz in day time and 6MHz in night time. It reflects back the electromagnetic waves of frequency upto 30 MHz but cannot reflect back the electromagnetic waves of frequency 40MHz or more. It exists in day as well as night time

3. The electromagnetic waves of frequency ranging from a few kilo hertz to a few hundered mega hertz are called radio waves.

The various frequency ranges used in radio waves or micro wave communication system are as follows:

(a) Medium frequency band (M.F) 300 to 3000kHz.

(b) High frequency band (H.F) 3 to 30 MHz

(c) very high frequency band (V.H.F) 30 to 300 MHz.

(d) Ultra high frequency band (U.H.F) 300 to 3000MHz

(e) Super high frequency band (S.H.F) 3000 to 30,000 MHz.

(f) Extra high frequency band (E.H.F) 30 to 300 GHz

The radio waves emitted from a transmitter antenna can reach the receiver antenna by the following mode of operation.

- **KENNELY HEAVISIDE LAYER :**

i) At 110 km above the surface of earth the concentration of electrons is very large. This layer is called Kennely Heaviside layer.

ii) The thickness of this layers is about of few km.

iii) Beyond this layer the electron concentration decreases upto 250 km

iv) From 250 km to 400 km, a layer of large concentration of electrons called Apple ton layer exists.

v) Above appleton layer, ie above Ionosphere the temperature is $927.6^0 C$.

	Ground Wave Propagation	Sky Wave Propagation	Space Wave Propagation
Channel	Ground	Layers of atmosphere	Line of sight communication
Method	Wave glides over the surface of earth diffraction effect	Due to reflection of radio waves from the layers having higher electron density	The radio waves travel from transmitting antenna to receiving antenna along a straight line
Frequency	Depends on power and frequency Less than 2 MHz	3 MHz to 30 MHz	Greater than 40 MHz
Uses	In medium wave Broad casting	Short wave broadcasting	FM broadcasting and microwave links
Range	Depends on height of the Antenna and Curvature of earth	Depends on the angle of incidence on the ionosphere. 150 km to 3000 km	Due to curvature of the earth the waves are blocked at a point
Attenuation	Attenuation increases with frequency		

Different layers of atmosphere and their interaction with the propagating electromagnetic waves :

Name of the layer	Approximate height	Exists during	Frequencies most affected
Troposphere	10 km	Day and night	VHF (up to several GHz)
Parts of Ionosphere : D (part of Stratosphere)	65-75 km	Day only	Reflects LF, absorbs MF and HF to some degree
E (Part of Stratosphere)	100 km	Day only	Helps surface waves reflects HF
FI (Parts of Mesophere)	170-190 km	Day time, merges with F2 at night	Partially absorbs HF waves yet allowing them to reach F2
F2 (Thermosphere)	300 km at night 250-400 km during day time	Day and night	Efficiently reflects HF waves, particularly at night

- **Ground wave propagation :** In this method, the radio waves are guided along the surface. The wave induces charges on the earth. These charges travel with the wave and this forms a current. Now the earth behaves like a leaky capacitor in carrying the induced current. The wave loses some energy, as energy is spent due to flow of charge through the earth's resistance. The wave also looses energy due to diffraction as it glides along the ground. The loss of energy increases as the frequency increases. Thus ground propagation is suitable upto 2MHz. As they loose energy they cannot go to long distances on the ground. Maximum range of the ground wave can be increased by increasing the power of the transmitter.
- **Sky wave propagation :** Above 2MHz and upto 30MHz, long distance communication takes place through ionosphere. The ionosphere reflects the radio waves back to the earth. This method is called sky wave propagation. It is used for shortwave broad casting services. Ionosphere is a thick blanket of 65 km to 400 km above the earth's surface. UV rays and other higher energy radiation coming from space results in the ionization of air molecules. The ionosphere is further divided into serval layers as shown in table below. It should be understood that degree of ionization changes with height. This is because the density of atmosphere decreases with height. At great heights, the radiation is intense, but the molecules available are few. On the other hand, near the earth's surface the molecular concentration is high but the intensity of radiation is low and thus again the ionization is low. Logically, the peak of ionization density occurs at some intermediate heights. The ionosphere acts as a mirror (reflector) for

frequencies of 3-30 MHz. Electromagnetic waves of frequences greater than 30 MHz pass through the atmosphere and skip.

The process of bending of EM waves is similar to total internal reflection in optics. The bending of waves can be easily explained on the basis of variation of refractive index of the ionosphere with change in electron density. Suppose that a radio wave enters the ionosphere from the underlying unionized medium. Since the refractive index of ionosphere decreases from D layer to F_2 layer, consequently, the incident ray will move away from the normal drawn at the point of incidence following the ordinary laws of refraction

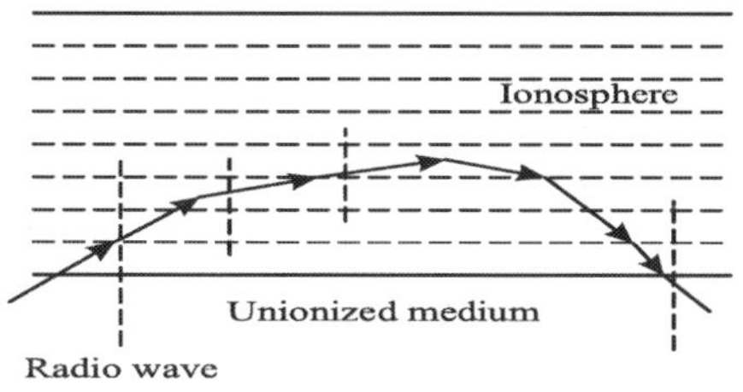

During the propagation in ionosphere the angle of refraction gradually increases and the ray goes on bending more and more till at some point, the angle of refraction becomes 90^0 and the wave travels parallel to the earth surface. This point is called point of reflection. Then the ray tends to move in the down ward direction and comes back to earth because of symmetry. Super high frequency (SHF) waves propagate as sky waves taking reflection at satellite.

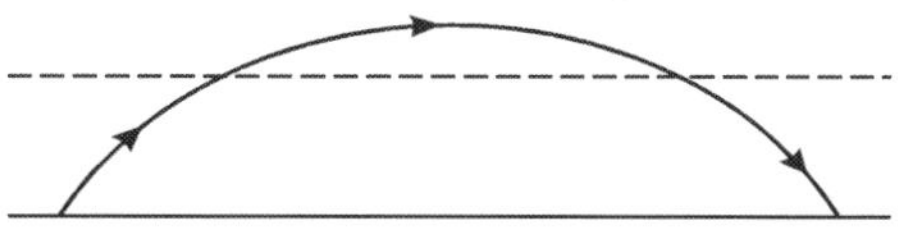

The sky wave propagation can cover a very long distance and so round the globe communication is possible.

(c) The sky waves being electromagnetic in nature, changes the dielectric constant and refractive index of the ionosphere. The effective refractive index of the ionosphere is

$$n_{eff} = n_0\left[1-\frac{Ne^2}{\varepsilon_0 m\,\omega^2}\right]^{1/2} = n_0\left[1-\frac{80.5N}{f^2}\right]^{1/2}$$

Where n_0 = refractive index of free space, N = electron density of ionosphere, ε_0 = permittivity of free space, e = charge on electron, m = mass of electron w = angular frequency of EM wave.

(d) As we go deep into the ionosphere, N increases so n_{eff} decreases. The refractions or bending of the beam will continue and finally it reflects back.

(e) The highest frequency of radio wave, which gets reflected to earth by the ionosphere after having been sent straight to it is

- ***Critical frequency (f_c)***

If maximum electron density of the ionosphere is N_{max} per m^3, then $f_c \approx 9(N_{max})^{1/2}$. Above f_c, a wave will penetrate the ionosphere and is not reflected by it.

(f): The highest frequency of radio waves which when sent at some angle of incidence, towards the ionosphere, get reflected and return to the earth is ***Maximum usable***

➢ ***frequency (MUF)*** $MUF = \frac{f_c}{\cos\theta}$

(g) The smallest distance from a transmitter along the earth's surface at which a sky wave of a fixed frequency but more than f_c is sent back to the earth is ***Skip distance.***

(h) The fluctuation in the strength of a signal at a receiver due to interference of two waves is ***fading***. Fading is more at high frequencies. It results into errors in data transmission and retrieval.

➢ **Wave propagation :** This method is used for line-of-sight [LOS] communication and also for satellite communication. At frequencies above 40MHz, communication is mainly by LOS method. At such frequencies, relatively smaller antenna can be erected above the ground. Because of LOS propagation, the direct waves get blocked, at some point due to the curvature of the earth as shown in the figure.

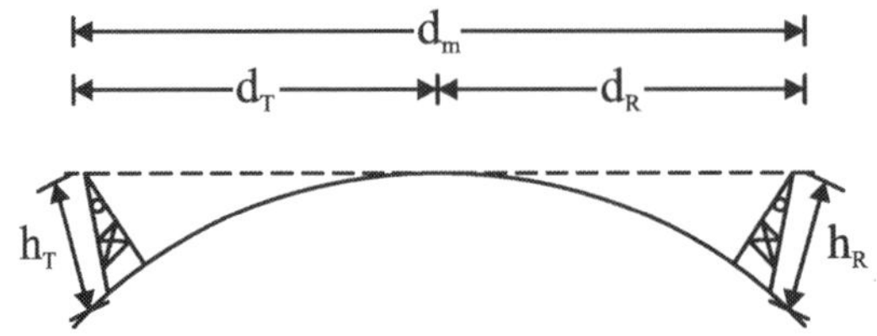

For the signal to be received beyond the horizon, the receiving antenna must be high enough to intercept the LOS waves. If the transmitting antenna is at a height h_T then it can be shown that the distance to the horizon d_T is given by $d_T = \sqrt{2Rh_T}$ where 'R' is the radius of earth. Similarly if the receiving antenna is at a height h_R, the distance to the horizon d_R is $d_R = \sqrt{2Rh_R}$

$\therefore$ The maximum distance d_M between the two antennas is $d_M = \sqrt{2Rh_T} + \sqrt{2RH_R}$ where

R = Radius of the earth.

h_T = height of the transmitting antenna and

h_R = height of the receiving antenna.

If the Population density around the tower is given, the number of persons covered by the transmitting tower

= (Area covered by the tower) × Population density.

$\therefore$ No. of persons $= \pi d^2 \times$ covered population density (Here d = radius of the area covered by single transmitting tower of height h_T)

Television broadcast, microwave and satellite communications are a few examples of communication systems that use space wave propagation. The figure below illustrates the various modes of wave propagation.

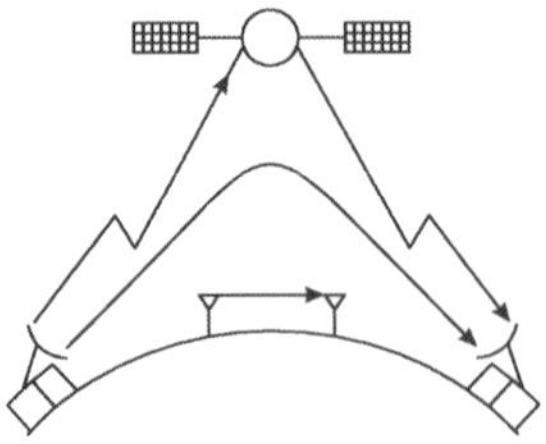

➢ **Range of TV transmission :**

As the frequency range of TV signals is 100-200 MHz, such signal transmission via ground waves is not possible. In such situations, we use line of sight transmission.

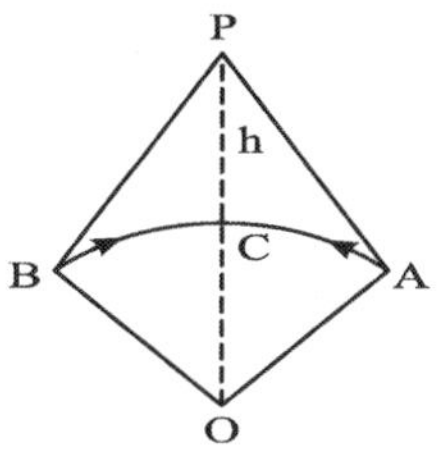

Let CP be the TV tower on the earth's surface. It's antenna is at P. Let PC = h. When TV broadcast is made, the signal can reach the earth upto A to B. There will be no reception of the signal beyond A and B. Arc length CA and CB is the range of TV transmission. If O is the centre of the earth, OA = OB = R is the radius of the earth, from right angled triangle OAP

$OP^2 = OA^2 + PA^2$

$(h+R)^2 = R^2 + PA^2$

$PA = PB = d$

$(h+R)^2 = R^2 + d^2$

$h^2 + R^2 + 2Rh = R^2 + d^2$

As $h << R$ we can ignore h^2

$d^2 = 2Rh$ and $d = \sqrt{2Rh}$

Range of TV transmission depends upon the height of the transmission antenna. Broadcasts are made from tall transmitting antenna.

- A **repeater** is a combination of a receiver and a transmitter. A repeater, picks up the signal from the transmitter, amplifies and retransmits it to the receiver sometimes with a change in carrier frequency. Repeaters are used to extend the range of a communication system as shown in figure. A communication satellite is essentially a repeater station in space. Use of repeater station to increase the range of communication

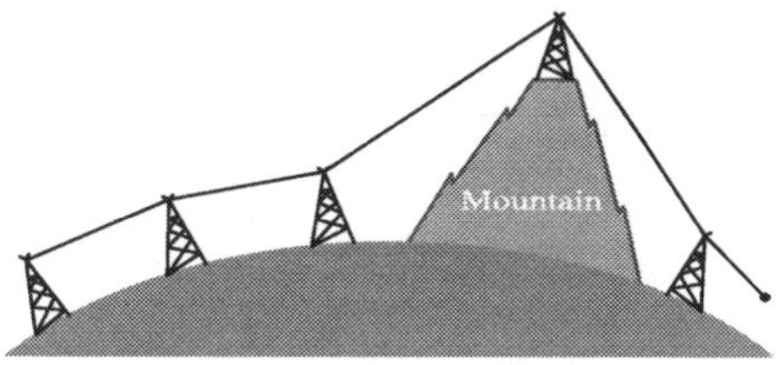

These problems are solved by using geostationary satellite as a communication satellite.

- **Satellite Communication:** Long distance communication beyond 10 to 20 MHz was not possible before 1960 because all the three modes of communication discussed above failed (ground waves due to conduction losses, space wave due to limited line of sight and sky wave due to the penetration of the ionosphere by the high frequencies beyond f_c).

Ionosphere behaves as a rarer medium by which carrier wave is reflected back if its frequency $f\,(\le f_c)$

where f_c is called a "critical frequency" and is given by $f_c = f_0\left(N_{max}\right)^{\frac{1}{2}}$

[N_{max} = maximum electron density.]

Satellite communication made this possible. The basic principle of satellite communication is shown in figure. A communication satellite is a spacecraft placed in an orbit around the earth. The frequencies used in satellite communication lie in UHF/ microwave regions. These waves can cross the ionosphere and reach the satellite.

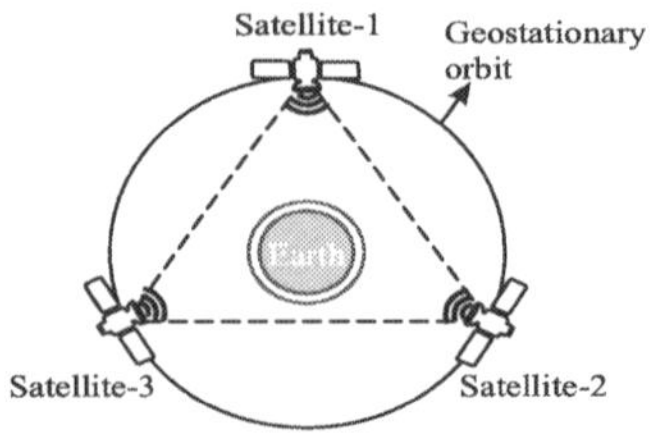

- A geostationary satellite has the same time period of revolution of earth. It locates at the height of 36000 *km* above the earth's surface (well above the ionosphere).
- A communication satellite is a spacecraft placed in an orbit around the earth which carriers a transmitting and receiving equipment called radio transponder. It amplifies the microwave signals emitted by the transmitter from the surface of earth and send to the receiving station on earth.
- The transmitted signal is UP-LINKED and received by the satellite station which DOWN-LINKS it with the ground station through its transmitter.

  The up-link and down-link frequencies are kept different (both frequencies being in the regions of UHF/ microwave).
- At least three geo-stationary satellites are required which are 120° apart from each other to have the communication link over the entire globe of earth.
- Satellite technology is very useful in collecting information about various factors of the atmosphere which governs the weather and climatic conditions.

  The satellite communication can be used for establishing mobile communication with great use the communication satellites are now being used in Global Positioning System (GPS). The ordinary users can find their positions within accuracy of $100m$.

  There are two types of satellites used for long distance transmission.

  (i) Passive satellite: It acts as reflector only for the signals transmitted from earth. Moon the natural satellite of earth is a passive satellite.

  (ii) Active satellite: It carries all the equipment used for receiving signals sent from the earth, processing them and then re-transmitting them to the earth. Now a days active satellites are in use.
- The Indian communication satellites INSAT-2B and INSAT-2C are positioned in such away in the outer space that they are accessible from any place in India.

REMOTE SENSING AND APPLICATION OF SATELLITE COMMUNICATION.

Remote sensing is the technique to obtain information about an object by observing it from a distance and without coming to actual contact with it.

- There are two types of remote sensing instruments: active and passive. Active instruments provide their own energy to illuminate the object of interest, as radar does. They send an energy pulse to the object and then receive and process the pulse reflected from the object. Passive instruments sense only radiations emitted by the object or solar radiation reflected from the object.
- The remote sensing is done through a satellite. The satellite used in remote sensing should move in an orbit around the earth in such a way that it always passes over the particular area of the earth at the same local time.

  The orbit of such a satellite is known as sun-synchronous orbit. A remote sensing orbit can be circular polar orbit or in highly inclined elliptical orbit.

- A remote sensing satellite takes, photographs of a particular region which nearly the same illumination every time it passes through that region.
- The most useful remote sensing technology is that it makes possible the repetitive surveys of vast areas in a very short time, even if the areas are inaccessible.
- Space based remote sensing is a new technology. It has high potential for applications in nearly all aspects of resource management.
- The Indian remote sensing satellites are

  IRS-1A, IRS-1B, and IRS-1C.
- Remote sensing is applied in (i) Meteorology (ii) Climatology (iii) Oceanography

  (iv) Archaeology, geological surveys. (v) Water resource surveys, (vi) Urban land use surveys. (vii) Agriculture and forestry and natural disaster.

  (viii) To detect movements of enemy army. (ix)To locate the place where underground nuclear explosion has carried out.
- **2.LINE COMMUNICATION**

  Line communication means interconnection of two points with the help of wires for exchange of information. There are three principal types

  (i)Two Wire Transmission Line

  (ii) Coaxial wire lines (coaxial cables)

  (iii) optical fibers
- **Two Wire Transmission Linc**

  The most commonly used two wire lines are: Parallel wire, twisted pair wires and co-axial cable.

(1) ***Parallel wire line:*** In a two wire transmission line, two metallic wires (may be hard or flexible) are arranged parallel to each other inside a protective insulation coating . Commonly used to connect an antenna with TV receiver. Such wires can suffer from interferences and losses.

(2) ***Twisted pair wire:*** It consists of two insulated copper wires twisted around each other at regular intervals to minimize electrical interference (to connect telephone systems). Used to connect telephone systems. It works well up to small distances. They cannot transmit signals over very large distances. They transmit both, the analog and digital signals. They are easy to install and cost effective.

- **Coaxial wire lines:** It consists of a central copper wire (which transmits surrounded by a PVC insulation over which a sleeve of copper mesh (outer conductor) is placed. The outer conductor is normally connected to ground and thus it provides an electrical shield to the signals carried by the central conductor. The outer conductor is externally covered with a polymer jacket for protection.

(a) Co-axial line wires can be used for microwaves and ultra-high frequency waves.

(b) The communication through co-axial lines is more efficient than through a twisted pair wire lines.

(c) Co-axial cables can be gas filled also. To reduce flash over between the conductor handling high power, N_2-gas is used in the cable.

- ***Impedance of Line***

  (1)Each portion of the transmission line can be considered as a small inductor, resistor and capacitor. As a result each length of transmission line has characteristic impedance.

(2) In case of co-axial cable, the dielectric can be represented by a shunt resistance G.

(3) When co-axial cable is used to transmit a radio frequency signal, X_L and X_C are large as compared to R and G respectively. Hence R and G can be neglected.

(4) In co-axial cable, R is zero, so no loss of energy and hence no attenuation of frequency signal occurs when transmitted along it. That's why co-axial cables are specially used in cable TV network.

➢ ***Characteristic impedance (Z_0)*** : It is defined as the impedance measured at the input of a line of infinite length.

a)For parallel line $Z_0 = \frac{276}{\sqrt{k}} \log \frac{2s}{d}$

d = Diameter of each wire

s = Separation between the two wires

k = Dielectric constant of the insulating medium

b) For co-axial line wire $Z_0 = \frac{138}{\sqrt{k}} \log \frac{D}{d}$

d = Diameter of inner conductor

D = Diameter of outer conductor

c) At radio frequency $Z_0 = \sqrt{\frac{L}{C}}$

The usual range of characteristic impedance for parallel wire lines is 150W to 600 W and for co-axial wire it is 40 W to 150 W.

➢ **Velocity factor of a line (*v. f.*)** : It is the ratio of reduction of speed of light in the dielectric of the cable

$$v.f. = \frac{v}{c} = \frac{\text{Speed of light in medium}}{\text{Speed of light in vacuum}} = \frac{1}{\sqrt{K}}$$

For a line *v.f.* is generally of the order of 0.6 to 0.9.

➢ **TELEPHONE LINKS**

(1) A telephone (the most common means of communication) link can be established with the help of co-axial cables, ground waves, sky waves, microwaves or optical fiber cables.

(2) Simultaneous transmission of a number of messages over a single channel without their interfering with one another is called multiplexing.

(3) Twisted pair wire lines provide a band width of 2 *MHz*, while co-axial cable provides a band width of 20 *MHz*. For further increase in band width, we use (i) microwave link (ii) communication satellite link.

➢ **OPTICAL COMMUNICATION**

The use of optical carrier waves for transmission of information from one place to another is called optical communication.

The information carrying capacity $\propto$ bandwidth $\propto$ frequency of carrier wave. Because of high frequency (10^{12} *Hz* to 10^{16} *Hz*) optical communication is better than others. (radio and microwave frequencies, 10^6 *Hz* – 10^{11} *Hz*).

Basic optical communication link is a point to point link having transmitter at one end, receiver at the other end and consists of three components namely

(1) Optical source and modulator

(2) Optical signal detector or photodetector (3) Optical fibre cable through which optical signal is transmitted.

➢ **Optical sources for communication links**

Light emitting diodes (LED) and diode lasers are preferred for optical source. LEDs are used for small distance transmission while diode laser is used for very large distance transmission.

For optical communication, light is to be **modulated** with the information signal. The frequency and intensity of light is sensitive to temparature changes,which is to be avoided. So suitable arrangement isrequired to obtain **thermal stability.**

- **Optical signal detector or photodetector:**

The optical signal reaching the receiving end has to be detected by a detector which converts light intoelectrical signals, So that the transmitted information may be decoded.

The optical detector should have

(i) size compatible with the fibre

(ii) High sensitivity at the desired optical wavelength

(iii) High response for fast speed data transmission/reception.

Semiconductor based photo-electors are used because they fullfill the above criteria

- **MODULATION AND ITS NECESSITY :**

Message signals are also called base band signals. Which essentially designate the band of frequencies representing the original signal, as delivered by the source of information. No signal, is a single frequency sinusoid, but it spreads over a range of frequencies called the signal bandwidth to transmit an electric signal frequency less than 20 kHz) over a long distance directly. It is clear that low frequency waves, can not travel long distances. Hence, to transmit low frequency wave over long distance, we take the help of high frequency waves called carrier wave. The low frequency wave is superposed over high frequency carrier wave. This process is called the modulation. The low frequency wave is called the modulating wave and the high frequency wave is called the carrier wave, and the resultant wave is called modulated wave. In this section we will discuss in detail about modulation. What is it? What is the need of modulation or how modulation is done etc.

No signal in general, is a single frequency but it spreads over a range of frequencies called the signal bandwidth. Suppose we with to transmit an electronic signal in the audio frequency (20Hz 20kHz) range over a long distance. Can we do it? No it cannot because of the following problems.

- **Size of antenna :** For transmitting a signal we need an antenna. This antenna should have a size comparable to the wavelength of the signal. For an electromagnetic wave of frequency 20kHz, wave length is 15km. Obviously such a long antenna is not possible and hence direct transmission of such signal is not practical.

The linear size of the antenna must be the order of the wave length and for effective transmission its length

must be $h = \frac{\lambda}{4}$

so that antenna properly senses the time variation of the signal.

Example1: For an electromagnetic wave of

$f = 20$ kHz, $\lambda = 15$ km Obviously, such a long antenna is not possible to construct and operate.

Hence direct transmission of such baseband signals is not practical.

Example2: If $f = 1$ MHz, then $\lambda = 300$ m

$h = 75$ m

Therefore, there is a need of translating the information contained in our original low frequency baseband signal into high or radio frequencies before transmission.

- ***Additional Information:***

(a) The distance between transmitting antenna and the horizon, $D_t = \sqrt{2Rh_t}$.

Where h_t = height of transmitting antenna R = Radius of the earth

(b) The distance between receiving antenna and the horizon, $D_r = \sqrt{2Rh_r}$.

Where h_r = height of receiving antenna

(c) The maximum distance between the transmitting antenna and receiving antenna D_m.

$D_m = D_r + D_t$

$D_m = \sqrt{2Rh_r} + \sqrt{2Rh_t}$.

Where R is the radius of earth.

$h_r > h_t$ so then the receiving antenna intercepts the line of sight waves.

- **Single antenna**

(d) The radius "d" of the area covered by a single transmitting tower of height h is given by $d = \sqrt{2R_e h}$. Where R_e is the radius of the Earth.

(e) If the Population density around the tower is given, the number of persons covered by the tower is = (Area covered by the tower) x Population density No. of persons covered = πd^2 x Population-density.

- **Effective power radiated by an antenna**

Power radiated by an antenna is proportional to $\left(\frac{l}{\lambda^2}\right)$. Where l is length of the antenna .

For a good transmissmission,high powers are required, hence low wavelength i.e high frequency transmissions are needed.

- **Mixing up of signals from different transmitters**

Suppose many people are talking at the same time or many transmitters are transmitting baseband information signals simultaneously. All these signals will get mixed up and there is no simple way to distinguish between them. This points out towards a possible solution by using communication at high frequencies and alloting a band of frequencies to each mesage signal for its transmission.

In doing so, we take the help of a high frequency signal, known as the carrier wave, and a process known as **modulation** which attaches information to it.

- **Modulation:** The process of superimposing information contained in the low frequency , message signal on a high frequency carrier wave, near transmitter is known as **modulation**.

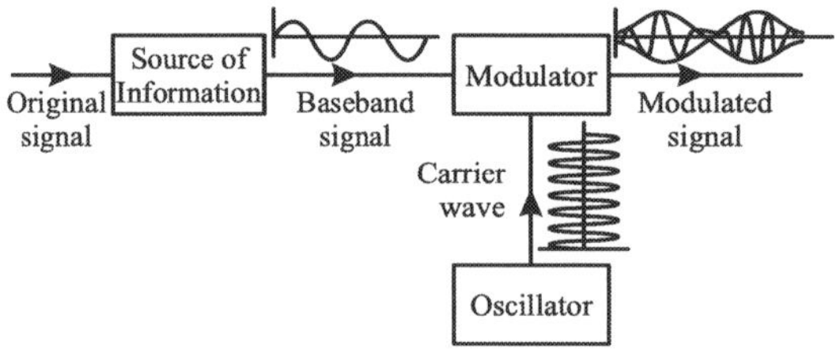

- **TYPES OF MODULATION:**

The carrier wave may be continuous (sinusoidal) or in the form of pulses as shown in figure(2).

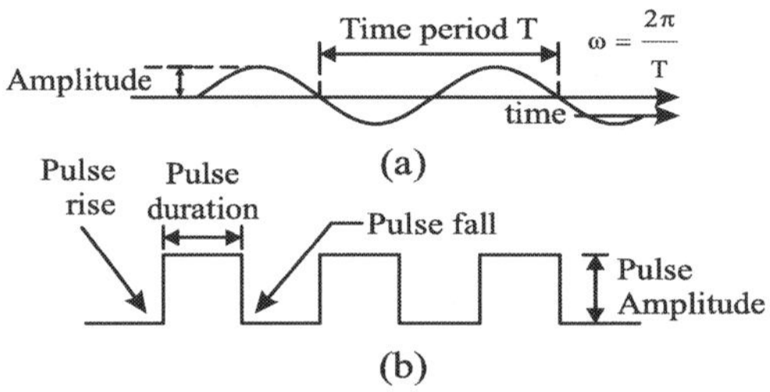

Therefore depending upon the specific characteristicof carrier wave which is being varied in accordance with the message signal, modulation can basically be differentiated as

(i) continuous wave modulation; and

(ii) pulse wave modulation.

- **According to the type of modulation**

For sinusodial continuous carrier waves

i) Amplitude Modulation (AM)

ii) Frequency Modulation (FM)

iii) Phase Modulation

- **For pulsed carrier waves**

i) Pulse Amplitude Modulation (PAM)

ii) Pulse Time Modulation (PTM)

a) Pulse Position Modulation (PPM)

b) Pulse Width Modulation (PWM) or Pulse Duration Modulation (PDM)

iii) Pulse Code Modulation (PCM)

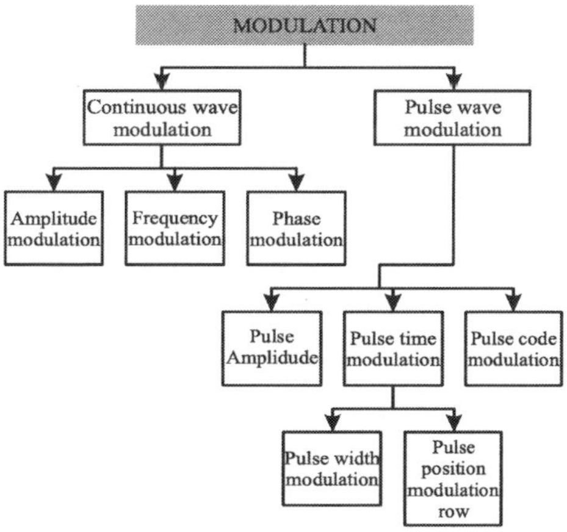

I) Continuous Wave Modulation

Eqution representing sinusoidal carrier wave can

$c(t) = A_c \sin(\omega_c t + \phi)$ ----(1)

where c(t) is the signal strength (voltage or current),

A_c is the amplitude

$(\omega_c t + \phi)$ is called argument of Phase angle of the carrier wave

$\omega_c (= 2\pi f_c)$ is the angular frequency

ϕ is the initial phase of the carrier wave. During the process of modulation, any of the two parameters, viz amplitude or phase angle, of the carrier wave can be controlled by the message or information signal. This results in two **types of modulations:**

i) Amplitude modulation (AM)

ii) Angle modulation

Angle modulation again can be of two types. They are

i) Frequency modulation (FM)

ii) Phase modulation (PM)

As shown in figure.

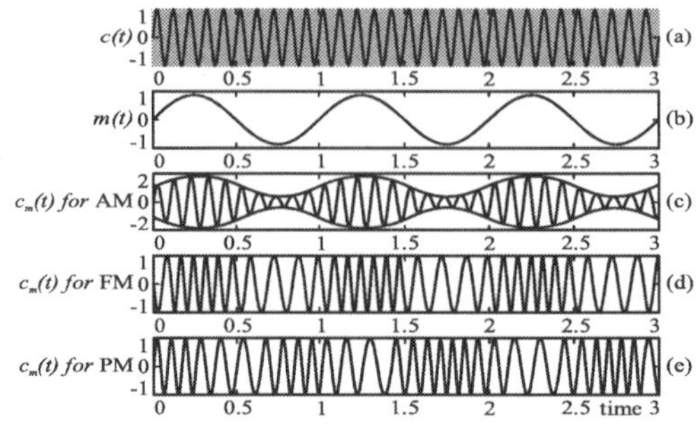

II) Pulse Wave Modulation.

The significant characteristics of a pulse are: i) Pulse amplitude

ii) Pulse duration or pulse width

iii) pulse position (denoting the time of rise or fall of the pulse amplitude) as shown in figure (3).

Types of pulse modulation:

a) pulse amplitude modulation (PAM),

b) pulse duration modulation (PDM) or pulse width modulation (PWM)

c) pulse position modulation (PPM).

I Continuous Wave Modulation:

1) Amplitude Modulation:

The method in which the amplitude of carrier is varied in accordance with the modulating signal keeping the frequency and phase of carrier wave constant is called amplitude modulation (AM).

Here we explain amplitude modulation process using a sinusoidal signal as the modulating signal.

Let $m(t) = E_m \sin \omega_m t$ represent the message or the modulating or base band signal. Here $\omega_m = 2\pi f_m$ is the angular frequency of the message signal.

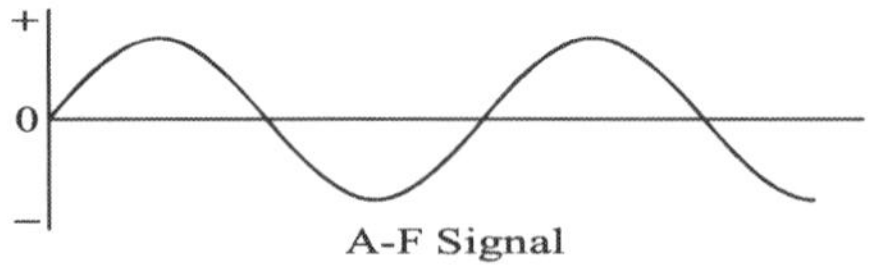

A-F Signal

$c(t) = E_c \sin \omega_c t$ represent carrier wave. Here $\omega_c = 2\pi f_c$ is the angular frequncy of the carrier signal

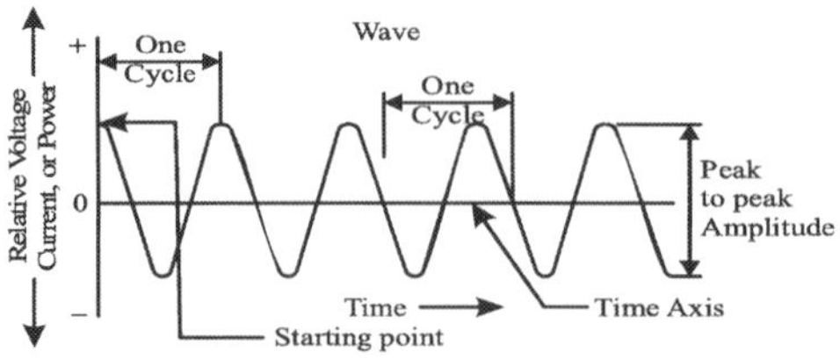

The modulated signal $c_m(t)$ can be written as

$$c_m(t) = (E_c + E_m \sin\omega_m t)\sin\omega_c t \text{ ----(1)}$$

$$c_m(t) = A_c\left(1 + \frac{E_m}{E_c}\sin\omega_m t\right)\sin\omega_c t \text{ -----(2)}$$

$$c_m(t) = E_c \sin\omega_c t + \mu E_c \sin\omega_m t \sin\omega_c t \text{ --(3)}$$

Using the trignometric relation

$\sin A \sin B = \frac{1}{2}(\cos(A-B) - \cos(A+B))$, we can write $c_m(t)$ of equation (3) as

$$c_m(t) = A_c \sin\omega_c t + \frac{\mu A_c}{2}\cos(\omega_c - \omega_m)t - \frac{\mu A_c}{2}\cos(\omega_c + \omega_m)t \text{ -----(4)}$$ Here

$\omega_c - \omega_m = 2\pi(f_c - f_m)$ = Lower side band frequency (LSB)

$\omega_c + \omega_m = 2\pi(f_c + f_m)$ = Upper side band frequency (USB)

Here (m or $\mu = E_m/E_c$) is the modulation index; (or) modulating factor.

In practice, μ is kept ≤ 1 to avoid distorition.

Depth of modulation $= \frac{A_m}{A_C} \times 100 = \mu \times 100$

Depth of modulation interms of E_{max} and E_{min}

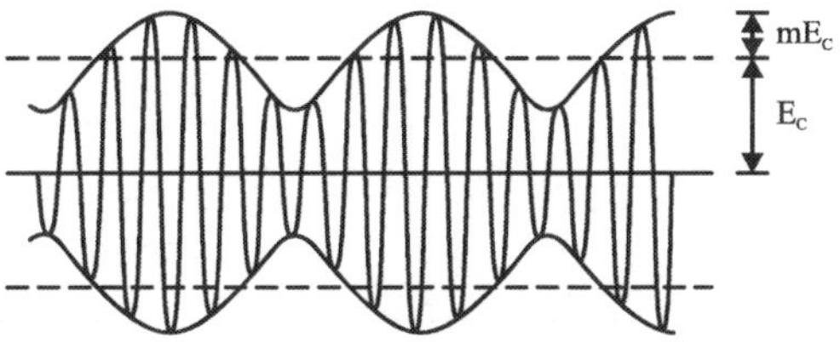

A.M. Wave

$$E_{max} = E_c + E_m = E_c(1+m)$$

$$E_{min} = E_c - E_m = E_c(1-m)$$

$$\frac{E_{max}}{E_{min}} = \frac{E_c + E_m}{E_c - E_m} = \frac{E_c(1+m)}{E_c(1-m)}$$

$$\boxed{\frac{E_{max} - E_{min}}{E_{max} + E_{min}} = m_a}$$

The Band width of AM wave is " $2f_m$ "

The modulated signal now consists of the carrier wave of frequency ω_c plus two sinusoidal waves each with a frequency slightly different from ω_c, known as side bands. The frequency spectrum of the amplitude modulated signal is shown in

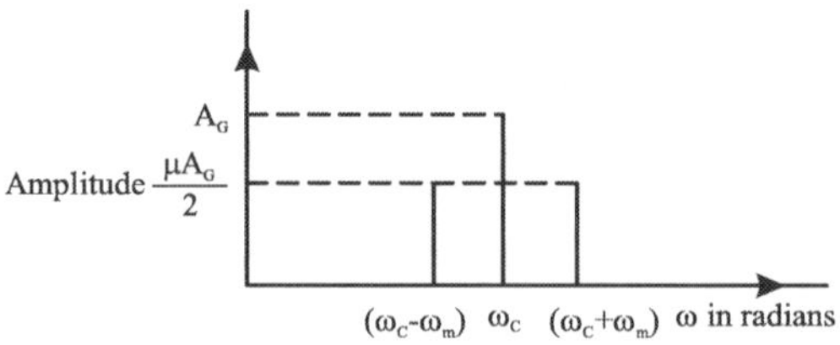

figure (5) A plot or amplitude versus ω for an amplitude modulated signal

As long as the broadcast frequencies (carrier waves) are sufficiently spaced out so that sidebands do not overlap, different stations can operate without interfering with each other.

Special cases of Amplitude modulation:

CaseI: In the absence of signal.

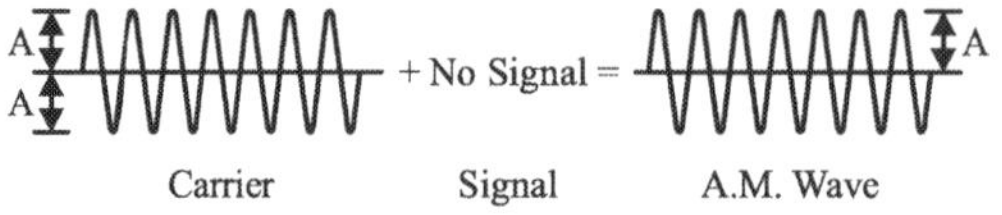

Modulation factor $m_a = \frac{0}{A} \times 100 = 0\%$

Case-II: When the signal amplitude is equal to CW wave.

Amplitude varies from 2A to zero.

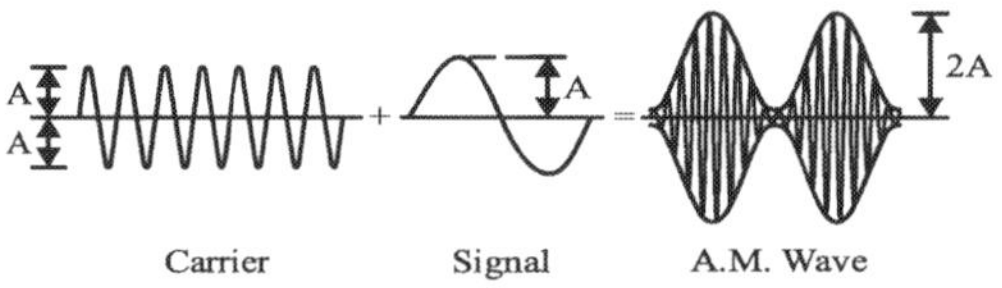

$$\frac{\text{Amplitude change in carrier wave}}{\text{Amplitude of CW}} = \frac{2A - A}{A} = 100\%$$

Case-III: When the amplitude of the signal is half of that of CW.

Amplitude of CW changes from A to $\left(A + \frac{A}{2}\right) = 1.5\,A$

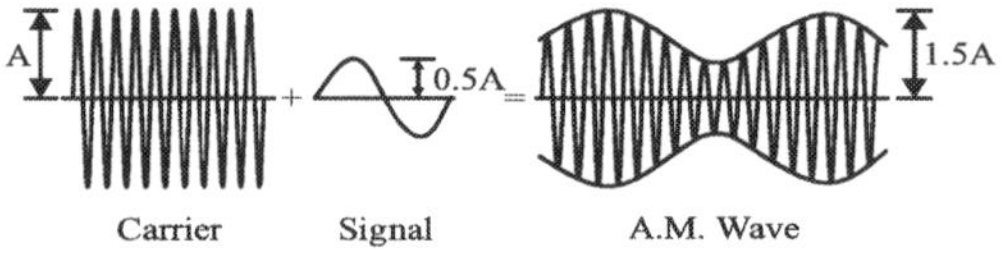

Modulation factor $= \frac{0.5\,A}{A} = 0.5$

$= 50\%$

Case-IV: When the amplitude of signal is 1.5 times that of the CW.

Amplitude of the modulated wave changes from 2.5 A to A

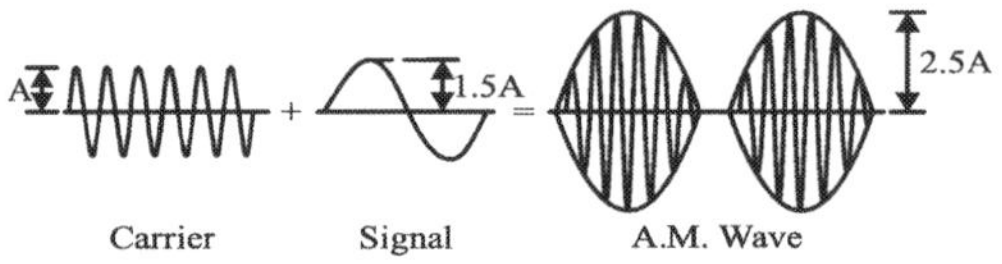

Modulation factor $m_a = \frac{2.5A - A}{A} = 1.5 = 150\%$.

In this case the quality of signal is lost

Note: A carrier wave is modulated by a number of sine waves with modulation indices m_1, m_2 and m_3. The total modulation index of the wave is

$$m = \sqrt{m_1^2 + m_2^2 + m_3^2}$$

- **Power out put in AM wave**

$$P_t = P_c + P_s$$

where P_t is power transmitted

P_c is power of carrier wave

P_s is total power of side bands

The equation of a carrier wave $Y_c = A_c \sin(w_c t + \phi)$

Power of carrier wave $P_c = \frac{[A_{rms}]^2}{R} = \frac{\left(\frac{A_c}{\sqrt{2}}\right)^2}{R} = \frac{A_c^2}{2R}$

The power of side bands = The power of lower side band + the power of upper side band

$$P_s = \frac{\left[\frac{(\mu A_c)}{2}/\sqrt{2}\right]^2}{R} + \frac{\left(\mu\frac{A_c}{2}/\sqrt{2}\right)^2}{R}$$

$$= \frac{\mu^2 A_c^2}{4R} = \frac{\mu^2}{2} P_c$$

$$P_t = P_c + P_s \quad = P_c + \frac{\mu^2}{2} P_c$$

$$= P_c\left(1 + \frac{\mu^2}{2}\right)$$

$$\therefore P_t = P_c\left(1 + \frac{\mu^2}{2}\right) ----(1)$$

$$\Rightarrow \frac{P_t}{P_c} = 1 + \frac{\mu^2}{2} ----(2)$$

$$\Rightarrow \left(\frac{i_t}{i_c}\right)^2 = 1 + \frac{\mu^2}{2} ----(3)\left[\because P \alpha i^2\right]$$

Example: If the modulation factor is 1 ie 100 % modulation then the useful power is 1/3 of the total power radiated. The remaining 2/3 power is contained by carrier wave

$$\frac{P_S}{P_T} = \frac{m_a^2}{2+m_a^2} = \frac{1}{3} \quad \text{and} \quad \frac{P_c}{P_T} = \frac{2}{2+m_a^2} = \frac{2}{3}$$

Transmission Efficiency $\eta = \dfrac{m^2}{2+m^2}$

PRODUCTION OF AMPLITUDE MODULATED WAVE

Amplitude modulation can be produced by a variety of methods. A conceptually simple method is shown in the block diagram of Fig.

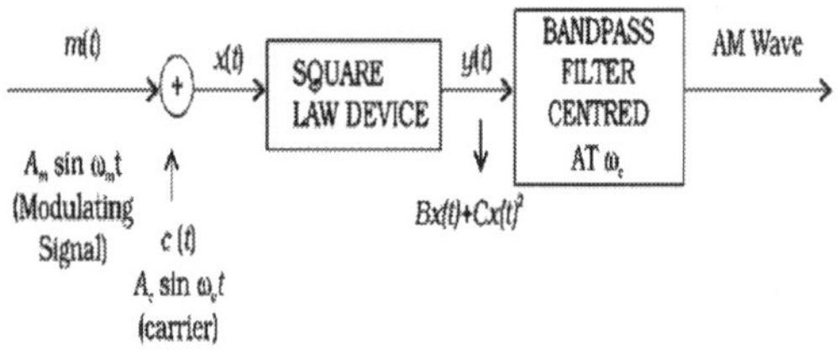

Here the modulating signal $A_m \sin \omega_m t$ is added to the carrier signal $A_c \sin \omega_c t$ to produce signal

$$x(t) = A_m \sin \omega_m t + A_c \sin \omega_c t$$

This signal is passed through a square law device which is a non linear device that can give tha output

$y(t) = Bx(t) + Cx^2(t)$ where B and C are constants

Thus,

$$y(t) = BA_m \sin \omega_m t + BA_c \sin \omega_c t$$

$$+C\left[A_m^2 \sin^2 \omega_m t + A_c^2 \sin^2 \omega_c t + 2A_m A_c \sin \omega_m t \sin \omega_c t\right] \text{-(6)}$$

$$= BA_m \sin \omega_m t + BA_c \sin \omega_c t + \frac{CA_m^2}{2} + A_c^2 - \frac{CA_m^2}{2}\cos 2\omega_m t - \frac{CA_c^2}{2}\cos 2\omega_c t$$

$$+CA_m A_c \cos(\omega_c - \omega_m)t - CA_m A_c \cos(\omega_c + \omega_m)t \text{ -(7)}$$

Where the trignometric relations $\sin^2 A = (1-\cos 2A)/2$ and the relation for sinA sinB mentioned earilier are used.

In equation, there is a dc term $c/2\left(A_m^2 + A_c^2\right)$ and sinusoids of frequencies

$\omega_m, 2\omega_m, \omega_c, 2\omega_c, \omega_c - \omega_m$ and $\omega_c + \omega_m$. As shown in figure this signal is passed through a band pass filter which rejects dc and the sinusoids of frequencies $\omega_m, 2\omega_m$ and $2\omega_c$ and retains the frequencies $\omega_c, \omega_c - \omega_m$ and $\omega_c + \omega_m$. The output of the band pass filter therefore is of the same form as equation(4) and is therefore an AM wave.

It is to be mentioned that the modulated signal cannot be transmitted as such. The modulator is to be followed by a power amplifier which provides the necessary power and then the modulated signal is fed to an antenna of appropriate size for radiation as shown in figure(7).

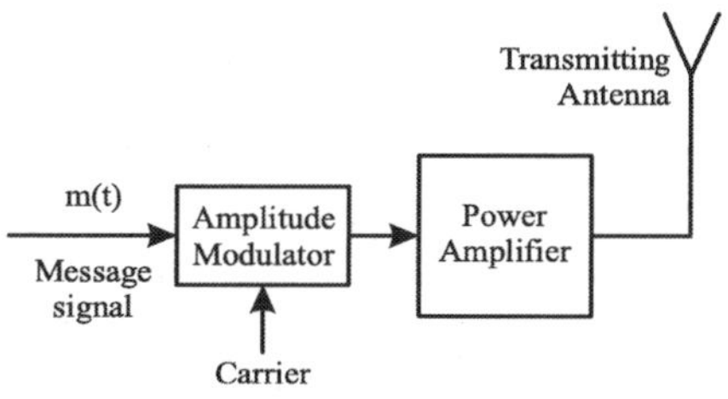

Figure(7)

- **A-M Transmitter.**

  In the block diagram of the AM transmitter the r-f section consists of an oscillator feeding a buffer, which in turn feeds a system of frequency multipliers and/or intermediate power amplifiers. If frequency multiplication is unneccessary, the buffer feeds directly into the intermediate power amplifiers which, in turn, drive the final power amplifier. The input to the antenna is taken from the final power amplifier.

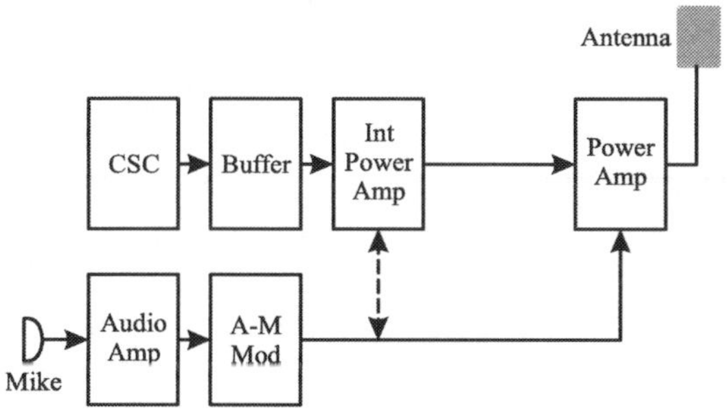

Amplitude-Modulated Transmitter

- **Detection of amplitude modulated wave**

  The transmitted message gets attenuated in propagating through the channel. The receiving antenna is therefore to be followed by an amplifier and a detector. In addition, to facilitate further processing, the carrier frequency is usually changed to a lower frequency by what is called an intermediate frequency (IF) stage preceding the detection. The detected signal may not be strong enough to be made use of and hence is required to be amplified. A block diagram of a typical receiver is shown in figure.

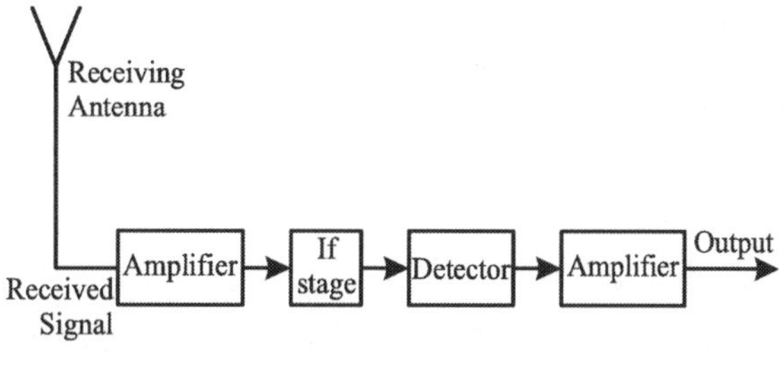

- **Simple demodulator circuit :**

  1) A diode can be used to detect or demodulate an amplitude modulated (AM) wave.

  2) A diode basically acts as a rectifier *i.e.* it reduces the modulated carrier wave into positive envelope only.

  3) The AM wave input is shown in figure. It appears at the output of the diode across PQ as a rectified wave (since a diode conducts only in the positive half cycle). This rectified wave after passing through the RC network does not contain the radio frequency carrier component. Instead, it has only the envelope of the modulated wave.

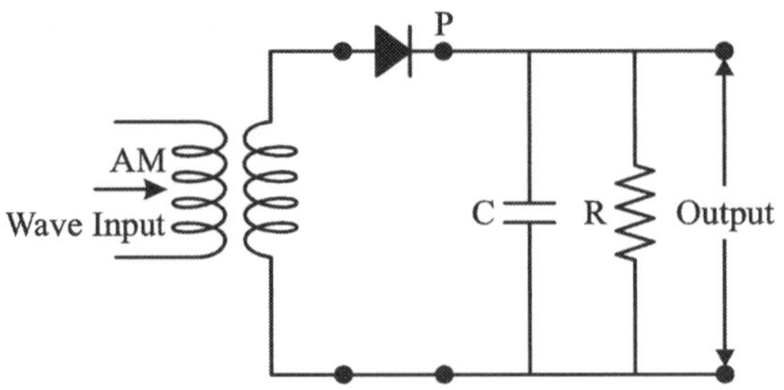

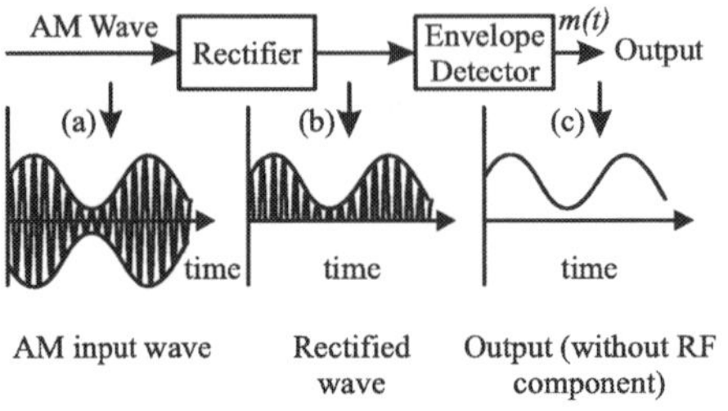

The capacitor connected in parallel with resistance R provides very low impedance at the carrier frequency and a much higher impedance at the modulating frequency. As a result capacitor effectively shorts or filters out the carrier, there by leaving the original modulating signal

In the actual circuit the value of RC (The time constant, t=RC) is chosen such that $\frac{1}{f_c} << RC$; where f_c = frequency of carrier signal.

Distortion in diode detectors: There are two types of distortions in diode detectors. Namely

a) Negative peak clipping

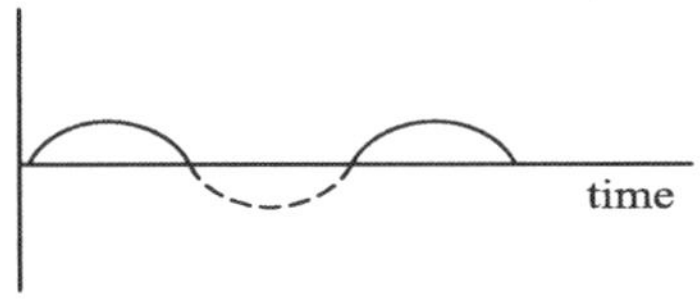

Figure shows the negative peak missing in the output message We know that

Modulation Index(m)= $\frac{E_m}{E_C} = \frac{I_m}{I_c}$ where $I_m = \frac{E_m}{Z_m}$ and $I_c = \frac{E_c}{R_c}$

Here z_m is audio load resistance of diode and R_c is dc diode resistance while $z_m < R_c$

hence $I_m > I_c$. This makes the modulation index in the demodulated wave higher than it was in modulated wave applied at the detector. In turn there is an increase in the chance of over modulation for modulation index nearer to 100% Due to this over modulation there is a negative clipping of the detector wave.

b) Diagonal clipping:

1. Diode ac load may no longer be purely resistive, it can contain reactive component.
2. At high modulation depths current will be changing so fast that the time constant of the load may be too slow to follow the changes. As a result current will decay exponentially. Hence output voltage follows the discharge law of the CR circuit

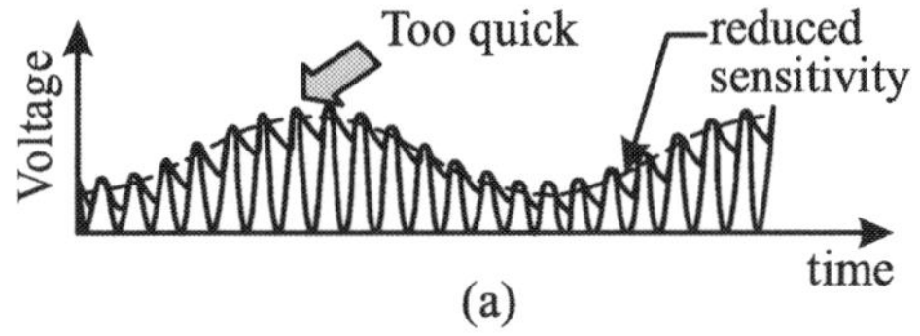

(a)

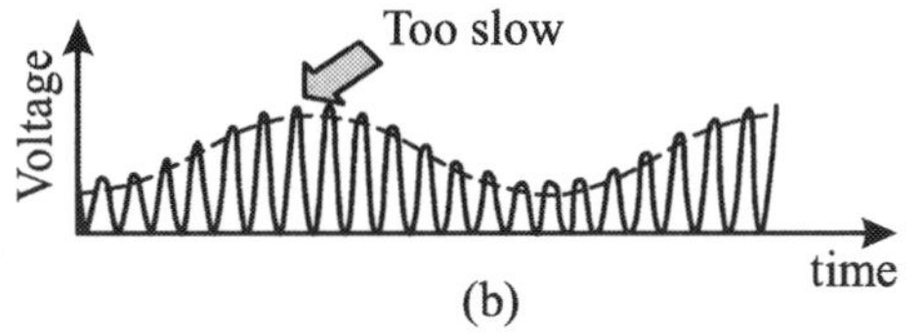

(b)

3. Condition necessary for avoiding distortion of this type is as follows.

$$I_m = \frac{E_m}{Z_m} = \frac{mE_c}{Z_m} \text{ and } I_c = \frac{E_c}{R_c}$$

$$m_d = \frac{I_m}{I_c} = \frac{mE_c/Z_m}{E_c/R_c} = \frac{mR_c}{Z_m}$$

Maximum value of $m_{dMaximum} = 1$

So maximum permissible transmitted modulation index will be $M_{maximum} = m_d \frac{Z_m}{R_c} = 1 \times \frac{Z_m}{R_c}$

W.E-1: ***The load on an Am diode detector consists of a resistance of 50 KΩ in parallel with a capacitor of 0.001 μF. Determine the maximum modulation index that the detector can handle without distortion when modulation frequency is (i) 1 kHz (ii) 5 kHz.***

Sol:

$$Z_m = R_c \| C = \frac{1}{\sqrt{\frac{1}{(R_c)^2} + \frac{1}{(X_c)^2}}}$$

$$M_{\max} = \frac{Z_m}{R_c}$$

$$= \frac{1}{R_c\sqrt{\frac{1}{(R_c)^2} + \frac{1}{(X_c)^2}}}$$

$$= \frac{1}{\sqrt{1 + (2\pi fCR_c)^2}}$$

For f=1 kHz

$$M_{\max} = \frac{1}{\sqrt{1 + 0.098696}} = 0.945$$

for f= 5 kHz

$$M_{max} = = \frac{1}{\sqrt{1+2.4674}} = 0.537$$

- **Limitation of amplitude modulation**

  (i) Noisy reception

  (ii) Low efficiency

  (iii) Small operating range

  (iv) Poor audio quality

2) Angle modulation:

The angle of the carrier wave is varied according to the base band signal while the amplitude is maintained constant. This method provides better discrimination against NOISE and INTERFERENCE then the Amplitude modulation

There are two ways of varying the angle of the carrier.

$\theta_i(t)$ is the angle of moduilated sinusoidal carrier assumed to be a function of the message signal.

Equation of the angle modulated wave is $S(t) = A_c \cos(\theta_i(t))$

where A_c is amplitude of the carrier.

Instantaneous frequency of the angle modulated wave is $f_i(t) = \frac{1}{2\pi}\frac{d\theta_i(t)}{dt}$

here $\frac{d\theta_i(t)}{dt}$ is the angular velocity of the PHASOR of length A_c

a) By varying the frequency ω_c, Frequency Modulation.

b) By varying the phase ϕ_c, Phase Modulation.

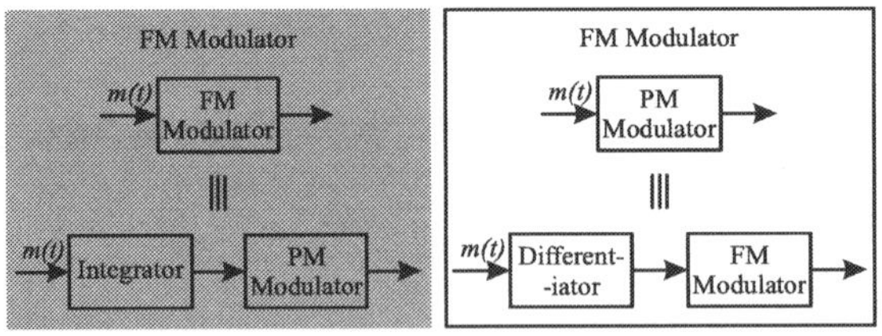

A comparison of FM and PM modulators.

i)Frequency modulation (FM)

The method in which the frequency of carrier is varied in accordance to the modulating signal, keeping the amplitude and phase of the carrier the same is called **Frequency modulation** (FM)

a) Audio quality of AM transmission is poor. There is need to eliminate amplitude sensitive noise. This is possible if we **eliminate amplitude variation.**

b) In FM the overall **amplitude** of FM wave remains constant at all times.

c) In FM, the total transmitted **power** remains constant.

d) ***Frequency deviation***: The maximum change in frequency from mean value (f_c) is known as frequency deviation. This is also the change or shift either above or below the frequency f_c and is called as frequency deviation.

$\delta = (f_{max} - f_c) = f_c - f_{min} = k_f \cdot \frac{E_m}{2\pi}$

k_f = Constant of proportionality.

It determines the maximum variation in frequency of the modulated wave for a given modulating signal.

e) Carrier swing (CS): The total variation in frequency from the lowest to the highest is called the carrier swing *i.e.* $CS = 2\ \delta f$

f) Frequency modulation index (m_f-) :

The ratio of maximum frequency deviation to the modulating frequency is called modulation index.

$$m_f = \frac{\delta}{f_m} = \frac{f_{max} - f_c}{f_m} = \frac{f_c - f_{min}}{f_m} = \frac{k_f E_m}{f_m}$$

g) Frequency spectrum: FM side band modulated signal consist of infinite number of side bands whose frequencies are

$$(f_c \pm f_m), (f_c \pm 2f_m), (f_c \pm 3f_m).......$$

The number of side bands depends on the modulation index m_f.

Band width $= 2n \times f_m$; where n = number of significant side band pairs

h) **Deviation ratio:** The ratio of maximum permitted frequency deviation to the maximum permitted audio frequency is known as deviation ratio. $(= \frac{(\Delta f)_{max}}{(f_m)_{max}})$

i) Percent modulation: The ratio of actual frequency deviation to the maximum followed frequency deviation

$$m = \frac{(\Delta f)_{actual}}{(\Delta f)_{max}}$$

j) Frequency spectra of fm waves under various conditions.

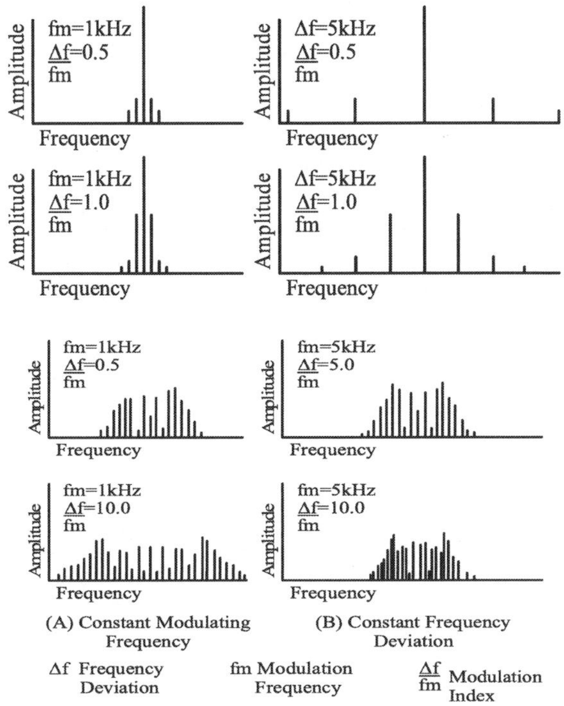

W.E-2: ***The carrier frequency is 500 kHz. The modulating frequency is 15 kilohertz and the deviation frequency is 75 kilohertz. Find***

a) modulation index b) Number of side bands c) Band width

Sol: $MI = \dfrac{\Delta f}{f_m} = 5$

We can have that there are 16 significant sidebands for a modulation index of 5.

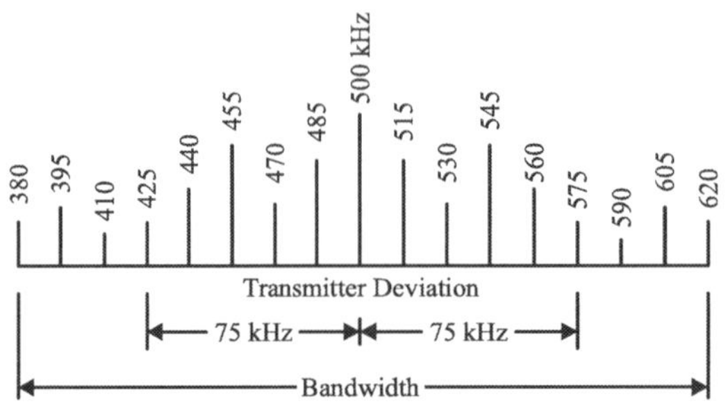

To determine total bandwidth for this case,

we use: $bW = f_m \times (\text{number of sidebands})$

bW = 15 x 16=240 kHz

ii) Phase Modulation (PM), phase of carrier is varied in accordance with modulating signal keeping namplitude and frequency constant. We use the term phase shift to characterize such changes. If phase changes after cycle k, the next sinusoidal wave will start slightly later than the time at which cycle k completes.

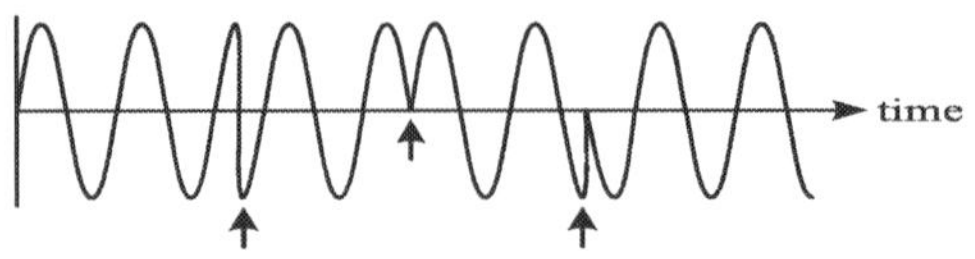

II) pulse wave modulation.

Here the carrier wave is in the form of pulses. Pulse modulation is an analog process as the modulating signal is analog.

The common pulse modulating systems are:

i) Pulse amplitude modulation(PAM):

The amplitude of the pulse varies in accordance with the modulating signal.

ii)Pulse time modulation (PTM):

Or

Pulse duration modulation(PDM)

a) Pulse width modulation (PWM):

The pulse duration varies in accordance with the modulating signal, or the width of the unmodulated signal is constant.

b) Pulse position modulation (PPM):

In PPM, the position of the pulses of the carrier wave train is varied in accordance with the instantaneous value of the modulating signal.

iii) Pulse code modulation (PCM):

The pulse amplitude, pulse width and pulse position modulations are not completely digital. A completely digital modulation is obtained by pulse code modulation (PCM) by following three operations.

(a) Sampling: It is the process of generating pulses of zero width and of amplitude equal to the instantaneous amplitude of the analog signal. The number of samples taken per second is called sampling rate.

(b) Quantization: The process of dividing the maximum amplitude of the analog voltage signal into a fixed number of levels is called quantization.

(c) **Coding:** The process of digitizing the quantised pulses according to some code is called coding.

➢ **Digital Communication And Quantisation Of Message Signal**

(Data Transmission and Retrieval)

Data means facts, concepts or instructions suitable for communication, interpretation or processing by human beings or by automatic means. Data in most cases consists of pulse type of signals.

In digital communication the modulating signals are discrete and are coded as represen tation of message signals to be transimitted. There are a number of *encoding steps in* digital communication, which makes its circutary complicated. Digital communication is error free and noise free.

The source encoder converts the informa tion into binery code. Encoder first digitise the analog waveform. Some times an additional encoding called *channel encoding* is carried out. In the final step, before transmission,the channel codes modulate a continuous wave form.

The pulse code modulated (PCM) signal is a series of 1's and 0's. The following three modulation techniques are used to transmit a PCM signal.

(1) Amplitude shift keying (ASK):

Two different amplitudes of the carrier represent the two binary values of the PCM signal. This method is also known as on-off keying (OOK)

1: Presence of carrier of same constant amplitude.

0: Carrier of zero amplitude.

(2) Frequency shift keying (FSK): The binary values of the PCM signal are represented by two frequencies.

1: Increase in frequency

0: Frequency unaffected

(3) Phase shift keying (PSK): The phase of the carrier wave is changed in accordance with modulating data signal.

1: Phase changed by p

0: Phase remains unchanged.

The analog signal is sampled by the sampler. The sampled pulses are then quantized. The encoder codes the quantized pulses according to the binary codes. After modulating the PCM signal (by ASK, FSK or PSK method)the modulated signal is, then transmitted into free space in the form of bits.

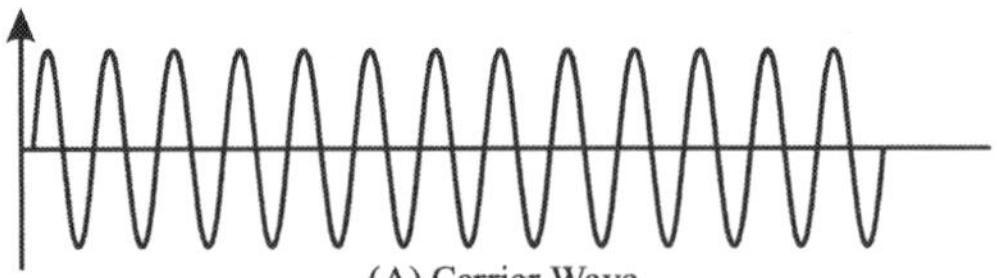

(A) Carrier Wave

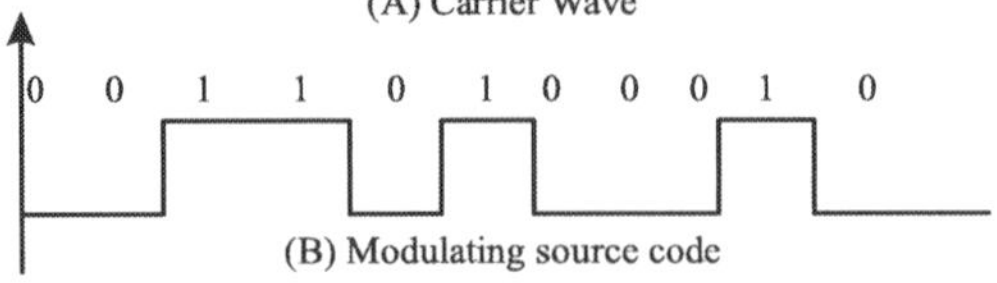

(B) Modulating source code

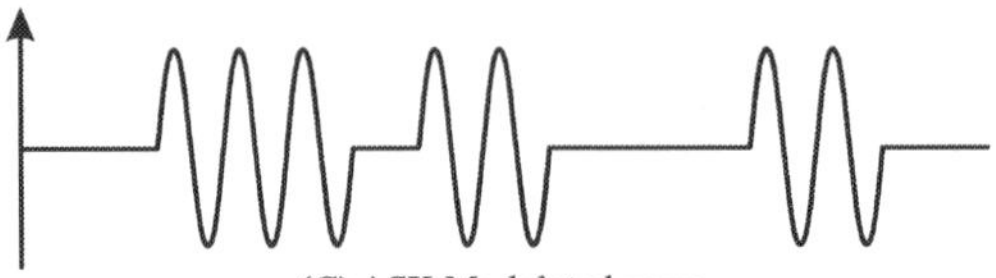

(C) ASK Modulated wave

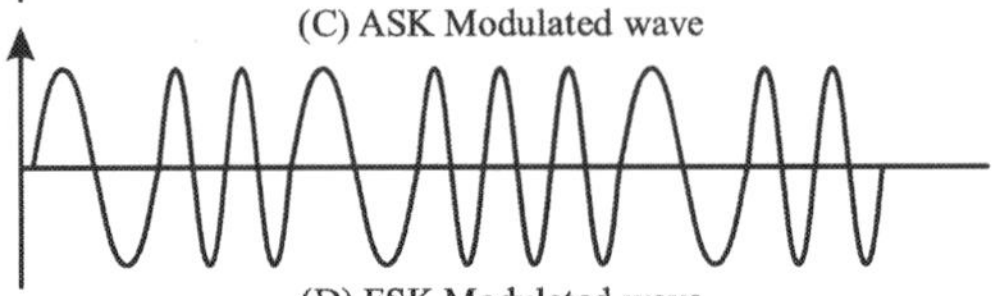

(D) FSK Modulated wave

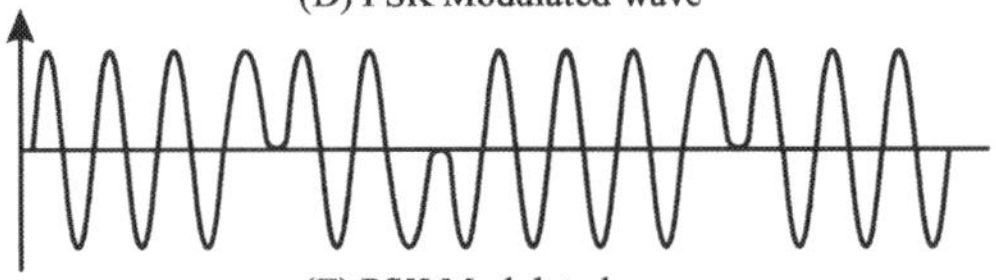

(E) PSK Modulated wave

➢ **MODEM AND FAX**

(1) **Modem:** Modems are used to interface two digital sources/receivers.

i) Word modem implies MODulator and DEModulator. Both the functions (modulation and demodulation) are included in a signal unit.

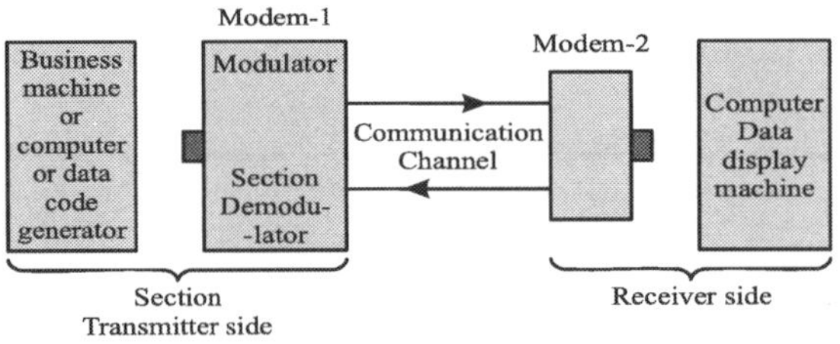

ii) Modems are placed at both ends of the communication circuit.

iii) The modem at the transmitting station changes the digital output from a computer to an analog signal, which can be easily sent via communication channel. While the receiving modem reverses the process.

iv)There are three modes of operation of a modem.

(a) Simplex mode: Data is transmitted in only one direction.

(b) Half duplex: Data is transmitted between the transmitter and the receiver in both direction, but only in one direction at a time.

(c) Full duplex: Data are transmitted between the transmitter and receiver in both directions at the same time.

(2) Fax (Facsimile transmission): The electronic reproduction of a document at a distance place is known as facsimile transmission (FAX).

The original written document is converted into transmittable codes and is converted in to electrical signals, which are then modulated and transmitted on to the receiving end.

- **The Internet :** It is a system with billions of users worldwide. It permits communication and sharing of all types of information between any two or more computers connected through a large and complex network. It was started in 1960's and opened for public use in 1990's. With the passage of time it has witnessed tremendous growth and it is still expanding its reach. Its applications include.
- **Email:** It permits exchange of text/ graphic material using email software. We can write a letter and send it to the recipient through ISP's (internet Service Providers) who work like the dispatching and receiving post offices.
- **File transfer:** A FTP (File Transfer Programmes) allows transfer of files/software from one computer to another connected to the internet.
- **World Wide Web (WWW):** Computers that store specific information for sharing with others provide websites either directly or through web service providers. Government departments, companies, NGO's (Non-Government Organizations) and individuals can post information about their activities for resticted or free use on their websites. This information becomes accessible to the users. Several search engines like Google, Yahoo!etc. help us in finding information by listing the related websites. Hypertext is a powerful feature of the web that automatically links relevant information from one page on the web to another using HTML (hypertext markup language)
- **E-commerce:** Use of the internet to promote business using electronic means such as using credit cards is called E-commerce. Customers view images and receive all the information about various products or services of companies through their websites. They can do on-line shopping from home/office. Goods are dispatched or services are provided by the company through mail/courier.
- **Chat:** Real time conversation among people with common interests through typed messages is called chat. Everyone belonging to the *chat group* gets the message instantaneously and can respond rapidly.
- **D.Mobile telephony :** The concept of mobile telephony was developed first in 1970's and it was fully implemented in the following decade. The central concept of this sytem is to divide the service area into a suitable number of *cells* centered on an office called MTSO (Mobile Telephone Switching Office). Each cell contains a low-power transmitter called a base station and caters to a large number of mobile receivers (popularly called cell phones). Each cell could have a service area of a few square kilometers or even less depending upon the number of customers. When a mobile receiver crosses the coverage area of one base station, it is necessary for the mobile user to be transferred to another base staion. This process is called *handover or handoff*. This process is carried out very rapidly, to the extent that the consumer does not even notice it. Mobile telephones operate typically in the UHF range of frequencies (about 800-950 MHz)

EXAMPLE PROBLEMS

W.E-3: ***How many AM broadcast stations can be accommodated in a 100 kHZ bandwidth if the highest modulating frequency of carrier is 5 kHZ?***

Sol. Any station being modulated by a 5 kHz singal will produce an upper side frequency 5 kHz above its carrier and a lower side frequency 5 kHz below its carrier, thereby requiring a bandwidth of 10 kHz. Thus, Number of stations accommodated

$$\frac{\text{Total bandwidth}}{\text{Bandwidth per station}}=\frac{100}{10}=10$$

W.E-4: ***How many 500 kHz waves can be on a 10km transmission line simultaneously?***

Sol: Let λ be the wavelength of 500 kHz signal. Then, $\lambda=\frac{c}{f}=\frac{3.0\times10^8}{5.0\times10^5}m=600m$

The number of waves on the line can be found from,

$$n=\frac{d}{\lambda}=\frac{10\times10^3}{600}=16.67$$

W.E-5: ***A two wire transmission line has a capacitance of 20 pF/m and a characteristic impedance of*** 50Ω

a) What is the inductance per metre of this cable?

b) Determine the impedance of an infinitely long section of such cable.

Sol: a) The characteristic impedance. $Z=\sqrt{L/C}$ $L=(Z^2)(C)=(50)^2(20\times10^{-12})H=0.05H/m$

b) The characteristic impedance of a transmission line is the impedance that an infinite length of line would present to a power supply at the input end of the line. Thus, $Z_\infty=Z_0=50\Omega$

W.E-6: ***T.V. transmission tower at a particular station has a height of 160m.***

a) What is the coverage range?

b) How much population is covered by transmission, if the average population density around the tower is 1200 per km^2 ***?***

c) What should be the height of tower to double the coverage range

Sol: a) Coverage range $d=\sqrt{2Rh}$

$=\sqrt{2\times6400\times10^3\times160m}$

$=45.254km$

b) Population covered

$=(\textit{population density})\times(\textit{area}\text{ covered})$

$=(1200)\times(\pi d^2)$

$=(2400\pi Rh)=2400\times3.14\times6.4\times10^3\times0.16$

$=77.17\,lac$

c) Coverage range $\propto\sqrt{h}$

Therefore coverage range can be doubled by making height of the tower four times to 640m. So, height of the tower should be increased by 480 m.

W.E-7: ***An audio signal given by*** $e_s=15\sin2\pi(200t)$ ***modulates a carrier wave given by*** $e_s=60\sin2\pi(100{,}000t)$. ***If calculate***

a) Percent modulation

b) Frequency spectrum of the modulated wave.

Sol: a) Signal Amplitude, $B=15$

Carrier amplitude, $A=60$

$$m=\frac{B}{A}=\frac{15}{60}=0.25$$

$\therefore$ Percentage modulation $=0.25\times100=25\%$

b) By comparing the given equations of signal and carrier with their standard form

$e_s=E_s\sin\omega_s t=E_s\sin2\pi f_s t$ and

$e_c=E_c\sin\omega_c t=E_c\sin2\pi f_c t$

we have signal frequency $f_s = 2000Hz$ and carrier frequency $f_c = 100,000Hz$

The frequencies present in modulated wave

i) $f_c = 100,000Hz = 100kHz$ ii) $f_c - f_s = 100,000 - 2000 = 98kHz$

iii) $f_c + f_s = 100kHz + 2kHz = 102kHz$

Therefore, frequency spectrum of modulated wave extends from 98kHz to 102 kHz is called band width.

W.E-8: ***The antenna current of an AM transmitter is 8A when only the carrier is sent but it increases to 8.93A when the carrier is modulated. Find percent modulation.***

Sol. The modulated or total power carried by AM wave $P_T = P_C\left(1+\frac{m^2}{2}\right)$. If R is load resistance. I_m is the current when carrier is modulated and I_c the current when unmodulated, then

$$\frac{P_T}{P_C} = \frac{I_m^2 R}{I_c^2 R}\ ; \qquad \therefore I + \frac{m^2}{2} = \frac{I_m^2 R}{I_c^2 R}$$

Given $I_m = 8.93A, I_c = 8A$

$$\therefore m^2 = 2\left[\left(\frac{8.93}{8.0}\right)^2 - 1\right] \therefore m = 0.7$$

Therefore, percentage modulation = 70%

W.E-9: ***A sinusoidal carrier voltage of 80 volts amplitude and 1 MHz frequency is amplitude modulated by a sinusoidal voltage of frequency 5kHz producing 50% modulation. Calculate the amplitude and frequency of lower and upper side bands.***

Sol: Amplitude of both LSB and USB are equal and given by

$$= \frac{mE_c}{2} = \frac{0.5 \times 80}{2} = 20volts$$

Now frequency of LSB = $f_c - f_s = (1000-5)kHz = 995kHz$

Frequency of USB = $f_c + f_s = (1000+5)kHz = 1005kHz$

W.E-10: ***The load current in the transmitting antenna of an unmodulated AM transmitter is 6 Amp. What will be the antenna current when modulation is 60%.***

Sol: Total power carried by AM wave

$$P_T = P_C\left(1+\frac{m^2}{2}\right)......(1)$$

Where P_c is the power of carrier component and m is the modulation factor. If R is the resistance, I_m the antenna load current when modulation is 60% and I_c is the antenna load current when un modulated, then

$$\frac{P_T}{P_C}=\frac{I_m^2R}{I_c^2R}, \therefore 1+\frac{m^2}{2}=\frac{I_m^2}{I_c^2} \text{ using (1)}$$

or $I_m = I_c\sqrt{\left\{\left(1+\frac{m^2}{2}\right)\right\}}$

Given $I_c = 6Amp, m = 0.6$

$$I_m = 6\left[1+\frac{(0.6)^2}{2}\right]^{1/2} = 6[1.086] = 6.52Amp$$

W.E-11: ***A carrier wave of 1000 W is subjected to 100% modulation. Calculate (i) Power of modulated wave, (ii) power is USB, (iii) power is LSB***

Sol: i) Total power of modulated wave

$$P_T = P_C\left(1+\frac{m^2}{2}\right);\ =1000\left(1+\frac{1^2}{2}\right)=1500watt$$

ii) Power in USB $=\frac{1}{2}P_{SB}$

Where power carried by side bands is given by amplitude modulation and detection

$$P_{SB} = P_C\left(\frac{m^2}{2}\right);\ =1000\left(\frac{1^2}{2}\right)=500watt$$

$$P_{USB} = \frac{1}{2}P_{SB} = \frac{1}{2}\times 500 = 250watt$$

iii) Since power in LSB = Power in USB

$P_{LSB} = P_{USB} = 250watt$

W.E-12: ***A transmitting antenna at the top of a tower has a height 32m and the height of the receiving antenna is 50m. What is the maximum distance between them for satisfactory communication in LOS mode? Given radius of earth*** $6.4\times10^6 m$.

Sol: $d_m = \sqrt{2\times64\times10^5\times32}+\sqrt{2\times64\times10^5\times50m}$

$= 64\times10^2\times\sqrt{10}+8\times10^3\times\sqrt{10}m$

$= 144\times10^2\times\sqrt{10}m = 45.5km$

W.E-13: ***A message signal of frequency 10 kHz and peak voltage of 10 volts is used to modulate a carrier of frequency 1 MHz and peak voltage of 20 volts. Determine***

a) modulation index

b) the side bands produced.

Sol: a) modulation index = 10/20 = 0.5

b) The side bands are at

$(1000+10)kHz = 1010\ kHz \& (1000-10kHz) = 990kHz$. are frequency.

CONCEPTUAL

1. **A transducer used at the transmitting end, serves the purpose of converting**
   1) electrical signal to sound form 2) sound signal to electrical form
   3) electrical signal to magnetic form 4) sound signal to magnetic form
2. **In a communication system, noise is most likely to affect the signal**
   1) at the transmitter 2) in the medium of transmission
   3) information source signal 4) at the destination
3. **Device that converts one form of energy into another is called**
   1) transmitter 2) transducer 3) receiver 4) channel
4. **The part of communication system that extracts the signal at the output of the channel is**
   1) transducer 2) transmitter 3) receiver 4) receiver or transmitter
5. **The attenuation of a signal is compensated by**
   1) rectifier 2) oscillator 3) modulator 4) amplifier

BAND WIDTH OF SIGNALS & BAND WIDTH OF RANSMISSION

6. **Modern communication systems use**
   1) analog circuits 2) digital circuits
   3) combination of analog & digital circuits 4) radio circuits
7. **The audio signal**
   1) can be sent directly over the air for large distance
   2) can not be sent directly over the air for large distance
   3) possesses very high frequency 4) possesses very low frequency
8. **A digital signal possess**
   1) continuously varying values 2) only two discrete values
   3) only four discrete values 4) constant values
9. **Digital signals**
   1) provide continuous set of values 2) represent values as randomly
   3) Utilise Decimal code system 4) Utilise binary code system
10. **Digital signals**
    i) do not provide a continuous set of values. ii) represents values as descrete steps.
    iii) can utilize binary system
    iv) can utilize decimal as well as binary system.
    The true option is.
    1) (i) & (ii) only 2) (ii) & (iii) only 3) (i), (ii) & (iii) only 4) (i),(ii),(iii)& (iv)

11 **A digital signal**
   1) is less reliable than analog signal 2) is more reliable than analog signal
   3) is equally reliable as the analog signal 4) Not at all reliable

12. **The band width required for transmiting video signal is**
    1) 50 KHz 2) 1 MHz 3) 4.2 MHz 4) 6 MHz
13. **Band width of an optical fiber is**
    1) more than 100 GHz 2) few kHz 3) less than 1MHz 4) less than 1GHz
14. **The short wave Radio broadcasting band is**
    1) 7 MHz to 22 MHz 2) 88 MHz to 108 MHz
    3) 30 KHz to 300 KHz 4) 3 GHz to 30 GHz

15. The FM Radio broad casting band is

1) 5 MHz to 30 MHz
2) 88 MHz to 108 MHz
3) 30 KHz to 300 KHz
4) 3 GHz to 30 GHz

16. The TV broad casting bands are

1) MF and HF bands
2) VHF and UHF bands
3) UHF and SHF bands
4) SHF and EHF band

17. A: Satellite communication uses different frequency bands for uplink and downlink
B: Bandwidth of video signals is 4.2 MHz

1) A is true but B is false
2) A is false but B is true
3) A and B are false
4) A and B are true

18. A: The frequency band of VHF is greater than UHF of TV transmission
B: Optical fiber transmission has frequency band of 1 THz to 1000 THz

1) A is true but B is false
2) A is false but B is true
3) A and B are false
4) A and B are true

19. The higher frequency TV broad casting bands range is

1) 54 - 72 MHz and 76 to 88 MHz
2) 174 - 216 MHz and 420 to 890 MHz
3) 896 to 901 MHz and 840 to 935 MHz
4) 5.925 to 6.425 GHz and 3.7 to 4.2 GHz

20. Frequency ranges for micro waves are :

1) 3×10^{9} to 3×10^{4} Hz
2) 3×10^{13} to 3×10^{9} Hz
3) 3×10^{14} to 3×10^{9} Hz
4) 3×10^{11} to 3×10^{9} Hz

21. The frequency band used for radar relay systems & T.V is

1) UHF 2) VLF 3) VHF 4) EHF

22. For TV transmission the frequency range employed

1) 30 - 300 MHz 2) 30 - 300 GHz 3) 30 - 300 KHz 4) 30 - 300 Hz

23. The frequency which is not part of AM broadcast

1) 100 kHz 2) 700 kHz 3) 600 kHz 4) 1500 kHz

24. Cellular Mobile works in the frequency range of

1) 840 to 935 MHz 2) 3.7 to 4.2 GHz 3) 420 to 890 MHz 4) 30 to 300 GHz

25. Frequency range used in down linking in satellite communication is

1) 0.896 to 0.901 GHz 2) 0.420 to 0.890 GHz 3) 5.925 to 6.425 GHz 4) 3.7 to 4.2 GHz

26. In the satellite communication, the uplinking frequency range is

1) 0.896 to 0.901 GHz
2) 0.420 to 0.890 GHz
3) 5.925 to 6.425 GHz
4) 3.7 to 4.2 GHz

27. In a communication system, noise is most likely to affect the signal

1) At the transmitter
2) In the channel or in the transmission line
3) In the information source
4) At the receiver

28. The frequency of a FM transmitter without signal input is called

1) Lower side band frequency
2) Upper side band frequency
3) Resting frequency
4) None of these

29. Indicate which one of the following system is digital

1) Pulse position modulation
2) Pulse code modulation
3) Pulse width modulation
4) Pulse amplitude modulation

30. Television signals on earth cannot be received at distances greater than 100 km from the transmission station. The reason behind this is that

1) The receiver antenna is unable to detect the signal at a distance greater than 100 *km*
2) The TV programme consists of both audio and video signals
3) The TV signals are less powerful than radio signals
4) The surface of earth is curved like a sphere

31. Audio signal cannot be transmitted because

1) The signal has more noise
2) The signal cannot be amplified for distance communication
3) The transmitting antenna length is very small to design
4) The transmitting antenna length is very large and impracticable

PROPAGATION OF EM WAVES IN THE ATMOSPHERE

32. A signal emitted by an antenna from a certain point can be received at another point of the surface in the form of

1) sky wave 2) ground wave 3) sea wave 4) both 1 and 2

33. An antenna is a device

1) That converts electromagnetic energy into radio frequency signal
2) That converts radio frequency signal into electromagnetic energy
3) That converts guided electromagnetic waves into free space electromagnetic waves and vice-versa
4) None of these

34. An antenna

1) Converts AF wave to RF wave 2) RF signal into electromagnetic energy
3) Converts the guided EM waves into free space EM waves and vice versa
4) Super imposes AF wave on RF wave

35. An antenna behaves as a resonant circuit only when the length is

1) equal to $\lambda/4$ 2) equal to $\lambda/2$
3) equal to the integral multiples of $\lambda/2$ 4) equal to $3\lambda/4$

36. If audio signal is transmitted directly into space, the length of the transmitting antenna required will be

1) extremely small 2)extremely large 3) infinitely large 4) moderate

37. The height of the antenna

a) limits the population covered by the transmission

b) limits the ground wave propagation

c) effectively used in line of sight communication

1) a & b are true 2) b & c are true 3) c & a are true 4) a, b,c are true

38. Statement A: If the antenna is vertical the vertically polarised EM wave is radiated

Statement B: The vertically polarised EM wave has electrical variations in the vertical plane

1) A is true but B is false 2) A is false but B is true
3) A and B are false 4) A and B are true

39. A: It is necessary for transmitting antenna must be at same height as that of receiving antenna for line of sight communication.

B: EM waves of frequency beyond 40 MHz, propagate as space waves.

1) both A and B are correct 2) both A and B are wrong
3) only A is correct 4) only B is correct

40. Broadcasting antennas are generally

1) Omnidirectional type 2) Vertical type
3) Horizontal type 4) None of these

MODULATION

41. The process of superimposing signal frequency (*i.e.* audio wave) on the carrier wave is known as

1)Transmission 2) Reception 3)Modulation 4) Detection

42. In frequency modulation

1) The amplitude of modulated wave varies as frequency of carrier wave
2) The frequency of modulated wave varies as amplitude of modulating wave
3) The amplitude of modulated wave varies as amplitude of carrier wave
4) The frequency of modulated wave varies as frequency of modulating wave

43. For transmitting audio signal properly

1) it is first superimposed on high frequency carrier wave
2) it is first superimposed on low frequency carrier wave
3) It is sent directly without superimposing on any wave
4) it is superposed with carrier wave of high velocity

44. The process of recovering the audio signal from the modulated wave is known as

1) amplification 2) rectification 3) modulation 4) demodulation

45. The most commonly employed analog modulation technique in satellite communication is the

1) amplitude modulation 2) frequency modulation
3) phase modulation 4) amplitude & phase modulation

46. The need for doing modulation is

1) to increase the intensity of audio signal 2) to decrease the intensity of audio signal
3) to transmit audio signal to large distances 4) to increase the frequency of the audio signal

47. The type of modulation is employed in India for radio transmission is

1) pulse modulation 2) frequency modulation
3) amplitude modulation 4) phase modulation

48. Modulation is used to

1) reduce the bandwidth 2) to seperate the transmission of different users
3) to ensure that intelligence may be transmitted to long distances
4) to allow the uses of practical antenna

49. The process of translating the information contained by the low base band signal to high frequencies is called

1) Detection 2) Modulation 3) Amplification 4) Demodulation

50. During the process of modulation the RF wave is called

1) Modulating wave 2) Modulated wave 3) Carrier wave 4) Audio wave

51. Modulation is required to

a) distinguish different transmissions
b) ensure that the information may be trans mitted over long distances
c) allow the information accessible for different people

1) a & b are true 2) b & c are true 3) c & a are true 4) a, b & c are true

52. The physical quantities of the wave used for modulation

1) Amplitude only
2) Amplitude and frequency
3) Amplitude, frequency and phase
4) Only frequency

53. In Amplitude modulation

1) The amplitude of the carrier wave varies in accordance with the amplitude of the modulating signal
2) The amplitude of carrier wave remains constant, frequency changes in accordance with the modulating signal
3) The amplitude of carrier wave varies in accordance with the frequency of the modulating signal
4) The amplitude changes in accordance with the wave length of the modulating signal

54. Amplitude modulation is used for broad casting because

1) it is more noise immune
2) it requires less transmitting power
3) it has simple circuit
4) it has high fidelity (faithful reproduction)

55. In amplitude modulation, carrier wave frequencies are .... than that compared to those in frequency modulation

1) lower
2) higher
3) same
4) lower or higher

56. AM is used for broad casting because,

1) it is more noise immune than other modulating systems
2) it requires less transmitting power compared with other systems
3) its use avoids receiver complexity
4) no other modulation system can provide the necessary bandwidth, faithful transmission.

57. Draw backs of Amplitude modulation

1) During transmission extreneous noise creeps in.
2) Most of the transmitting power is wasted, as it does not contain useful information.
3) The reception is not clear in the case of weak signals due to noise
4) The receiver set is complex

58. In amplitude modulation

1) only amplitude is changed but frequency remains same
2) both amplitude & frequency changes equally
3) both amplitude & frequency changes unequally
4) only frequency changes but amplitude remains constant.

59. The limitation of amplitude modulation is

1) clear reception
2) high efficiency
3) small operating range
4) good audio quality

60. In frequency modulation

1) Frequency of CW remains constant but amplitude changes in accordance with modulating wave frequency
2) Frequency of CW changes in accordance with the modulating wave frequency but the amplitude also changes.
3) Frequency of CW changes in accordance with the frequency of modulating wave frequency but the amplitude remains constant.
4) Frequency of CW changes in accordance with the amplitude of modulating wave amplitude

61. In T.V. broadcasting both picture and sound are transmitted simultaneously. In this

1) audio signal is frequency modulated and video signal is amplitude modulated
2) both audio and video signals are frequency modulated
3) audio signal is amplitude modulated and video signal is frequency modulated
4) both audio and video signals are amplitude modulated

62. Effective power radiated by an antenna is

1) Proportional to the square at the length of the antenna
2) inversely proportional to the wavelength
3) inversely proportional to the square of the wavelength
4) proportional to the wavelength

63. The concepts of communication are

a) mode of communication b) need for modulation
c) types of modulation d) detection of modulated wave
1) a, b, c are true 2) b, c, d are true 3) c, d, a are true 4) a, b, c & d are true

64. The difference between phase and frequency modulation

1) practically they are same but theoretically they differ
2) lies in the poorer audio response of phase modulation
3) lies in the poorer audio response of frequency modulation
4) lies in the definitions of modulation and their modulation index

65. Basically, the product modulator is

1) An amplifier 2) A mixer
3) A frequency separator 4) A phase separator

66. Which of the following is the disadvantage of FM over AM

1) Larger band width requirement 2) Larger noise
3) Higher modulation power 4) Low efficiency

67. Audio signal cannot be transmitted as such because

1) the signal has more noise
2) the signal cannot be amplified for distance communication
3) the transmitting antenna length is very small to design
4) the transmiting antenna length is very large and impracticable

68. The examples of broadcast are

A) radio B) television C) telephony D) internet
1) A & B 2) A, B & D 3) A,B & C 4) B & D

69. The waves relavent to telecommunications are

1) visible light 2) infrared 3) ultraviolet 4) microwave

70. While tuning in a certain broad cast station with a receiver, we are actually

1) varying the local oscillator
2) varying the resonant frequency of the circuit for the radio signal to be picked up
3) tuning the antenna 4) varying the current of receiver set

71. Long distance short-wave radio broadcasting uses

1) Ground wave 2) Ionospheric wave 3) Direct wave 4) Sky wave

72. Radio waves of constant amplitude can be generated with

1) filter 2) rectifier 3) FET 4) oscillator

73. Advantage of HF transmission is

A) that the length of antenna is small
B) that the antenna can be mounted at larger heights
C) that the power radiated is more for a given length of antenna
1)a & b 2) b& c 3) a & c 4) a,b & c

74. The intensity of the ground waves decrease with increase of distance due to

1) Interference 2) Diffraction 3) Polarization 4) Due to unknown reason

75. High frequency waves are

1) absorbed by F layer
2) reflected by the E layer
3) capable of use for long distance transmission
4) affected by the solar cycle

76. As the e.m. waves travel in free space

1) absorption takes place
2) attennuation takes place
3) refraction takes place
4) reflection takes place

77. The electromagnetic waves of frequency 80 MHz and 200 MHz

1) can be reflected by troposphere
2) can be reflected by ionosphere
3) can be reflected by mesosphere
4) cannot be reflected by any layer of earth's atmosphere

78. The better propagation mode to propagate television frequency and radar signals is

1) satellite communication
2) ground propagation
3) polarized communication
4) skywave communication

79. Micro wave link repeaters are typically 50 km apart

1) because of atmospheric attenuation
2) because of the earths curvature
3) to ensure that signal voltage may not harm the repeater
4) to reduce the interference of microwaves

80. Attenuation of ground waves is due to

1) Diffraction effect
2) Radio waves induce currents in the ground because of the polarisation

1) a & b are true 2) Only a is true 3) Only b is true 4) Both a & b false.

81. The ground wave eventually disappears, as one moves away from the transmitter, because of

1) interference from the sky wave
2) loss of line of signal condition
3) maximum single - hop distance limitation
4) diffraction effect causing tilting of the wave

82. The range of ground wave transmission can be increased by

1) increasing the power of transmitter with the use of HF
2) increasing the power of transmitter with the use of VLF
3) decreasing the power and increasing the frequency of radio waves
4) decreasing both power and frequency of radio waves

83. Space wave propagation is used in

a) microwave communication
b) satellite communication
c) TV transmission

1) Only a 2) Both a & b 3) Both b & c 4) a ,b & c

84. Frequencies in the UHF range normally propagate by means of:

1) Ground waves 2) Sky waves. 3) Surface waves 4) Space waves.

85. When a sky wave is reflected onto the ground

1) frequency of the reflected wave is different to that of incident wave
2) there is a phase difference introduced to the reflected wave
3) the reflected wave is out of phase with incident wave and reach the receving antenna along with the direct wave from transmitting antenna causing interference.
4) the waves are not reflected by the ground.

86. The electromagnetic waves of frequency 2 MHz to 30 MHz are

(1)In ground wave propagation (2)In sky wave propagation

(3)In microwave propagation (4)In satellite communication

87. Among the following frequencies one will be suitable for beyond-the horizon communication using sky waves is

1) 10 kHz 2) 10 MHz 3) 1GHz 4)1000GHz

88. Among the following, the waves which can penetrate the ionosphere are

1)10GHz 2)10MHz 3)20MHz 4) 25 MHz

89. A: At great heights from surface of earth and close to earth ionisation of air molecules is low.
B: EM waves of frequencies beyond 30 MHz penetrate ionosphere and escape.

1) both A and B are correct 2) both A and B are wrong

3) only A is correct 4) only B is correct

90. Through which mode of propagation, the radio waves can be sent from one place to another

1) Ground wave propagation 2) Sky wave propagation

3) Space wave propagation 4) All of them

91. The frequency at which communication will not be reliable is

1) 100 KHz 2)1 MHz 3) 10 GHz 4)100 GHz

92. The frequency above which radiation of electrical energy is practical is

1) 0.2 kHz 2) 2 kHz 3) 20 kHz 4) 2Hz

93. The radio waves of frequency 300 MHz to 3000 MHz belong to

1) High frequency band 2) Very high frequency band

3) Ultra high frequency band 4) Super high frequency band

94. Coaxial cable is an example of

1) Optical fibre 2) Free space 3) Wire medium 4) Sea medium

95. Optical fibre communication is generally preferred over general communication system because

1) it is more efficient 2) of signal security 3) both (1) & (2) 4) it is easily available

96. The attenuation in optical fibre is mainly due to

1)Absorption 2)Scattering 3)Neither absorption nor scattering 4)Both 1 and 2

97. Consider telecommunication through optical fibres. Which of the following statements is not true

(1)Optical fibres may have homogeneous core with a suitable cladding

(2)Optical fibres can be of graded refractive index

(3)Optical fibres are subject to electromagnetic interference from outside

(4)Optical fibres have extremely low transmission loss

98. The phenomenon by which light travels in an optical fibres is

1)Reflection 2)Refraction 3)Total internal reflection 4) Transmission

99. A laser is a coherent source because it contains

1) Many wavelengths 2) Uncoordinated wave of a particular wavelength

3) Coordinated wave of many wavelengths 4) Coordinated waves of a particular wavelength

100. In which of the following remote sensing technique is not used

1)Forest density 2)Pollution 3)Wetland mapping 4)Medical treatment

CONCEPTUAL - KEY

1) 2	2) 2	3) 2	4) 3	5) 4	6) 2	7) 2	8) 2	9) 4	10) 3
11) 2	12) 3	13) 1	14) 1	15) 2	16) 2	17) 4	18) 2	19) 2	20) 2
21) 1	22) 1	23) 1	24) 1	25) 4	26) 3	27) 2	28) 3	29) 2	30) 4
31) 4	32) 4	33) 3	34) 3	35) 1	36) 2	37) 4	38) 4	39) 4	40) 2
41) 3	42) 2	43) 1	44) 4	45) 2	46) 3	47) 3	48) 1	49) 2	50) 3
51) 4	52) 3	53) 1	54) 3	55) 1	56) 3	57) 2	58) 1	59) 3	60) 3
61) 1	62) 3	63) 4	64) 1	65) 2	66) 1	67) 4	68) 2	69) 4	70) 3
71) 3	72) 4	73) 4	74) 2	75) 2	76) 2	77) 4	78) 1	79) 2	80) 1
81) 4	82) 2	83) 4	84) 4	85) 3	86) 2	87) 2	88) 1	89) 1	90) 4
91) 1	92) 3	93) 3	94) 3	95) 3	96) 4	97) 3	98) 3	99) 4	100) 4

LEVEL - I - 1

1. **An optical communication system is operating at a wavelength of 800 nm, it's optical source frequency is**

   1) 3.8×10^{14} Hz 2) 3.8×10^{12} Hz 3) 3.8×10^{10} Hz 4) 3.8×10^{9} Hz

2. **A carrier wave of 1000 kHz is used to carry the signal, the length of the transmitting antenna will be equal to**

   1) 3 m 2) 30 m 3) 300 m 4) 3000 m

3. **A transmitting antenna is at a height of 40 m and the receiving antenna is at a height of 60 m. The maximum distance between them for satisfactory communication is nearly**

   1) 22.5 km 2) 27.5 km 3) 50 km 4) 25 km

4. **A T V transmitting antenna is 80 m tall. If the receiving antenna is on the ground the service area is**

   1) 12 sq km 2) 3215 sq km 3) 144sq km 4) 32 sq km

5. **The maximum distance upto which TV transmission from a TV tower of height 'h' can be received is proportional to**

   1) $h^{1/2}$ 2) h 3) $h^{3/2}$ 4) h^2

6. **In short wave communication waves of which of the following frequencies will be reflected back by the ionospheric layer, having electron density 10^{11} per m^3**

   1) 2 *MHz* 2) 10 *MHz* 3) 12 *MHz* 4) 18 *MHz*

7. **In an amplitude modulated wave for audio frequency of 500 *cycle/second*, the appropriate carrier frequency will be**

   1) 50 *cycles/sec* 2) 100 *cycles/sec* 3) 500 *cycles/sec* 4) 50,000 *cycles/sec*

8. **The modulation index of an FM carrier having a carrier swing of 200 *kHz* and a modulating signal 10 *kHz* is**

   1) 5 2) 10 3) 20 4) 25

9. **If a number of sine waves with modulation indices n_1, n_2, n_3 .... Modulate a carrier wave, then total modulation index (n) of the wave is**

   1) $n_1 + n_2 .... + 2(n_1 + n_2.....)$ 2) $\sqrt{n_1 - n_2 + n_3..........}$

   3) $\sqrt{n_1^2 + n_2^2 + n_3^2.........}$ 4) $n_1 + n_2 ....$

10. **A sky wave with a frequency 55 *MHz* is incident on *D*-region of earth's atmosphere at 45°. The angle of refraction is (electron density for *D*-region is 400 *electron/cm*^{3})**

1)60° 2)45° 3) 30° 4)15°

11. **What should be the maximum acceptance angle at the aircore interface of an optical fibre if n_1 and n_2 are the refractive indices of the core and the cladding, respectively**

1) $\sin^{-1}(n_2/n_1)$ 2) $\sin^{-1}\sqrt{n_1^2-n_2^2}$ 3) $\left[\tan^{-1}\frac{n_2}{n_1}\right]$ 4) $\left[\tan^{-1}\frac{n_1}{n_2}\right]$

12. **The characteristic impedance of a coaxial cable is of the order of**

1)50 Ω 2)200 Ω 3)270 Ω 4)None of these

13. **The velocity factor of a transmission line *x*. If dielectric constant of the medium is 2.6, the value of *x* is**

1)0.26 2)0.62 3)2.6 4)6.2

14. **For television broadcasting, the frequency employed is normally**

1)30-300 *MHz* 2)30-300 *GHz* 3)30-300 *KHz* 4)30-300 *Hz*

15. **Because of tilting which waves finally disappear**

1) Microwaves 2)Surface waves 3)Sky waves 4)Space waves

16. **A radar has a power of 1kW and is operating at a frequency of 10 GHz. It is located on a mountain top of height 500m. The maximum distance upto which it can detect object located on the surface of the earth (Radius of earth $6.4\times10^6 m$) is**

1) 80 km 2) 16 km 3) 40 km 4) 64km

LEVEL - I -1- KEY

1) 1	2) 3	3) 3	4) 2	5) 1	6) 1	7) 4	8) 2	9) 3	10) 2
11) 2	12) 3	13) 2	14) 1	15) 2	16) 1				

LEVEL - I -1- HINTS

1. $c = n\lambda \Rightarrow n = c/\lambda$
2. length of antenna $\approx \lambda$

Here $l = \lambda = \frac{C}{v} = \frac{3\times10^8}{1000\times10^3} = 300m$

3. distance between transmitter and receiver is $s = \sqrt{2Rh_T} + \sqrt{2Rh_R}$
4. area covered, $A = \pi d^2$ where $d = \sqrt{2Rh}$
5. $d = \sqrt{2Rh} \Rightarrow d \alpha \sqrt{h}$
6. By using $f_c \approx 9(N_{max})^{1/2} \Rightarrow f_c \approx 2\ MHz$
7. Carrier frequency > audio frequency
8. CS = 2 ´ Δf or $\Delta f = CS/2$

$\Delta f = \frac{200}{2} = 100\ kHz$; Now $m_f = \frac{\Delta f}{f_m} = \frac{100}{10} = 10$

10. $n_{eff} = n_0\sqrt{1-\left(\frac{80.5\,N}{v^2}\right)} = 1\sqrt{1-\frac{80.5\times(400\times10^6)}{(55\times10^6)^2}} \approx 1$

Also $n_{eff} = \frac{\sin i}{\sin r} \Rightarrow \sin r = \sin i \Rightarrow r = i = 45°$

11. Core of acceptance angle $\theta = \sin^{-1}\sqrt{n_1^2 - n_2^2}$

12. conceptual

13. $v.f. = \frac{1}{\sqrt{k}} = \frac{1}{\sqrt{2.6}} = 0.62$

14. VHF (Very High Frequency) band having frequency range 30 *MHz* to 300 *MHz* is typically used for TV and radar transmission.

15. conceptual

16. Range of radar on earth surface (optical distance, for space wave, i.e., line of view)

$$Range = \sqrt{(R+h)^2 - R^2} = \sqrt{2Rh + h^2} = \sqrt{2Rh} = \sqrt{2\times6400\times\frac{1}{2}km} = 80km$$

LEVEL - I -2

1. An optical communication system is operating at a wavelength of 600 nm, it's optical source frequency is

1) 3.8×10^{14} Hz 2) 3.8×10^{12} Hz 3) 3.8×10^{10} Hz 4) 5×10^{14} Hz

2. A carrier wave of 2000 kHz is used to carry the signal, the length of the transmitting antenna will be equal to

1) 3 m 2) 30 m 3) 150 m 4) 3000 m

3. A transmitting antenna is at a height of 25 m and the receiving antenna is at a height of 64 m. The maximum distance between them for satisfactory communication is nearly

1) 22.5 km 2) 46.5 km 3) 50 km 4) 25 km

4. A T V transmitting antenna is 100 m tall. If the receiving antenna is on the ground the service area is

1) 12 sq km 2) 4000sq km 3) 144sq km 4) 32 sq km

5. The maximum distance upto which TV transmission from a TV tower of height 'h' can be received is proportional to (R earth radius)

1) $(Rh)^{\frac{1}{2}}$ 2) h 3) $h^{3/2}$ 4) h^2

6. In short wave communication waves of which of the following frequencies will be reflected back by the ionospheric layer, having electron density 10^{12} per m^3

1) 2 *MHz* 2) 9 *MHz* 3) 12 *MHz* 4) 18 *MHz*

7. In an amplitude modulated wave for audio frequency of 1000 *cycle/second*, the appropriate carrier frequency will be

1) 50 *cycles/sec* 2) 100 *cycles/sec* 3) 500 *cycles/sec* 4) 40,000 *cycles/sec*

8. The modulation index of an FM carrier having a carrier swing of 300 *kHz* and a modulating signal 10 *kHz* is

1) 15 2) 10 3) 20 4) 25

9. **A sky wave with a frequency 64.4 *MHz* is incident on *D*-region of earth's atmosphere at 45°. The angle of refraction is (electron density for *D*-region is 400 *electron/cm³*)**

1)60° 2) 90° 3) 45° 4)15°

10. **The velocity factor of a transmission line *x*. If dielectric constant of the medium is 1.8, the value of *x* is**

1)0.26 2)0.62 3) 0.74 4)6.2

11. **A radar has a power of 1kW and is operating at a frequency of 10 GHz. It is located on a mountain top of height 700m. The maximum distance upto which it can detect object located on the surface of the earth (Radius of earth $6.4\times10^6 m$) is**

1) 80 km 2) 16 km 3) 40 km 4) 92 km

LEVEL - II -2 - KEY

1) 4 2) 3 3) 2 4) 2 5) 1 6) 2 7) 4 8) 1 9) 3 10) 3
11) 4

LEVEL - II -2-HINTS

1. $C = n\lambda$; $n = \frac{C}{\lambda}$

2. Length of antenna $\approx \lambda, \nu = 2000KHz$

here $l = \lambda = \frac{C}{\nu} = \frac{3\times10^8}{2000\times10^3} = 150m$

3. Distance between transmitter and receiver is $d = \sqrt{2Rh_T} + \sqrt{2Rh_R}$

$= \sqrt{2\times6400\times10^3\times25} + \sqrt{2\times6400\times10^3\times64}$ = 46.5 km

4. Area covered $A = \pi d^2$; where $d = \sqrt{2Rh}$

5. $d = \sqrt{2Rh}$; $d \propto \sqrt{Rh}$; $d \propto (Rh)^{\frac{1}{2}}$

6. By using $f_c \approx 9(N_{max})^{\frac{1}{2}}$; $f_c \approx 9(10^{12})^{\frac{1}{2}}$

$f_c \approx 9MHz$

7. Carrier frequency > audio frequency

8. $CS = 2\Delta f$

$\Delta f = \frac{CS}{2} = \frac{300KHz}{2} = 150KHz$

$m_f = \frac{\Delta f}{f_m} = \frac{150KHz}{10KHz} = 15$

9. $n_{eff} = n_0\sqrt{1-\frac{80.5N}{v^2}}$

$= 1\sqrt{1-\frac{80.5\times400\times10^6}{\left(64.4\times10^6\right)^2}}$; ≈ 1

Also $n_{eff} = \frac{\sin i}{\sin r} \Rightarrow \sin i = \sin r$; $\Rightarrow i = r = 45^0$

10. $V.f = \frac{1}{\sqrt{k}}$; $= \frac{1}{\sqrt{1.8}} = 0.74$

11. Range of Radar on earth surface (optical distance, for space wave, i.e line of view)

Range $= \sqrt{(R+h)^2 - R^2} = \sqrt{2Rh + h^2}$

$= \sqrt{2Rh} = \sqrt{2\times6400\times0.7}$

$[700m = 0.7km] = 92$ km

LEVEL - II -3

1. **The maximum distance between the transmitting and receving TV towers is 72 km . If the ratio of the heights of the TV transmitting tower to receiving tower is 16 : 25, the heights of the transmitting and receiving towers are**

   1) 51.2 m ; 80 m 2) 40 m ; 80 m 3) 80 m ; 125 m 4) 25 m ; 75 m

2. **The maximum distance between the transmitting and receiving TV towers is D. If the heights of both transmitting and receiving towers are doubled then the maximum distance between them becomes**

   1) 2D 2) $\sqrt{2}$ D 3) 4D 4) D/2

3. **A T.V. tower has a height of 10 m is in a region of average population density 100π /km2. The number of people that can recieve the transmission is nearly**

   1) 1,28,000 2) 64,000 3) 2,56,000 4) 32,000

4. **At certain distance from a transmitting tower a receiver tower of height 20 m is used to receive direct signal. Another tower is installed beyond the first along the same line of sight to receive the signals from the same transmitter. Its height is 44% more than the first receiving tower. Then the separation between the two receiving towers is**

   1) 6.4 km 2) 3.2 km 3) 1.6 km 4) 0.8 km

5. **The TV signals have a band width of 6MHz. The number of TV channels that can be accommodated in a band width 12 GHz is**

   1) 2 2) 20 3) 200 4) 2000

6. **A carrier wave of peak voltage 12 V is used to transmit a message signal. The peak voltage of the modulating signal in order to have a modulation index of 75% is**

   1) 6 V 2) 9 V 3) 4 V 4) 15 V

7. **A modulating signal is a square wave as shown in the figure**

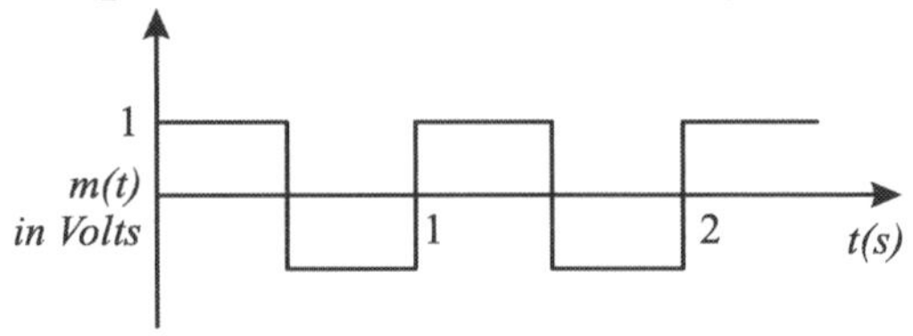

The carrier wave is given by $c(t) = 2\sin(8\pi t)$ *volts*.

The modulation index is

1) 2 2) 0.75 3) 0.5 4) 1.5

8. **A audio signal** $15\sin 2\pi(1500t)$ **amplitude modulates** $60\sin 2\pi(10^5)t$. **The depth of modulation is**

1) 25% 2) 20% 3) 50% 4) 40%

9. **An AM wave is given by** $V = 1500[1+0.5\sin 12560t]\sin(5.26\times10^5 t)$.

The modulating frequency is

1) 2.0 kHz 2) 1.0 kHz 3) 12.5 kHz 4) 50 kHz

10. **The amplitude modulated current is given by** $i = 125[1+0.6\sin 2900t]\sin(5.50\times10^5 t)$.

The RMS value of carrier current will be

1) $\frac{125}{\sqrt{2}}A$ 2) $\frac{100}{\sqrt{2}}A$ 3) $\frac{75}{\sqrt{2}}A$ 4) $\frac{50}{\sqrt{2}}A$

11. **An audio signal** $25\sin 2\pi(1400t)$ **amplitude modulates** $80\sin 2\pi(10^5 t)$. **The two side band frequencies are**

1) 98.6 kHz, 101.4 kHz 2) 92.5 kHz, 105.5 kHz

3) 94 kHz, 102.5 kHz 4) 96 kHz, 106 kHz

12. **If** f_c **and** f_m **are the frequencies of carrier wave and signal, then the band width is**

1) f_m 2) $2f_m$ 3) f_c 4) $2f_c$

13. **The number of AM broadcast stations that can be accommodated in a 300kHz band width for the highest modulating frequency 15 kHz will be**

1) 10 2) 5 3) 7 4) 12

14. **A 1000 kHz carrier is simultaneously modulated with 300Hz, 800 Hz and 2 kHz audio waves. The frequencies present in the output are**

1) 999.7 kHz, 1000.3kHz, 999.2kHz 2) 1000.8kHz, 998 kHz, 1002kHz

3) 1002.8 kHz, 996kHz, 1106 kHz 4) Both (1) and (2)

15. **Depth of modulation in terms of** E_{max} **and** E_{min} **is**

1) $m_a = \frac{E_{max} + E_{min}}{E_{min}}$ 2) $m_a = \frac{E_{max} - E_{min}}{E_{max}}$

3) $m_a = \frac{E_{max} - E_{min}}{E_{max} + E_{min}}$ 4) $m_a = \frac{E_{max} + E_{min}}{E_{max} - E_{min}}$

16. **If V_c is amplitude of carrier wave in AM transmitter where modulation factor is m then amplitude of side bands can be**

1) $\frac{V_c}{2m}$ 2) $\frac{m}{2}V_c$ 3) mV_c 4) $\frac{V_c}{2}$

17. **If a carrier wave of amplitude 10mV is modulated by an audio signal of amplitude 2mV, then amplitude of LSF or USF is given by**

1) 0.2 mV 2) 0.5 mV 3) 1 mV 4) 2 mV

18. **The amplitude modulated current is given by $i = 125[1 + 0.6\sin 2900t]\sin 10^6 t$. The depth of modulation is**

1) 60% 2) 6% 3) 36% 4) 66%

19. **If minimum voltage in an AM wave was found to be 2V and maximum voltage 10V. The % modulation index**

1) 80% 2) 66.67% 3) 64.25% 4) 76.25%

20. **Sinusoidal carrier voltage of frequency 1.5 *MHz* and amplitude 50 *V* is amplitude modulated by sinusoidal voltage of frequency 10 *kHz* producing 50% modulation. The lower and upper side-band frequencies in *kHz* are**

1) 1490, 1510 2)1510, 1490 3) $\frac{1}{1490}, \frac{1}{1510}$ 4) $\frac{1}{1510}, \frac{1}{1490}$

21. **A transmitter supplies 9 *kW* to the aerial when unmodulated. The power radiated when modulated to 40% is**

1) 5 *kW* 2) 9.72 *kW* 3) 10 *kW* 4) 12 *kW*

22. **The total power content of an AM wave is 1500 *W*. For 100% modulation, the power transmitted by the carrier is**

1) 500 *W* 2) 700 *W* 3) 750 *W* 4) 1000 *W*

23. **The bit rate for a signal, which has a sampling rate of 8 *kHz* and where 16 quantization levels have been used, is**

1) 32000 *bits/sec* 2) 16000 *bits/sec* 3) 64000 *bits/sec* 4) 72000 *bits/sec*

24. **The antenna current of an AM transmitter is 8 *A* when only carrier is sent but increases to 8.96 *A* when the carrier is modulated sinusoidally. The percentage modulation is**

1)50% 2)60% 3)65% 4)71%

25. **In a diode AM-detector, the output circuit consist of $R = 1k$W and $C = 10\,pF$. A carrier signal of 100 *kHz* is to be detected. Is it good**

1) Yes 2) No
3) Information is not sufficient 4) None of these

26. **In an FM system a 7 *kHz* signal modulates 108*MHz* carrier so that frequency deviation is 50 *kHz*. The carrier swing is**

1) 7.143 2) 8 3) 0.71 4) 350

27. **Maximum useable frequency (MUF) in *F*-region layer is *x*, when the critical frequency is 60 *MHz* and the angle of incidence is 70°. Then *x* is**

(1)150 *MHz* 2)170 *MHz* (3)175 *MHz* 4)190 *MHz*

28. Suppose that the modulating signal is $m(t)=2\cos(2\pi f_m t)$ and the carrier signal is $x_c(t)=A_c\cos(2\pi f_c t)$ which one of the following is a conventional AM signal without over modulation

1) $x(t)=A_c m(t)\cos(2\pi f_c t)$
2) $x(t)=A_c(1+m(t))\cos(2\pi f_c t)$
3) $x(t)=A_c\cos(2\pi f_c t)+\dfrac{A_c}{4}m(t)\cos(2\pi f_c t)$
4) $x_c(t)=A_c\cos(2\pi f_m t)+\cos(2\pi f_c t)+A_c\sin(2\pi f_m t)\sin(2\pi f_c t)$

29. $C(t)$ and $m(t)$ are used to generate an AM signal. The modulation indes of generated AM signal is 0.5. Then the quantity $\dfrac{P_{Total\ SB}}{P_{Carrier}}=$

1) 1/8
2) 1/4
3) 2/3
4) 9/8

30. Which of one of the following modulated signals are recovered up to a scaling factor using envelope detedtor

1) $20\cos 200\pi t+30m(t)\cos 200\pi t$
2) $20\cos 200\pi t+16m(t)\cos 200\pi t$
3) $10m(t)\cos 400\pi t$
4) None

31. A message signal $m(t)=4\cos 2000\pi t$ modulates the carrier $C(t)=\cos 2\pi f_c t$ where $f_c=1MHz$ to produce an AM signal. For demodulation using envelope detector the time constant RC should satisfy

1) $0.5ms<RC<1ms$
2) $1\mu s \ll RC \ll 0.5ms$
3) $RC \gg 1\mu s$
4) $RC \gg 1\mu s$

32. A 1 MHz carrier signal is modulated by a symmetrical sinusoidal wave for period of 100 μs in a nonlinear (square law device). Which of the following frequencies will not be present in the modulated signal

1) 990 kHz
2) 1010 kHz
3) 1020 kHz
4) 1030 kHz

33. A given AM transmitter develops an unmodulated power out put of 1kW across 50 Ω resistance. When a message signal of amplitude 5 V is applied on it then the side bands carry 40% of power of carrier. Amplitude of the carrier signal used is

1) 505.952 V
2) 126.488
3) 252.976
4) 316.22

34. Let $m(t)=\cos(4\pi\times10^3 t)$ and $C(t)=5\cos(2\pi\times10^6 t)$ are the message and carrier signals modulation index is 0.5. What is the efficiency achieved?

1) 8.33%
2) 11.11%
3) 20 %
4) 25%

LEVEL– II -3- KEY

1) 3	2) 2	3) 1	4) 2	5) 4	6) 2	7) 3	8) 1	9) 1	10) 1
11) 1	12) 2	13) 1	14) 4	15) 3	16) 2	17) 3	18) 1	19) 2	20) 1
21) 2	22) 4	23) 1	24) 4	25) 2	26) 1	27) 3	28) 3	29) 1	30) 2
31) 2	32) 3	33) 4	34) 2						

LEVEL - II -3-HINTS

1. $\frac{h_t}{h_r}=\frac{16}{25}$; $d=\sqrt{2Rh_t}+\sqrt{2Rh_r}$

$72\times10^3=\sqrt{2R\frac{16}{25}h_r}+\sqrt{2Rh_r} \Rightarrow h_r=125m \,\&\, h_t=80m$

2. s= $\sqrt{2Rh_T}+\sqrt{2Rh_R}$ when both h_T and h_R are doubled then s becomes $\sqrt{2}$ times.

3. Number of people that can receive
$=$ area $\times$ population density
$=\pi d^2 \times$ population density

4. distance between transmitter and first receiver is x and between transmitter and second receiver is y.

$x=\sqrt{2Rh_T}+\sqrt{2Rh_1}$

$y=\sqrt{2Rh_T}+\sqrt{2R(1.44h_1)}$

find y -x

5. Number of channels $=\frac{12\times10^9}{6\times10^6}=2000$

6. $\mu=\frac{A_m}{A_c}\Rightarrow\frac{3}{4}=\frac{A_m}{12}$

7. $\mu=\frac{A_m}{A_c}=\frac{1}{2}=0.5$

8. $\mu=\frac{A_m}{A_c}\times100=\frac{15}{60}\times100=25\%$

9. $C_m(t)=A_c(1+\mu\sin\omega_m t)\sin\omega_c t$

$f_m=\frac{\omega_m}{2\pi}=\frac{12560}{2\pi}=2khz$

10. $I_{rms}=\frac{I_{peak}}{\sqrt{2}}=\frac{125}{\sqrt{2}}$

11. $LSB=f_c-f_m=10^5-1400=98.6khz$

$USB=f_c+f_m=10^5+1400=101.4khz$

12. band width $=(f_c+f_m)-(f_c-f_m)=2f_m$

13. $n=\frac{\text{band width}}{\text{signal band width}}=\frac{300}{2\times15}=10$

14. frequencies in output are $f_c-f_m \,\&\, f_c+f_m$

15. $E_{max}=A_c+A_m$

$E_{min}=A_c-A_m\Rightarrow\mu=\frac{A_m}{A_c}=\frac{E_{max}-E_{min}}{E_{max}+E_{min}}$

16. Amplitude of LSB = USB = $\mu\frac{A_c}{2}$

17. Amplitude of LSB = USB = $\mu\frac{A_c}{2}=\frac{A_m}{A_c}\times\frac{A_c}{2}=\frac{A_m}{2}$

18. $\mu=0.6\times100=60\%$

19. $\mu=\frac{10-2}{10+2}\times100$

20. (1) Here, $f_c=1.5\ MHz=1500\ kHz,\ f_m==10\ kHz$

\ Low side band frequency

$=f_c-f_m=1500\ kHz-10\ kHz=1490\ kHz$

Upper side band frequency

$=f_c+f_m=1500\ kHz+10\ kHz=1510\ kHz$

21. (2) $P_t=P_c\left[1+\frac{m^2}{2}\right]=9\left[1+\frac{(0.4)^2}{2}\right]$

$=9\left[1+\frac{0.16}{2}\right]$ (Q $m=40\%=0.4$)

$=9\ (1.08)=9.72\ kW$

22. (4) $\frac{P_t}{P_c}=1+\frac{m^2}{2}$ or $P_c=P_t\left[\frac{2}{2+m^2}\right]$

\ $P_c=1500\left[\frac{2}{2+1}\right]$ $\therefore\ m=100\%=1=1000\ W$

23. (1) If n is the number of bits per sample, then number of quantisation level = 2^n Since the number of quantisation level is 16

$\Rightarrow\ 2^n=16$ Þ $n=4$

bit rate = sampling rate × no. of bits per sample = 8000 × 4 = 32,000 *bits/sec.*

24. (4) We know that $\left(\frac{I_t}{I_c}\right)^2=1+\frac{m^2}{2}$

Here, $I_t=8.96A$ and $I_c=8A$

$\left(\frac{8.96}{8}\right)^2=1+\frac{m^2}{2}$ or $1.254=1+\frac{m^2}{2}$

or $\frac{m^2}{2}=0.254$ or $m^2=0.508$ or $m=0.71=71\%$

25. (2) For demodulation $\frac{1}{f_c}<<RC$

$\frac{1}{f_c}=\frac{1}{100\times10^3}=10^{-5}\ s$

$RC=10^3\times10\times10^{-12}\ s=10^{-8}\ s$

We see that $\frac{1}{f_c}$ here is not less than RC as required by the above condition. Hence, this is not good.

26. (1) Carrier swing

$$=\frac{\text{Frequency deviation}}{\text{Modulating frequency}}=\frac{50}{7}=7.143$$

27. (3) $MUF=\frac{f_c}{\cos\theta}=\frac{60}{\cos 70^\circ}=175\ MHz$

28. Convention AM signal withour over modulation means m<1

$$x(t)=A_c\left(1+m(t)\right)\cos\left(2\pi f_c t\right)$$

For option C $x(t)=A_c\left(1+\frac{1}{4}m(t)\right)\cos\left(2\pi f_c t\right);\quad \therefore m=\frac{1}{4}$

29. $\frac{P_{Total\ SB}}{P_{Carrier}}=\frac{P_c\left(m^2/2\right)}{P_c}=\frac{1}{8}$

30. For option 1 $m=\frac{30}{20}=1.5$ m>1

For option 2 $m=\frac{16}{20}=0.8$ m<1

For option 3 signal contains only side bands so message cannot be detected.

31. $f_c=1\ MHz$ and $f_m=2000Hz$

$$\frac{1}{f_c}\ll RC\ll\frac{1}{f_m}$$

32. The frequency components of modulation signal are from nonlinear device are

$f_c,\ f_c\pm f_m,\ f_c\pm 3f_m\ldots.$

So the frequencies present in the AM are 100 kHz, 990 kHz, 1010 kHz, 970 kHz, 1030 kHz

33. $\frac{A_C^{\ 2}}{2R}=1kW\Rightarrow\frac{A_c^{\ 2}}{2\times 50}=10^3$

34. $\eta=\frac{m^2}{2+m^2}$

LEVEL - II -4

1. The maximum distance between the transmitting and receving TV towers is 64 km . If the ratio of the heights of the TV transmitting tower to receiving tower is $4:9$, the heights of the transmitting and receiving towers are

1) 51.2 m ; 80 m 2) 1280 m ; 2880 m 3) 80 m ; 125 m 4) 25 m ; 75 m

2. The maximum distance between the transmitting and receiving TV towers is D. If the heights of both transmitting and receiving towers are halved then the maximum distance between them becomes

1) 2D 2) $D/\sqrt{2}$ 3) 4D 4) D/2

3. A T.V. tower has a height of 5 m is in a region of average population density 100 π /km2. The number of people that can recieve the transmission is nearly

1) 1,28,000 2) 64000 3) 2,56,000 4) 32,000

4. **At certain distance from a transmitting tower a receiver tower of height 180 m is used to receive direct signal. Another tower is installed beyond the first along the same line of sight to receive the signals from the same transmitter. Its height is 21% more than the first receiving tower. Then the separation between the two receiving towers is**

1) 6.4 km 2) 3.2 km 3) 1.6 km 4) 4.8 km

5. **The TV signals have a band width of 4MHz. The number of TV channels that can be accommodated in a band width 16 GHz is**

1) 2 2) 20 3) 200 4) 4000

6. **A carrier wave of peak voltage 10 V is used to transmit a message signal. The peak voltage of the modulating signal in order to have a modulation index of 80% is**

1) 6 V 2) 9 V 3) 8 V 4) 15 V

7. **A modulating signal is a square wave as shown in the figure**

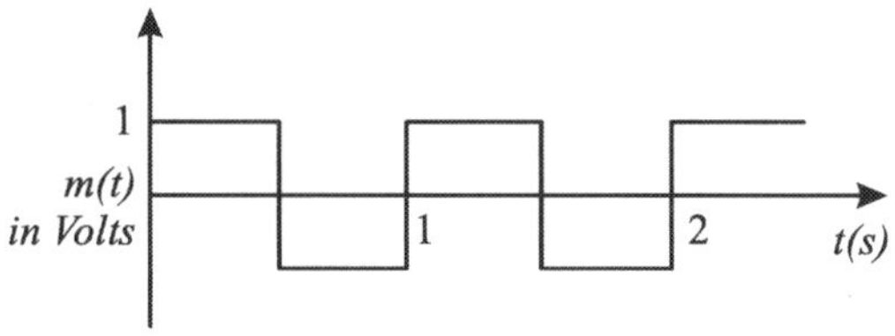

The carrier wave is given by . $c(t)=1.33\sin(8\pi t)$ *volts* **The modulation index is**

1) 2 2) 0.75 3) 0.5 4) 1.5

8. **A audio signal** $20\sin 2\pi(1500t)$ **amplitude modulates** $40\sin 2\pi(10^5)t$**. The depth of modulation is**

1) 25% 2) 20% 3) 50% 4) 40%

9. **An AM wave is given by** $V=1500[1+0.5\sin 62800t]\sin(5.26\times10^5 t)$.

The modulating frequency is

1) 20 kHz 2) 10 kHz 3) 12.5 kHz 4) 50 kHz

10. **The amplitude modulated current is given by** $i=75[1+0.3\sin 2003t]\sin(7\times10^5 t)$.

The RMS value of carrier current will be

1) $\frac{125}{\sqrt{2}}A$ 2) $\frac{100}{\sqrt{2}}A$ 3) $\frac{75}{\sqrt{2}}A$ 4) $\frac{50}{\sqrt{2}}A$

11. **An audio signal** $5\sin 2\pi(1600t)$ **amplitude modulates** $20\sin 2\pi(2\times10^5 t)$**. The two side band frequencies are**

1) 98.4 kHz, 101.6 kHz 2) 92.5 kHz, 105.5 kHz
3) 32.1 kHz 31.59 kHz 4) 96 kHz, 106 kHz

12. **If** $2f_c$ **and** $3f_m$ **are the frequencies of carrier wave and signal, then the band width is**

1) f_m 2) $6f_m$ 3) f_c 4) $2f_c$

13. **The number of AM broadcast stations that can be accommodated in a 140kHz band width for the highest modulating frequency 10 kHz will be**

1) 10 2) 5 3) 7 4) 12

14. **A 2000 kHz carrier is simultaneously modulated with 600Hz, 800 Hz and 4 kHz audio waves. The frequencies present in the output are**

1) 2000.6 kHz,2000.8 kHz,1996 kHz
2) 1999.4 kHz, 2004 kHz, 1999.2kHz
3) 1002.8 kHz, 996kHz, 1106 kHz
4) Both (1) and (2)

15. **If a carrier wave of amplitude 20mV is modulated by an audio signal of amplitude 4mV, then amplitude of LSF or USF is given by**

1) 0.2 mV
2) 0.5 mV
3) 1 mV
4) 2 mV

16. **The amplitude modulated current is given by $i = 70\left[1 + 0.06\sin 2900t\right]\sin 10^6 t$. The depth of modulation is**

1) 60%
2) 6%
3) 36%
4) 66%

17. **If minimum voltage in an AM wave was found to be 1.1 V and maximum voltage 10V. The % modulation index**

1) 80.2%
2) 66.67%
3) 64.25%
4) 76.25%

18. **Sinusoidal carrier voltage of frequency 3 *MHz* and amplitude 50 *V* is amplitude modulated by sinusoidal voltage of frequency 10 *kHz* producing 50% modulation. The lower and upper side-band frequencies in *kHz* are**

1) 1490, 1510
2) 3010, 2990
3) $\frac{1}{1490}, \frac{1}{1510}$
4) $\frac{1}{1510}, \frac{1}{1490}$

19. **A transmitter supplies 6 *kW* to the aerial when unmodulated. The power radiated when modulated to 60% is**

1) 5 *kW*
2) 9.72 *kW*
3) 7.08 *kW*
4) 12 *kW*

20. **The total power content of an AM wave is 6000 *W*. For 100% modulation, the power transmitted by the carrier is**

1) 500 *W*
2) 700 *W*
3) 750 *W*
4) 4000 *W*

21. **The bit rate for a signal, which has a sampling rate of 8 *kHz* and where 32 quantization levels have been used, is**

1) 32000 *bits/sec*
2) 40000 *bits/sec*
3) 64000 *bits/sec*
4) 72000 *bits/sec*

22. **The antenna current of an AM transmitter is 10 *A* when only carrier is sent but increases to 10.1 *A* when the carrier is modulated sinusoidally. The percentage modulation is**

1)50%
2)60%
3) 200%
4)71%

23. **In a diode AM-detector, the output circuit consists of $R = 1k\Omega$ and $C = 10\,pF$. A carrier signal of 1000 *MHz* is to be detected. Is it good**

1) Yes
2) No
3) Information is not sufficient
4) None of these

24. **In an FM system a 9 *kHz* signal modulates 108*MHz* carrier so that frequency deviation is 60 *kHz*. The carrier swing is**

1) 7.143
2) 8
3) 6.67
4) 350

25. **Maximum useable frequency (MUF) in *F*-region layer is *x*, when the critical frequency is 60 *MHz* and the angle of incidence is 60°. Then *x* is**

1)150 *MHz*
2)170 *MHz*
3)120 *MHz*
4)190 *MHz*

LEVEL -II -4- KEY

1) 2	2) 2	3) 2	4) 4	5) 4	6) 3	7) 2	8) 3	9) 2	10) 3
11) 1	12) 2	13) 3	14) 4	15) 4	16) 2	17) 1	18) 2	19) 3	20) 4
21) 2	22) 3	23) 1	24) 3	25) 3					

LEVEL - II -4-HINTS

1. $d = \sqrt{2Rh_T} + \sqrt{2Rh_R}$; $\frac{h_T}{h_R} = \frac{4}{9}; d = 64km$

2. $D = \sqrt{2Rh_T} + \sqrt{2Rh_R}$

3. $A = \pi D^2 = \pi[2Rh_2]$

   $n = (Population\ density)A$

4. $d = \sqrt{2Rh_2} - \sqrt{2Rh_1}$; $h_2 = h_1 - 21\%$ of h_1

5. $n = \frac{16GHz}{4MHz}$

6. Modulation index

   $= \frac{\text{Amplitude of modulating signal}}{\text{Amplitude of carrier wave}}$

7. $m = \frac{a_m}{a_c}$

8. $\text{depth} = \frac{a_c - a_m}{a_c} \times 100$

9. $f_m = \frac{w_m}{2\pi}$

10. $i_{rms} = \frac{i_0}{\sqrt{2}}$

11. $LSB = f_c - f_m = 10^5 - 1600 = 98.4KHz$

    $USB = f_c + f_m = 10^5 + 1600 = 101.6KHz$

12. Frequencies in out put are $2f_c - 3f_m$ and $2f_c + 3f_m$

13. $n = \frac{band\ width}{signal\ band\ with} = \frac{140}{2 \times 10} = 7$

14. Frequencies in out put are $f_c - f_m$ and $f_c + f_m$

15. Amplitude of $L.S.B = U.S.B = \mu\frac{A_c}{2} = \frac{A_m}{A_c}.\frac{A_c}{2} = \frac{A_m}{2}$

16. Depth of modulation $= 0.06 \times 100 = 6\%$

17. $\mu = \frac{10 - 1.1}{10 + 1.1} \times 100 = 80.2\%$

18. $f_c - f_m$ and $f_c + f_m$

19. $P_t = P_c\left[1+\frac{m^2}{2}\right] = (6KW)\left[1+\frac{(0.6)^2}{2}\right]$

20. $P_c = P_t\left[\frac{2}{2+m^2}\right]$

21. No. of quantization levels = 2^n

bit rate = sampling power x no. of bits per sample

22. $\left(\frac{I_t}{I_c}\right)^2 = 1+\frac{m^2}{2}$

23. For demodulation $\frac{1}{f_c} << RC$

24. Carrier swing $= \frac{frequency\ deviation}{Modulating\ frequency}$

25. $MUF = \frac{f_c}{\cos\theta}$

LEVEL - III

1. **The tuned circuit of an oscillator in a simple AM transimitter employs a 250 micro henry coil and 1nF condenser. If the oscillator output is modulated by audio frequency upto 10KHz, the frequency range occupied by the side bands in KHz is**

1) 210 to 230 2) 258 to 278 3) 308 to 328 4) 118 to 128

2. **A TV tower has a height of 70m. If the average population density around the tower is $1000 km^{-2}$, the population covered by the TV tower**

1) 2.816×10^6 2) 2.86×10^9 3) 2.816×10^3 4) 2.816×10^{12}

3. **A carrier wave is modulated by a number of sine waves with modulation indices 0.1, 0.2, 0.3. The total modulation index (m) of the wave is**

1) 0.6 2) 0.2 3) $\sqrt{0.14}$ 4) $\sqrt{0.07}$

4. **The maximum peak-to-peak voltage of an AM wave is 16 mV and the minimum peak-to-peak voltage is 4mV. The modulation factor is equal to**

1) 0.6 2) 0.3 3) 0.8 4) 0.25

5. **An AM wave is exressed as $e = 10(1+0.6\cos 2000\pi t)\cos 2\times10^8\pi t$ volts, the minimum and maximum values of modulated carrier wave are**

1) 10 V, 20 V 2) 4 V, 8 V 3) 16 V, 4 V 4) 8 V, 20 V

6. **The audio signal voltage is given by $V_m = 2\sin 12\pi\times10^3 t$. The band width and LSB if carrier wave has a frequency $3.14\times10^6 rad/s$**

1) 12 KHz; 494 KHz 2) 6 KHz; 313 KHz
3) 6 KHz; 494 KHz 4) 18 KHz; 494 KHz

7. **A TV transmission tower at a particular station has a height of 160m. Radius of earth is 6400km**

i. The range it covers is 45255 m

ii. The population that it covers is77.42 lakhs. When population density is $1200 km^{-2}$

iii. The height of antenna should be increased by 480 m to double the coverage range

1) i and ii are true 2) ii and iii are true 3) i and iii are true 4) i, ii and iii are true

8. **The tuned circuit of the oscillator in a simple AM transmitter employs a 40 μH coil and 12 nanofarad(nF) capacitor. If the oscillator output is modulated by audio frequency of 5 kHz,Which of the following frequencies doesn't appear in the out put AM?**

1) $f_{USB} = 225$ KHz 2) $f_{USB} = 235$ kHz 3) f_c=230 kHz 4) $f_c = 235$kHz

9. **A 400 watt carrier is modulated to a depth of 80%. Caluculate the total power in the modulated wave.**

1) 528 W 2) 128 W 3) 256 W 4) 400 W

10. **Caluculate modulation index if carrier waves is modulated by three signals with modulation indices as 0.6, 0.3 and 0.4**

1) 1.0 2) 0.70 3) 0.78 4) 1.3

11. **A 1000 KHz carrier is simoultaneously modulated with f_{m1}= 300 Hz, f_{m2}= 800 Hz and f_{m3}= 1 KHz audio sine waves. What will be the frquencies present in the output?**

a) $f_{LSB1} = 999.7 KHz$, $f_{USB1} = 1000.3 KHz$

b) $f_{LSB2} = 999.2 KHz$, $f_{USB2} = 1000.8 KHz$

c) $f_{LSB3} = 999 KHz$, $f_{USB3} = 1001 KHz$

d) $f_{LSB3} = 990 KHz$, $f_{USB3} = 1010 KHz$

1) a only 2) b, c and d 3) a, b and c 4) a, c only

LEVEL-III - KEY

1) 3 2) 1 3) 3 4) 1 5) 3 6) 1 7) 4 8) 2 9) 1 10) 3
11) 3

LEVEL-III - HINTS

1. $f_c = \frac{1}{2\pi\sqrt{LC}} = 318\text{khz}, f_m = 10\text{khz}$

$LSB = f_c - f_m = 308$
$USB = f_c + f_m = 328$

2. Population = Population density × Area

= Population density × $\pi \times 2Rh$

3. $m = \sqrt{m_1^2 + m_2^2 + m_3^2}$

4. $\mu = \frac{(16-4)/2}{(16+4)/2}$

5. $E_{max} = (1+\mu)E_c$
$E_{min} = (1-\mu)E_c$

6. band width $= 2f_m$

$LSB = f_c - f_m$

7. $d = \sqrt{2Rh}$

Population = Population density $\times \pi d^2$

$d \alpha \sqrt{h}$

9. $P_{Total} = P_C\left(1+\frac{m^2}{2}\right) = 528$ watts

10. $m_t = \sqrt{m_1^2 + m_2^2 + m_3^2} = 0.78$

LEVEL - IV

1. Consider telecommunication through optical fibres. Among the following statements one is not true

1)optical fibers can be of graded refractive index

2)optical fibers are subjected to electromagnetic interference from out side

3)optical fibers have extremely low transmission loss

4)optical fibers have homogeneous core with a suitable cladding

2. A 600 W carrier is modulated to a depth of 75% by a 400Hz sine wave. The total antenna power is

1) 769 W 2) 796 W 3) 679 W 4)637.5 W

3. The antenna current of an AM transmitter is 8A when only the carrier is sent, but it increases to 8.93A when the carrier is modulated by a single wave. The percentage modulation is

1) 60.1% 2) 70.1% 3) 80.1% 4) 50.1%

4. The antenna current is 8A only when only carrier wave is transmitted and 9.6A when AM wave is transmitted. If carrier power is 10kw, the modulating power is

1) 14.4 kW 2) 1.2 kW 3) 2 kW 4) 4.4 kW

Passage - I

A message signal of frequency 10kHz and peak voltage 10V is used to modulate a carrier wave of frequency 1MHz and peak voltage 20volts.

5. The modulation index is

1) 0.2 2) 0.3 3) 0.5 4) 0.8

6. Among the following are USB and LSB are

1) 1010kHz, 990 kHz 2) 990 kHz, 1010kHz 3) 950 kHz, 1050kHz 4) 1050kHz, 990 kHz

7. The bandwidth of signal is

1) 10kHz 2) 20 kHz 3) 30 kHz 4) 40 kHz

Passage - II

8. When only carrier is transmitted antenna current observed is 8A when it is modulated with 500 Hz sine wave antenna current becomes 9.6A. The percentage of modulation is

1) 80% 2) 20% 3) 94.26% 4) 83.76%

9. **1% of 10^{12} Hz of a satellite link was used for telephony. The number of channels or subscribers if each channel is of 8 kHz are**

1) 2.5×10^{7} 2) 1.25×10^{6} 3) 2.5×10^{8} 4) 1.25×10^{8}

10. **A carrier is modulated simultaneously by 3 sine waves of modulation indices 0.3, 0.4 and 0.45 respectively. The net modulation index is**

1) 1.15 2) 0.67 3) 0.57 4) 0.84

11. **A 1 kW carrier is modulated to a depth of 80%. The total power in the modulated wave is**

1) 1.32 2) 1.56 3) 1.84 4) 1.96

Passage - III

The expression for an AM voltage is $e=5\left(1+0.6\cos1000\pi t\right)\cos5\times10^{6}\pi t$

12. **The peak value of carrier wave is**

1) 30 V 2) 6 V 3) 3 V 4) 5 V

13. **The percentage of modulation is**

1) 0.6% 2) 6% 3) 60% 4) 66%

14. **Frequency of audio signal is**

1) 1000 Hz 2) 5000 Hz 3) 500Hz 4) 250 Hz

15. **Frequency of carrier signal is**

1) $5\times10^{6}\,Hz$ 2) 1000Hz 3) $2.50\times10^{3}\,Hz$ 4) $2.50\times10^{6}\,Hz$

16. **Frequencies of sideband is**

1) 2.5005×10^{6} Hz, 2.4995×10^{6} Hz 2) 2.505×10^{6} Hz, 2.495×10^{6} Hz

3) 2.505×10^{6} kHz, 2.495×10^{6} kHz 4) 2.505 MHz, 2.495kHz

Passage - IV

An amplitude modulated voltage is expressed as $e=10\left(1+0.8\cos2000\pi t\right)\cos3\times10^{6}\pi t\ volt$

17. **The peak value of carrier wave is**

1) 5 V 2) 8 V 3) 10 V 4) 100 V

18. **The depth of modulation is**

1) 8 2) 0.8 3) 800 4) 80

19. **The frequency of carrier wave is**

1) $3\times10^{6}\,Hz$ 2) $1.5\times10^{3}\,Hz$ 3) $3\times10^{3}\,Hz$ 4) $1.5\times10^{6}\,Hz$

20. **The minimum and maximum value of modulated carrier wave is**

1) 2V, 18V 2) 4V, 16V 3) 2V, 8V 4) 4V, 12V

21. **The frequency of modulating signal is**

1) 1000 Hz 2) 100Hz 3) 10MHz 4) 100 MHz

22. **The bandwidth of modulated signal is**

1) 1000 Hz 2) 2000 Hz 3) 1500 Hz 4) 2500 Hz

ASSERTION AND REASON

Instructions for the following questions:

In each of the following questions a statement of Assertion (A) is given followed by corresponding statement of Reason (R) just below it. Of the statements mark the correct answers.

1) A and R are true and R is the correct explanation of A

2) A and R are true but R is not correct explanation of A

3) A is true but R is false | **4) A is false but R is true.**

23. **A:** It is not necessary for a transmitting antenna to be at the same height as that of receiving antenna for line-of sight communication

R: If the signal is to be received beyond the horizon then the receiving antenna must be high enough to intercept the line-of sight waves

24. **A:** Low frequency audio signals cannot be transmitted directly over long distances

R: To transmit low frequency audio signals over long distances, the audio signals are superimposed on a high frequency carrier signal

25. **A:** The microphone acts as a transducer.

R: In radio communication, microphone is used to convert sound signal into electrical signal.

26. **A:** A FM signal is less susceptible to noise than an AM signal.

R: In FM transmission, the message signal is in the form of frequency variations of carrier waves. During modulation process, the noise gets amplitude modulated.

27. **A:** Artifical satellites are not necessary for long distance TV transmission.

R: The TV signals are not reflected by the ionosphere. They are reflected back to the earth from an earth station by making use of artifical satellites.

28. **A:** Greater the height of a TV transmitting antenna, greater is its coverage.

R: The height of transmitting antenna is directly proportional to the square of the distance of TV coverage

29. **A:** Long distance radio broadcasts use short wave bands.

R: The short waves are absorbed by the earth but reflected by F layer in ionosphere. After reflection, they reach the surface of earth back only at a large distance from the transmitter.

30. **A:** TV signals cannot be transmitted via sky wave propagation.

R: Ionosphere is unable to reflect radio waves of frequencies greater than 40MHz.

MATRIX MATCHING

31. **Match the frequency band with the type of use**

Frequency Band	**Type of use**
a) LF	e) Radio Broad casting
b) HF	f) Marine and navigational aid
c) VHF	g) Satellite communication
d) SHF	h) TV Broad casting

1) a - e; b - f; c - h; d - g
2) a - f; b - e; c- g; d- h
3) a - f; b - e; c - h; d - g
4) a - e; b - g; c - f; d - h

32. **Match layer of ionosphere and height of the layer from surface of earth for atmosphere**

a) D	e) 250 - 400 km
b) E	f) 100 km

c) F
d) F_2
g) 65 - 75 km
h) 170 - 190 km

1) a-g; b-f; c-h; d-e 2) a-h; b-g; c-e; d-f 3) a-g; b-h; c-f; d-e 4) a-g; b-h; c-e; d-f

33. Match layer of ionosphere and frequencies most affected

a) D
b) E
c) F
d) F_2

e) helps surface waves and reflects HF
f) efficiently reflects HF waves
g) partially absorbs HF
h) reflects LF and absorbs MF

1) a-g; b-f; c-h; d-e
2) a-h; b-e; c-g; d-f
3) a-h; b-e; c-f;d -g
4) a-e; b-h; c-g; d-f

34. Match the frequency band with the type of use service frequency band

a) AM broadcast
b) satellite communication
c) FM broadcast
d) Cellular mobile

e) 88 - 108 MHz
f) 896 - 935 MHz
g) 540 - 1600 kHz
h) 3.7 - 6.5 GHz

1) a-g; b-h; c-f; d-e
2) a-h; b-g; c-e; d-f
3) a-f; b-h; c-e; d-g
4) a-g; b-h; c-e; d-f

35. Match List-1 with List-2

List-1	**List-2**
Communication mode	Example
a) Point to point communication	e) RADAR
b) broadcast communication	f) AM Radio
c) Line of sight communication	g) FM Radio
d) Satellite communication	h) Traditional telephony
	i) Mobile telephony
	j) TV

1) a-h; b-f,g,j; c-e;d-i,j
2) a-e; b-i,j; c-h;d-g,f
3) a-f ; b-h,i,j; c-g;d-e
4) a-g; b-f,g,j; c-e;d-i,j

36. Match List-1 with List-2

List-1	**List-2**
Name of device	**use**
a) Antenna	e) sends out information
b) Transmitter	f) picksup information
c) Receiver	g) converts energy in one form to another form
d) Transducer	h) radiates signal
	i) recieves signal

1) a-e; b-g,h;c-f;d-h 2) a-h,i;b-e;c-f;d-g 3) a-f; b-e;c-f,i;d-h 4) a-h;b-g,i;c-f;d-e

37. Match List-1 with List-2

List-1	List-2
Name	Frequencies most affected
a) Part of stratosphere(D)	e) VHF(upto several GHz)
b) Part of mesophere (F_1)	f) reflects LF
c) Part of thermosphere(F_2)	g) Partially absorbs HF
d) Troposphere	h) Efficiently reflects HF

1) a-e;b-f,g,h;c-i;d-h 2) a-f;b-f,g,h;c-f;d-e 3) a-f,g,h;b-g;c-i;d-e 4) a-g;b-f,g,h;c-f;d-e

38. Match List-1 with List-2

List-1	**List-2**
Type of propagation	Frequencies
a) sky waves	e) 1.5 MHz
b) space wave	f) 20 MHz
c) ground wave	g) 30MHz
d) micro wave	h) 50MHz
	i) 3 GHz

1) a-e,i;b-e;c-h;d-i 2) a-f,g;b-h;c-e;d-i 3) a-i,g;b-e;c-h;d-f 4) a-g;f-h;c-i,b;d-e

LEVEL - IV - KEY

1) 2	2) 1	3) 2	4) 4	5) 3	6) 1	7) 2	8) 3	9) 2	10) 2
11) 1	12) 4	13) 3	14) 3	15) 4	16) 1	17) 3	18) 2	19) 4	20) 1
21) 1	22) 2	23) 1	24) 1	25) 1	26) 1	27) 4	28)1	29) 1	30) 1
31) 3	32) 1	33) 2	34) 4	35) 1	36) 2	37) 3	38) 2		

LEVEL - IV - HINTS

2. $P_t = P_c\left(1+\frac{\mu^2}{2}\right) = 600\left[1+\frac{(0.75)^2}{2}\right] = 768.5W$

3. $\left(\frac{i_t}{i_c}\right)^2 = 1+\frac{\mu^2}{2} \Rightarrow \left(\frac{8.93}{8}\right)^2 = 1+\frac{\mu^2}{2}$

$\Rightarrow \mu = 0.701 = 70.1\%$

4. $\frac{P_t}{P_c} = \left(\frac{i_t}{i_c}\right)^2 \Rightarrow \frac{P_t}{10} = \left(\frac{9.6}{8}\right)^2 = 14.4kw$

$\therefore P_m = 14.4 - 10 = 4.4\ kW$

5. $\mu = \frac{10}{20}$

6. $USB = f_m + f_c$

$LSB = f_m - f_c$

7. $\text{bandwidth of signal} = 2f_m$

8. $\frac{i_t}{i_c} = \sqrt{1+\frac{\mu^2}{2}}$

9. 1% of 10^{12} Hz $= 10^{10}$ Hz

$\text{Number of channels} = \frac{10^{10}}{8\times10^3} = 1.25\times10^6$

10. $m = \sqrt{m_1^2 + m_2^2 + m_3^2}$

11. $P_t = P_c\left[1+\frac{\mu^2}{2}\right]$

12. eqn of AM voltage is given by

$e = A_c(1+\mu\sin\omega_m t)\sin\omega_c t$

$\therefore A_c = 5\,\text{v}$

13. $\mu = 0.6\times100 = 60\%$

14. $f_m = \frac{\omega_m}{2\pi} = \frac{1000\pi}{2\pi} = 500$

15. $f_c = \frac{\omega_c}{2\pi} = \frac{5\times10^6\pi}{2\pi}$

16. $USB = f_c + f_m$

$LSB = f_c - f_m$

17. eqn of AM voltage is given by

$e = A_c(1+\mu\sin\omega_m t)\sin\omega_c t$

$\therefore A_c = 10\,\text{V}$

18. $\mu = 0.8$

19. $f_c = \frac{\omega_c}{2\pi} = \frac{3\times10^6\pi}{2\pi}$

20. $E_{min} = (1-\mu)A_c, \quad E_{max} = (1+\mu)A_c$

21. $f_m = \frac{\omega_m}{2\pi} = \frac{2000\pi}{2\pi}$

22. band width $= 2f_m$

Made in the USA
Monee, IL
05 November 2026

67811022R10164